# Underutilized and Underexploited Horticultural Crops

**VOLUME : 2**



NIPA® GENX ELECTRONIC RESOURCES & SOLUTIONS P. LTD.
New Delhi-110 034

# Underutilized and Underexploited Horticultural Crops

**VOLUME : 2**

*Editor*

**Prof. K.V. Peter**

Professor of Horticulture and Former Vice-Chancellor
Kerala Agricultural University
Thrissur - 680 656s
Kerala (India)

**NIPA® GENX ELECTRONIC RESOURCES & SOLUTIONS P. LTD.**

New Delhi-110 034

**NIPA® GENX ELECTRONIC RESOURCES & SOLUTIONS P. LTD.**

101,103, Vikas Surya Plaza, CU Block
L.S.C. Market, Pitam Pura, New Delhi-110 034
Ph : +91 11 27341616, 27341717, 27341718
E-mail: newindiapublishingagency@gmail.com
web: www.nipabooks.com

***For customer assistance, please contact***
Phone: + 91-11-27 34 17 17
Fax: + 91-11- 27 34 16 16
E-Mail: feedbacks@nipabooks.com



ISBN: 978-93-58870-63-3



Composed and Designed by NIPA®.

**AVRDC**

**The World Vegetable Center**

**Regional Center for South Asia**

ICRISAT Campus, Patancheru 502 324, Hyderabad, A.P. India

## FOREWORD

Nature bestowed man with about 2.5 lakh plant species and perhaps the same number of unknown species. Out of this plant wealth, 80,000 plant species are edible but man exploited only 3,000 of these plant species for food, shelter and clothing. Commercial scale cultivation is limited to 150 species and consumption for food is limited to 20 different kinds of crops, the world over. Four crops rice, wheat, maize and potato alone supply about 60% of total calories intake. The food basket base is shrinking consequent to monocrop culture and reduced area under cropping. There is reduction in the consumption of organic compounds making ourselves vulnerable to stresses and blockage of immuno regulators in the body system. Awareness to enlarge food basket incorporating edible but so far underutilized and underexploited crops has come up as a positive development in food and nutrition. Traditional and ethnic foods are much valued for its presence of lesser known vegetables, fruits, spices, aromatic plants and tuber crops. A few of the underutilized vegetables contain antinutrient factors like free nitrates and oxalates. In fact, these vegetables are consumed only at negligible quantity. They supply micronutrients, essential amino acids, fatty acids, flavourants and pigments.

The Andaman and Nicobar islands sustain a large number of non traditional horticultural crops. Noni (*Morinda citrifolia*), alligator apple (*Anona glabra*), Custard apple (*Anona squamosa*), Sour sop (*Anona muricata*), Ramphal (*Anona reticulata*), Karonda (*Carissa carandus*), Carambola *(Averrhoa carambola*), Bilimbi (*Averrhoa bilimbi*), Wild mango (*Mangifera andamanica, Mangifera griffithi, Mangifera camptosperma*), wild amra (*Dracontomelum dao*), Amra (*Spondias cytheres*), Laseda (*Cordia obliques*), badam (*Terminalia catappa*), Chalta (*Dillenia indica*), Sapida (*Baccaurea ramiflora*), Mangosteen (*Garcinia mangostana*), West Indian cherry (*Malphigia glabra*), Jack fruit (*Artocarpus heterophyllus*), Monkey Jack (*Artocarpus lakoocha)*, Amjeer(*Ficus carica*), Wild jamun *(Syzygium claviflorum*), Jambose *(Syzygium jambolana),* Wild rose apple *(Syzygium samarangense),* Rose apple (*Syzygium aquem*), Kewra (*Pandanus andamansium*), Passion fruit(*Passiflora edulis*), Wild ber (*Zyzyphus glabra*), Bael (*Aegle marmelos*), Rambutan (*Nephelium lappaceum*) and Bakul(*Mimusops elengi*) are a few of under utilized fruits with considerable economic value in Andaman and Nicobar islands alone. Aonla (*Emblica officinalis)*, an indigenous fruit with high nutritive value and medicinal and therapeutic properties, is not clearly understood. Out of 110 genera and 640 species under Cucurbitaceae, 36 genera and 100 species occur in India. Cucumis has more than 30 species, out of which *C. sativus* and *C. melo* are grown commercially. Long melon, snap melon, winter melon, mango melon, pickling melon and kachri are potential melons still underutilized. Ash gourd (*Benincasa hispida*) has considerable medicinal value. Large number of cucurbits are still underexploited and underutilized. Mateera is an indigenous type of water melon (*Citrullus lanatus*) grown extensively in North-Western India particularly in arid and semi arid areas of Rajasthan. Mouth appeal of Mateera is attributed to sweetness, taste, aroma and texture of edible flesh(pulp). There are a number of carotene rich leaf vegetables and lesser known like alternanthera which grow under conditions of stresses- biotic and abiotic. Giant granadilla provides edible fruits tasting like pear on maturity and unripe fruits can be boiled and eaten as vegetable. There are a large number of under utilized root and tuber crops like Suran (*Amorphophallus sp.*), Kiriki kukri (*Arisaema totuosum)* and Samsfali (*Asparagus sp.),* which are rich sources of starch with possible uses in ethanol production for green energy. Ornamentals play soothening and aesthetic role in human activities. There are a large number of under exploited ornamentals like Liliums, Lotus and Cassias with potential values. Energy plantations are gaining public attention in view of depleting non renewable sources of petroleum based energy. Trees like Chironji and Mahua are so far not utilized for energy, though there is potential. The present volume II of "Underutilized and Underexploited Horticultural Crops" edited by Prof. K.V. Peter deals with root and tuber crops, ornamentals, fruits, vegetables, herbs and spices and energy crops having considerable economic value, but not fully exploited. I compliment Prof. K.V. Peter and all the contributors for the commitment and hard work in the compilation. I congratulate New India Publishing Agency, New Delhi for efficient and error free printing.


**M L CHADHA**

(Director)

AVRDC-The World Vegetable Center

Regional Center for South Asia

# DEDICATION

This volume is devoted to the memory of my uncle late Sri T.M. Joseph, A-7, Adayar Boat Club Road, Chennai who inspired me to be a good scientist of use to society and people.

# Acknowledgement

The Editor is grateful to Dr. M.L Chadha, Director, AVRDC-The World Vegetable Center, Regional Center for South Asia, ICRISAT, Hyderabad for the FOREWORD. I am also grateful to Dr. R Keshvachandran, Associate Professor and Co-ordinator Bioinformatics Centre, Kerala Agricultural University and Miss Alishya Joy JRF for the technical support and meticulous timings.

# INTRODUCTION

Horticultural Genetic Bio-resources are a few of biological assets available to resource- poor farmers to ensure sustainable production to improve livelihood options. Rapid loss of genetic diversity (genetic erosion) is of global concern and is particularly severe in cropped lands because of monocultures demanding immediate measures for conservation. Widening of food basket is a protective measure against natural and man-made calamities. Traditional and indigenous knowledge on nutritional curatives, protective and industrial uses of present underutilized and underexploited horticultural crops would add to the menu of health foods of today and tomorrow. The regional office of International Plant Genetic Resources Institute (IPGRI) at Serdang, Malaysia initiated *in situ* conservation of plant genetic resources. The Andaman and Nicobar groups of Islands sustain plant diversity in a large number of non traditional horticultural crops like noni, alligator apples, custard apple, soursop, ramphall, karonda, carambola, bilimbi, wild mango,wild amra, amara, laseda, badam, chalta, mangosteen, West Indian cherry, Jack fruit, monkey jack,fig, wild jamun, jamboza, wild rose apple, rose apple, kewra, passion fruit, wild ber, bael, rambutan and bakeral. Therapeutic values of underutilized fruits like aonla (goose berry) rich in vitamin C (500-1500 mg/ 100g), pectin, iron, calcium and phosphorus, are now understood. Aonla is the main ingrediant of Chavanprash and Triphal powder, used for curing different human ailments. Fruits contain chemical substances gallic acid and leucoanthocyanin having antioxidant properties. Ber is another minor fruit of India grown under marginal conditions, providing quality yields at lower cost. Fruits of ber are eaten afresh but other forms like dried, candied and squash can also be prepared from ber. India produces many deciduous fruits including pome fruits (apple and pear) and stone fruits (peach, plum, apricot and cherry) in considerable quantity. These are mainly grown in the North-Western Indian states of J & K, Himachal Pradesh, Uttarakhand and Uttar Pradesh. Development of high quality peach and nectarine (prunes) cultivars with low chilling needs has greatly expanded the potential for commercial peach and nectarine production in North and North Eastern India. Among vegetables for fibres, minerals, vitamins, and non-calorific value foods, cucurbits have multipurpose values. They are the modest sources of antioxidants, having antiglycymic properties. Large number of cucurbits are still underexploited and underutilized, which include Little gourd, Malabar gourd, Buffalo gourd, Sweet gourd, Fluted pumpkin, Stuffing cucumber, Oyster nut and Pointed gourd. Indian arid zone is one of the largest sub tropical deserts of the world, of which 20% is arid and the remaining semi-arid with varied habitats. Drier conditions favour development of quality vegetables particularly cucurbits, chillies and spices. Indigenous vegetables like mateera, snap melon and tinda are rescue crops, providing sustainable income to farmers. Among underexploited leaf vegetables in tropical climates, alternanthera is lesser known, but rich in protein (5g/100g of leaf), Vit A (1.95 mg), B3 (1.2 mg) and Vit C (17.0 mg). The family Cucurbitaceae consists of nearly 100 genera and over 750 species. Stuffing cucumber, chayotes, bittermelon, snake gourd, ivy gourd, buffalo gourd, long melon, round melon, snap melon and sweet gourd have considerable economic values. Giant granadilla is another viny crop belonging to Passion fruit family with potential uses as an ornamental, a fruit at maturity and vegetable at tender stage. The fruit is valued in tropics as antiscortutic and stomachic. Many underutilized root and tuber crops occur in Himalayan region. The genera Amorphophallus, Arisaema, Asparagus, Berium, Campanula, Chacrophyllum, Codonopsis, Crambles, Dioscorea, Helianthus, Inula, Lilium, Polygonatum, Pueraria and Typha carry several species of economic value. Liliums are grouped under high value flowers. Lily is a symbol of love and suited well as a cut flower. Lilies though ranking 10$^{th}$ among the top selling flowers in the world, are not exploited in India. Plants for renewable green energy are getting attention in view of depleting petroleum resources. Gaseous energy from carbon sources like chirongi and mahua is ecofriendly and less pollutive. Chirongi is an excellent tree of agro forestry and social forestry, especially in wasteland development and dry land horticulture. Fresh ripe fruit is very palatable and is largely eaten raw or roasted and oily kernels are the most important parts. Chirongi oil is non- repellant and non-toxic and is suitable for human use. Mahua (*Brassia latifolia*) is a multi purpose tree. Mahua flowers are rich sources of sugar, vitamins and minerals. Every part of Mahua yields an economic product of great potential value. Spices add flavour, colour and fragrance to food preparations. Long coriander is a herb introduced from S Mexico to Panama, Columbia, Bolivia, Brazil and Cuba to Trinidad. It is used as a substitute for coriander. Long coriander belongs to the same family as coriander, but the plant shape and form do not bear much resemblance. Among spices and herbs, curcumas have medicinal, aromatic and antioxidant properties. Turmeric powder and paste are used in a large number of ailments.

The second volume of Underutilized and Underexploited Horticultural Crops carries 20 chapters contributed by 33 working scientists from 16 Research Institutes/ Universities. The volume has been foreword Dr. M.L. Chadha, Director, Asian Vegetable Research and Development Centre (World Vegetable Centre), Taiwan, now based at ICRISAT, Hyderabad.

# Contents

## Underutilized Root and Tuber Crops



## Underutilized Flowers



## Underutilized Trees



## Underutilized Spices



■■■

# General

**Underutilized and Underexploited Horticultural Crops, Vol. 2, 2007**
*K.V. Peter (ed.), pp. 1-9*
New India Publishing Agency, New Delhi (India)
E-mail : newindiapublishingagency@gmail.com
Web: www.bookfactoryindia.com

# 1 *In situ* Conservation of Horticultural Crops

**BHUWON STHAPIT**

*In situ* conservation specialist, International Plant Genetic Resources Institute, Regional Office of Asia, the Pacific and Oceania (APO) Serdang, Malaysia

Genetic resources for food and agriculture are the biological basis of world food and nutrition security and they directly support the livelihood of people. Horticultural genetic resources are a few of biological assets available to resource-poor farmers to ensure sustainable production to improve livelihood options. Genetic diversity gives resource-poor farmers a low cost option to manage the shocks and vulnerability of production systems by its ability to adopt to changing environments. It is the foundation upon which plant breeding depends for the creation of new varieties and is, therefore, of critical public value for global food security. The rapid loss of genetic diversity (genetic erosion) is a worldwide concern and is particularly severe among crop landraces. Farmer's varieties or landraces may contain co-adapted gene complexes that have evolved over decades and are the most valued genetic resources as they provide options for a range of climatic and edaphic stresses. Therefore, agricultural biodiversity has a critical public value for global food security (FAO, 1978).

## What is *in situ* conservation?

Genetic resources can be conserved by two approaches: *in situ* (on-farm-in its place of origin) or *ex situ*, (off-site, outside its place of origin) as in botanical gardens, field genebanks and seed genebanks. The Convention on Biological Diversity (CBD) defined *in situ* conservation as means of conservation of ecosystems and natural habitats and the maintenance and recovery of viable populations of species in their natural surroundings and in the case of domesticates or cultivated species, in their surroundings where they have developed their distinctive properties (Article 2 CBD, UNCED, 1992). In a simple term, *in situ* (on-farm) conservation of landraces refers to plants or its wild relatives that are conserved in the very place where they developed their present-day characteristics. In *in situ* conservation, entire agro-ecosystem is conserved, including cultivated crops and their wild and weedy relatives that may be growing in nearby areas (Brush, 1995; Brush, 2000; Maxted *et al.*, 2004).

*In situ* conservation, which is a dynamic form of PGR management, allows the processes of natural and human selection to continue (Bellon *et al.*, 1996). Crop genetic resources are passed down to generations of farmers and are subject to different selection pressures. Therefore, *in situ* conservation is generally used to describe a management process by which farmers maintain traditional crop varieties which they develop in their local conditions and which they continue to manage and improve. Thus, the conservation of specific genotype(s) is secondary to the continuation of the processes that allow the material to evolve and change over time (Altieri and Merrick, 1987; Jarvis and Hodgkin, 2000). Traditionally, fruits are managed in a variety of production systems such as natural forest, protected areas, buffer zones, home gardens, semi-commercial and commercial orchards. Horticultural crop species-whether cultivated or wild-have generally been selected to suit the environment in which they are cultivated or selected naturally to satisfy particular needs of its growers and users, such as colour, flavour and taste. Wild fruit species which evolved in the region have developed specific adaptations through natural selection and have specific patterns of distribution. Although there may often be continued gene introgression with wild relatives near natural forest and home gardens, farmers' varieties are the products of domestication followed by constant development through farmers' selection. The sustainable management of genetic diversity of locally developed fruit tree species (farmer varieties), with associated wild relatives, by farmers or communities within traditional agricultural (*in situ*), horticultural (orchards) or agri-silvicultural systems (home gardens) is defined as the *in situ* conservation (Altieri and Merrick, 1987; IPGRI 1991; UNCED, 1992; Bellon *et al.*, 1996; Maxted *et al.*, 2002).

### *What factors shape the crop genetic diversity?*

Environmental, biological, cultural, socio-economic and policy factors influence a farmer's decision to select or replace or maintain a particular crop cultivar at any given time (Jarvis *et al.*, 2000; 2004). Although environment mainly determines the genetic diversity in the process of planting, managing and harvesting, the farmers also make crucial decisions that affect genetic diversity of the crop populations. For example, farmers search locally adapted crops/cultivars for the specific land types and farming systems, and select and maintain seeds they like. Farmer's preference for specific colour, food, taste and type also demand diverse crop varieties. Cultural identity also shapes the traditional food culture (tastes and preferences) and thus influences the crops and varieties grown (Smale *et al.*, 2004). Over time, a farmer may alter the genetic structure of a crop population and hence, the goal of *in situ* conservation is to encourage farmers to select, maintain, and exchange local crop diversity for benefit of mankind. This selection pattern differs whether farmer's livelihood depends upon commercial orchards or traditional homegarden /mixed farming systems.

### *Why in situ conservation?*

The majority of crop seed is from varieties developed by farmers over many generations of selection without direct inputs of formal plant breeding. The seed nurtured by farmers provides opportunities for continuous crop adaptation and selection. Thus, diversity in landraces, co-adapted to various biotic and abiotic stresses, is used as the primary breeding or rootstock material for modern varieties. Many horticultural crops and varieties were domesticated from natural forest or wild, and both farmers and plant breeders have used them to develop improved fruit cultivars. There is significant increase over time in the number of landraces that are used in the Consultative Group on International Agricultural Research (CGIAR), for example, Centro Internacional de Mejoraniento de Maiz Y Trigo (CIMMYT) and International Rice Research Institute (IRRI) (Witcombe, 1999). The most dramatic example is of wheat in CIMMYT, where the number of landraces used increased from less than ten in 1950 to more than 60 in 1990. The rice cultivars released by IRRI since 1980 have more than 37 landraces in their pedigree. The continuing use of landraces by farmers contributes to stable food production and income, especially in marginal environments where impacts of modern varieties have been limited. Therefore, the Convention on Biological Diversity (CBD) has recognized the continued maintenance of traditional varieties *in situ* as an essential component of sustainable agricultural development.

The potential threat that the loss of genetic diversity poses to the world's food supply has been recognized for many years and has led to the *ex situ* storage of genetic materials in genebanks. While this form of conservation remains no doubt useful, especially for immediate use in plant breeding, it has major drawbacks in terms of effectiveness and extensiveness. Firstly, an *ex situ* genebank freezes the natural evolutionary process, making the seeds more maladaptive for the future. Secondly, *ex situ* collections are more vulnerable to mismanagement and the transmission of seed-borne pathogens. Thirdly, *ex situ* conservation of genetic resources either as seed (orthodox seed crops) in cold stores or in field genebanks (fruit crops with recalcitrant seeds and vegetatively propagated crops) is costly.

In addition to allowing evolution to continue, *in situ* conservation contributes to the conservation of diversity at all levels - the ecosystem, the species, and the genetic diversity within species. It also empowers the farmers to exercise control over their plant genetic resources, the major biological asset and uses it to improve the livelihoods. *In situ* conservation strategies also promote a broader range of partnerships in conservation efforts involving a diversity of stakeholders to meet desired objectives. In addition, it is a powerful strategy to integrate a farming community into the national PGR system, particularly horticultural crops.

### *Who is responsible for in situ conservation?*

Recognizing these benefits, several international conventions and agreements, including the CBD, the Global Plan of Action (GPA) of the FAO have called for the conservation and sustainable use of agricultural biodiversity. Among other things, each of these instruments not only recognizes the countries' responsibilities to conserve and use their PGRFA, but also the importance of equitable sharing of the benefits derived from the use of resources and technologies (UNCED, 1992).

In most developing countries, farmers practice *de facto* conservation of landraces as an integral part of their traditional farming systems and have done so for centuries. However, this requires some level of institutionalization and integrating into national efforts. The international community has also recognized the critical role of local institutions of genetic resources whether they are identified as farmers, indigenous and local communities, notably in the preamble to the CBD, which has been ratified by 181 countries.

### *What basic information is needed to understand in situ conservation?*

In response to these challenges, in 1995, IPGRI and its national partners launched a project in nine countries, 'Strengthening the scientific basis of *in situ* conservation of agricultural biodiversity in nine countries including Nepal and Vietnam to understand four basic questions (Gauchan *et al.*, 2003; Jarvis *et al.*, 2004):

- What is the amount and distribution of the genetic diversity maintained by farmers over space and time?
- What are the processes (consciously or unconsciously) used to maintain the genetic diversity *in situ*?

- Who maintains genetic diversity within a community and how? (Subedi *et al.*, 2003)
- What factors influence farmers' decisions on maintaining traditional varieties?

Understanding the above mentioned questions provides the knowledge needed not only to better manage horticultural genetic resources on farm, but also to develop options for better livelihoods and income. Experience suggests that effective management and conservation of horticultural genetic resources *in situ* take place where the resources are valued and used to meet needs of local communities. If horticultural genetic resources are going to be conserved *in situ*, it must happen as a spin-off of farmer's production activities directed to his/her livelihood. This means *in situ* conservation efforts must be carried out within the framework of farmer's livelihood needs.

### *What are the criteria for selecting in situ conservation sites?*

A fundamental problem faced by any *in situ* conservation effort is locating crop populations to focus. It is essential to consider some generalized criteria for selection of sites; ecosystems, intra-specific diversity of target species, species adaptation, genetic erosion, diverse use values, and interests of farming community, partners and government agencies and logistics for monitoring (Jarvis *et al.*, 2000). One of the often cited disadvantages of *in situ* conservation is the difficulty of accessing the material conserved. This is mainly because the *in situ* conservation efforts to date have not been mainstreamed and not linked to national PGR efforts.

### *How feasible is in situ conservation?*

Survey carried out by FAO in recent years illustrated that *ex-situ* approach is the main focus of many national programmes in the APO region. It also showed that the role of *in situ* approach is increasingly appreciated as complementing *ex situ* conservation. It is important to understand how farmers value local crop diversity and how much they are willing to pay for such genetic resources. In general terms, horticultural biodiversity provides many goods and services of environmental, economic, social and cultural importance; these environmental goods and services also contribute to sustainable livelihoods in number of ways.

The economic concept that farmers' varieties embody both (i) 'private' values in the harvest the farmer enjoys, either directly as food or feed, or indirectly through the cash obtained by selling the seed/grain and purchasing other items, and (ii) 'public' values in its contribution to the genetic diversity from which future generations of farmers and consumers will also benefit (Smale *et al.*, 2004). The least cost conservation occurs in sites that are most highly ranked in terms of public benefits and where, because the private benefits farmers obtain from growing genetically diverse varieties are the greatest, the public interventions to encourage them to do so will be the least. Those crop genetic resources, which have both low farmer utility (current private value) and public value, will have difficulty to conserve *in situ* unless public interventions are made for adding benefits.

### *How in situ programme provides benefits to community?*

In order for local crop diversity to be maintained by farmers, the resources must have some value and/or be competitive with other options, a farmer might have. Benefits may be economic, ecological and socio-cultural for farmers, communities and society. Two options can be considered in adding benefits; the first through participatory plant breeding and the second through public awareness, better marketing, and policy incentives (Gauchan, Sthapit and Jarvis, 2003). The first option is to seek improved quality, disease resistance, high yield, better taste, and other preferred traits through breeding, seed networks and modified farming systems. The second option includes adding value to local crop resources so that the demand for the material or some derived product may be increased. These diverse options will emerge when the community, researchers, and developmental institutions are directly involved in management of traditional knowledge and genetic resources for biodiversity-based livelihood and income generation. This is only possible if the local capacity of farming communities and institutions are strengthened for making appropriate decisions and these partners also take up the responsibility of monitoring local crop diversity after developmental interventions (Sthapit and Jarvis, 2006).

### *How to implement in situ conservation programme?*

It is important to focus not only in scientific understanding of *in situ* management of agricultural biodiversity but also to develop institutional capacity to run internally driven *in situ* conservation programmes. It is realised that local communities and village-level opinion leaders need to understand how they can use their own local biodiversity and mobilize social and human capitals to generate financial resources for developing livelihood options and conservation actions. Once they realised importance of community management of biodiversity and other community resources, conservation efforts are much easier to tackle. Following steps are considered essential for community-based biodiversity management (CBM approach) of agricultural biodiversity *in situ* (Figure 1).

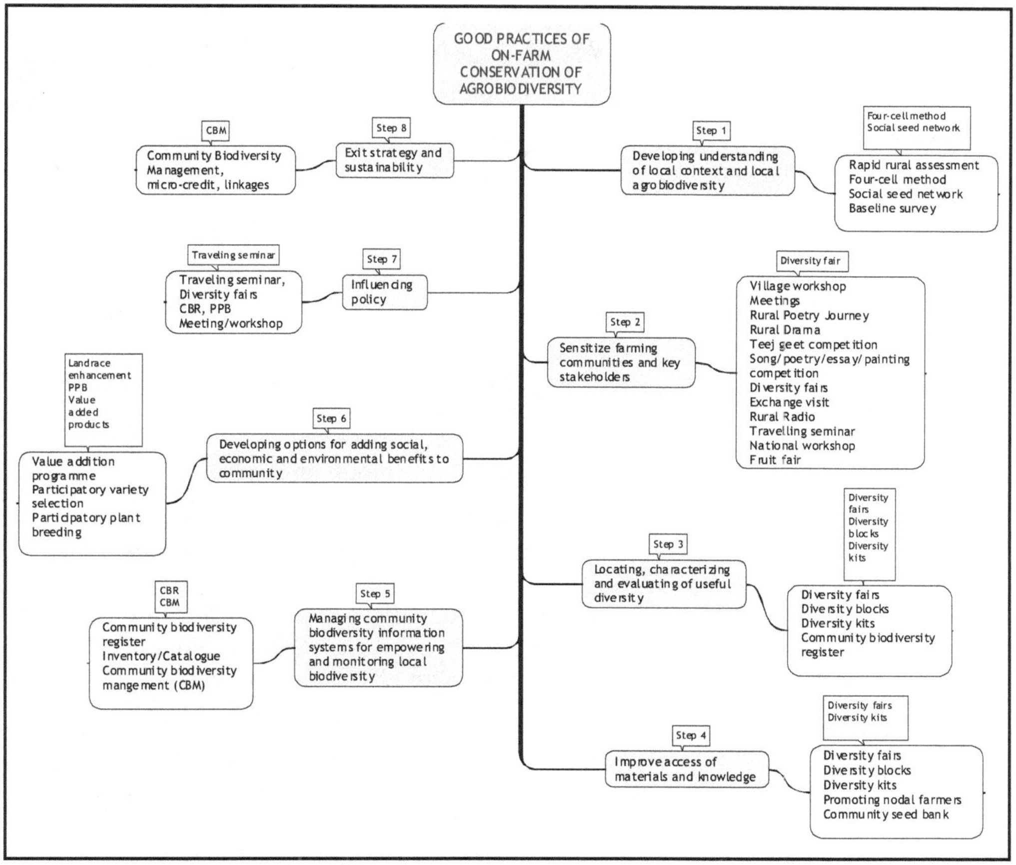


**Figure 1.** Good practices of on-farm conservation of agricultural biodiversity (Sthapit and Jarvis, 2006)

## Conclusion

Sustainable *in situ* conservation is possible only when farmers, communities, and national institutions perceive benefits in terms of social, economic, and environmental services. Once we understand that the farmer management of local crop diversity is primarily livelihood option for rural community, and then cost of *in situ* conservation is much cheaper than *ex situ*. In the process of farming, farmers not only derive social, economic and environmental benefits from local genetic resources but also the evolutionary potential of such genetic resources. It is also important to note that *in situ* conservation *per se* is not a panacea on its own as it is neither recommended as a universal practice nor a feasible method in all circumstances. It has a place and time, as *in situ* conservation can be transient and subject to change over time and that provides the major link with *ex situ* conservation.

## References

Altieri, M. A. and Merrick, L.C. 1987. *In situ* conservation of crop genetic resources through maintenance of traditional farming systems. *Economic Botany*, 41:86-96.

Bellon M. R. 1996. The dynamics of crop intra-specific diversity: A conceptual framework at the farmer level. *Economic Botany*, 50(1): 29-36.

Brush, S.B. 1995. *In situ* Conservation of landraces in Centers of crop diversity. *Crop Science*. 35: 346-354.

Brush S.B. 2000. Genes in the Field: On-Farm Conservation of Crop Diversity. IDRC, Canada/IPGRI, Rome, pp. 3-26.

FAO, 1998. The state of the world's plant genetic resources for food and agriculture. FAO, Rome, 510pp.

Gauchan, D, Sthapit B.R. and Jarvis D.I. 2003. Agrobiodiversity conservation *in situ*: Nepal's contribution to a scientific basis for policy recommendations. IPGRI, Rome.

IPGRI, 2001. On farm management of crop genetic diversity and the convention on biological diversity's programme of work on agricultural biodiversity. A synthesis paper prepared by the International Plant Genetic Resources Institute, Rome, Italy for Subsidiary Body on Scientific, Technical and Technological Advice of Convention on Biological Diversity, Montreal, 12-16 November 2001.

Jarvis D.I., and Hodgkin T. 2000. Farmer decision making and genetic diversity: linking multidisciplinary research to implementation on farm. Chapter 11. Pp. 261-278 (in) Genes in the field con-farm conservation of crop diversity (S.B. Brush, ed), IPGRI, IDRC and Lewis publishers, Washington DC.

Jarvis, D.I., Myer L., Klemick H., Guarino L., Smale S., Brown A.H.D., Sadiki M., Sthapit B.R. and Hodgkin T. 2000. A training guide for *in situ* conservation on-farm. Version I. IPGRI, Rome.

Jarvis D.I., Zoes V, Nares D and Hodgkin T. 2004. On-farm management of crop genetic diversity and the Convention on Biological Diversity programme. *Plant Genetic Resources Newsletter* 138: 5-17.

Maxted N, Guarino L, Myer L, and Chiwona E.A. 2002. Towards a methodology for on-farm conservation of plant genetic resources. *Genetic Resources and Crop Evolution* 49:31-46.

Smale M, Bellon M, Jarvis D and Sthapit B. 2004. Economic concepts for designing policies to conserve crop genetic resources on-farms. *Genetic Resources and Crop Evolution*, 51:121-135.

Sthapit B.R. and Jarvis D.I. 2006. Good practices of community-based on-farm management of agricultural biodiversity in Nepal: lessons learned. Pp. 91-111 (in) Sthapit BR, Shrestha, PK, Upadhyay, MP and Jarvis, D (eds) 2005. On-farm conservation of agricultural biodiversity in Nepal. Vol 2: Managing diversity and promoting its benefits. Proc. of the Second National Workshop, 25-27 August 2004, Nagarkot, Nepal

Subedi A, Chaudhary P, Baniya B, Rana R, Tiwari R.K., Rijal D.K., Jarvis D. and Sthapit B. 2003. Who maintains crop genetic diversity and how: Implications for on-farm conservation and utilization. *Culture & Agriculture,* 25(2):41-50.

UNCED 1992. Biodiversity Convention. United Nations Conference on Environment and Development (UNCED), Geneva.

# Underutilized Fruits

**Underutilized and Underexploited Horticultural Crops, Vol. 2, 2007**
*K.V. Peter (ed.), pp. 11-54*
New India Publishing Agency, New Delhi (India)
E-mail : newindiapublishingagency@gmail.com
Web: www.bookfactoryindia.com

# 2 Underutilized Fruit Crops of Andaman and Nicobar Islands

**D.R. SINGH**
Senior Scientist, Division of Horticulture and Forestry
Central Agricultural Research Institute, Port Blair - 744 101

The Andaman and Nicobar group of Islands, popularly known as the Bay islands, consists of 572 islands, islets and rocks, situated off the Eastern coast of India. The Islands are covered from coast to coast with an almost impenetrable growth of luxuriant tropical rain forests. The luxuriant growth of dense forests with tangled mass of climbers, lianes, canes and plant communities provide an additional source of income. Many species of bamboos, bushes, shrubs and stunted trees have tremendous values (Mondal, *et al* 2000). Most of the diversity is still collected from wild forest habitats largely composed of trees, shrubs and other perennial vegetations. Underutilized fruits provide food, nutrition and sustenance to the natives. Broadly, such underutilized diversity in edible fruits falls into four categories (Groot *et al* 1993).

1. Cultivated locally but are underutilized.
2. Commonly cultivated for which new uses can be found.
3. Cultivated in a country or region but their values are proven elsewhere under similar climatic conditions.
4. Those which are harvested from the wild.

Potential of a large number of world's fruits still remains to be explored and utilized. This situation exists largely because of very little research and developmental efforts made in the difficult and remote tropical and temperate habitats where most of such fruits occur. Lack of awareness about their potential uses and low and erratic bearing of most underutilized species have also contributed to their neglect. (Meenu Srivastava, 2004).

Recent years witnessed considerable improvement and attention on production of underutilized fruits. As alternate crops, these fruit crops help in diversification besides meeting multifarious human needs. To ensure sustainable utilization of these known resources, efforts are needed to conserve valuable germplasms and also to standardize utilization and processing technologies for their produces.

Andaman and Nicobar Islands, being rich in plant diversity, have a very large number of non traditional horticultural crops but information on nutritive values of these fruit crops are not available (Patel *et al* 2006). Apart from nutritive value, these fruit crops are particularly more important for medicinal properties. These fruits possess both restorative and curative principles. These could be aromatic, cooling, digestive, stomachic, stimulant, useful in seasoning, maturation and fermentation of culinary, processed foods and drinks. A few others have chemical principles responsible for their astringent, alternative antihelmintic, anodyne, emenagogue and emollient properties. In addition to their therapeutical values, they provide nutrition, strength and vigour to our body and restore loss of minerals and amino acids and thus protect against many deficiencies and diseases. These fruits provide not only nutritious food but also stable income to the tribals living in the forests and degraded wastelands and to the inhabitants of scarcity areas. Quite often, they serve as famine insurance during scarcity years by providing fruits, fire wood, fence materials, leaf fodder etc. Since many of these fruits are nutritionally rich and of great medicinal value, considerable knowledge on edible wild and unexploited fruits lie scattered. Indigenous and local knowledge on various usages of several useful, untapped fruits occurring wild, semi wild and cultivated in backyards in Andaman and Nicobar Islands were gathered from tribals and farmers living in the interior Islands and forests and from intermittent reports and random information by various workers (Sreekumar *et al.*, 1996) of which wild fruit species offer great potential for crop improvement breeding and selection. In Andaman and Nicobar Islands , marginal and wastelands of different kinds viz. sand dunes, ravines, acidic and marshy soils, which are not suitable for cultivation of high input demanding fruit crops,exist either due to poor quality of soil or lack of water resources. Such lands are suitable for growing low input demanding underutilized fruit crops to diversify pressure on present day agriculture, which is so inevitable in view of increasing population pressure and fast depletion of natural resources and changing human needs (Patel *et al.*, 2006).

## Underutilised and Underexploited Fruits

### Noni

*Morinda citrifolia*, **Common Name** - Noni, **Family** - Rubiaceae. is among the original "canoe plants" that Hawaii's Polynesian colonizers brought with them in their voyaging canoes. The voyagers valued the plant for its medicine and dyes.

#### *Habitat and Distribution*

Found wild in coastal areas, wastelands and fences all over South Andaman. The strong aroma of fruit is off-putting and, for this reason, the plant is not cultivated widely in gardens. It is however, useful for coastal erosion control. It is the best suited to tropical areas.

#### *Botanical Description*

*M.citrifolia* is a large shrub to medium tree varying from between 3 to 12 metres high. It has oval shaped leaves to about 300 mm long by 150 mm wide. The white flowers occur in the leaf axils in clusters mainly in summer and autumn. They are followed by succulent fruits, which fuse into a large compound structure as they ripen. The fruits are edible but have a very pungent aroma when ripe, apparently to attract fruit bats which are the dispersal agents for seeds. Each fruit contains about 150 seeds, brown in colour (Nelson, 2001).

#### *Propagation*

By seeds, grafting and hard wood and soft wood cuttings. (Singh *et al*, 2006).

#### *Uses*

The fruijuice has a range of medicinal properties. In fact, a small industry has built up around the fruit which is being marketed as "Indian Noni". The juice is high in vitamin C and there is high demand for it as an alternative medicine against a host of illnesses like arthritis, diabetes, high blood pressure, muscle aches and pains, menstrual difficulties, headaches both mild and severe, heart diseases, AIDS, cancers, gastric ulcers, sprains, mental depression, senility, poor digestion, atherosclerosis, blood vessel problems, drug addiction and more. Leaves are also eaten as vegetable and the fruits have antiseptic and antibiotic properties. Scientific evidence on the benefits of juice is limited but there is anecdotal evidence for successful treatment of colds and influenza. (Abbott, 1985 and Elkins, 1997).

## Alligator Apple

*Anona glabra* L.; **Common Name-** Jungli Sitaphal; **Family-** Anonaceae

### *Habitat and Distribution*

Found wild in South Andaman, opposite to Murugan temple, on east bank of Dilthaman tank, in swamps.

### *Botanical Description*

Small spreading tree, 3-8m. high. Bark cracked brown with pale lenticels. Roots with thick corky pith. Leaves ovate-oblong, obtuse or acuminate at apex, 6-20 x 3-8 cm, glabrous, fragrant, petioles 1-3 cm long or conical, 6-10 x 5-8 cm, smooth or very faintly reticulate towards base, glabrous, greenish-yellow with a creamy yellow, sweet scented pulp (smelling like pineapple or a semi ripened jack fruit), seeds numerous (counted up to 180/ fruit) and glabrous. Flowering/ fruiting is from July-Sept.

### *Propagation*

By smooth seeds, oblong or elliptic-long, sandal wood coloured, 1.8 x 0.8 cm rounded at both ends.

### *Uses*

Salt resistant, can be grown with associated mangrove plants like *Pandanus lerum*, suitable as root stock for custard apple, powder made out from different parts like seed, bark and leaf .It can be tried for snail control. It can be exploited for afforestation in Coastal areas. Pithy wood roots are used for making floats and corks.

## Custard Apple

*Anona squamosa* L.; **Common Name-** Sita Phal; **Family-** Anonaceae

### *Habitat and Distribution*

Custard apple is American in origin and was introduced to India by Portuguese. The species is maintained in several parts of India. In Andaman and Nicobar Islands, the custard Apple was introduced by settlers and at present is grown in backyards and also as semi wild in forests.

### *Botanical Description*

A woody shrub or small tree with fairly stout trunk, reaching a height of 5-6m, the branches bear alternate leaves which are 6-12cm long, oblong lanceolate, glabrous and pubescent beneath, pellucid-punetate. Flowers are solitary and greenish. Flowers are carried singly or in small groups and occupy an extra axillary position. They are greenish yellow, comprising three fused sepals, six petals, of which outer whorl of threads are about 2.5cm. long and inner and very much reduced and almost scale like. There are numerous stamens made up of fleshy filaments with two longitudinally splitting pollen sacs on outer surface and numerous fairly hairy sepals, which are quite free from each other in the centre. The fruit is formed by upgrowth of the receptacle and development of carpels, which fuse together into fleshy mass of white pulp. In the white pulp, black seeds are embedded or buried. Surface of fruit is covered by scale like structures and carpels retain their individuality in custard apple.

### *Propagation*

By seeds. Vegetative propagation gains popularity now. Stem cuttings, inarching and budding are common methods.

### *Uses*

As it is naturally cross pollinated and most of the plantations of this species are of seedling origin and produce fruits of variable quality, it offers ample scope for selection. The heterozygocity and nature of cross pollination bring wide variations in the seedling progenies. (Hiremath *et al,* 2006) .Almost entirely eaten as dessert fruit, although pulp of fruit may be mixed with milk to form a drink or made into ice-cream. The fruit is a good source of proteins, carbohydrates with 12.0 - 18.5% sugar and 0.26 - 0.65% acidity. Custard apple leaves have medicinal properties. Oil is extracted from custard apple seeds for making soap and the cake is used as manures.

## Sour Sop

*Anona muricata* L.; **Common Name**- Sour Sop; **Family**- Anonaceae

### *Habitat and Distribution*

It is American in origin and was introduced to India by Portuguese. The species is being maintained in several parts of India. In Andaman and Nicobar Islands, the Sour Sop was introduced by settlers and at present is grown in backyards and also as semi wild in forests.

### *Botanical Description*

Sour Sop is a small evergreen tree usually growing up to 5-8m height. Leaves are small oval shaped, which are leathery, glossy and have dark green colour. They are 1.5 - 2.5cm in length and have characteristic pungent odour when crushed. Flowers are large, hermaphrodite, thick and fleshy and grow on small axillary stems or often directly on trunk of tree. The fruit is syncarpous formed by fusion of pistils and receptacle into a large fleshy structure. Average fruit weight is 700g. Sour sop is irregular in shape but most frequently the fruit is ovoid or heart shaped. Length of fruit is 15-30cm. The skin is thick, leathery and very dark green with many short curved fleshy spines. The mesocarp contains numerous shiny dark brown seeds, ovoid in shape and measuring about 2 cm in length and 1 cm in breadth. The white cottony pulp is very juicy. Due to its distinctive aroma, it is highly relished. The sour sop tree bears fruit continuously after 3 to 4 years of age. The main fruiting season is June to August.

### *Propagation*

By seeds and by budding/ grafting.

### *Uses*

Sour sop can be consumed as a desert. However, it is more often consumed after extraction of pulp. It can be made into fruit jelly or used in preparation of beverages, sherbets, ice creams and syrups. A number of medicinal properties are attributed to leaves and juice of Sour Sop. It is a rich source of vitamin B complex.

## Ramphall

*Anona reticulata* L.; **Common Name-** Ramphall/ Bullock Heart; **Family-** Anonaceae

### *Habitat and Distribution*

Is American in origin and introduced to India by Portuguese. The species is maintained in several parts of India. In Andaman and Nicobar Islands, the Ramphall was introduced by settlers and at present is grown in backyards and also in semi wild forests.

### *Botanical Description*

A small tree, leaves 12-24cm long, glabrous, lateral nerves 15-18 pairing. Flowers are yellow green. Heart shaped fruit is slightly larger than *A.squamosa* and carpels are fused to form a rind marked with hexagonal areoles. It has the advantage of containing a few seeds but the pulp is less delicately flavoured. The sugar content is slightly less than the custard apple. Less resistant to cold than custard apple.

### *Propagation*

By seeds and budding/ grafting.

### *Uses*

Can be consumed as a desert. It is more often consumed after extraction of pulp. It can be made into fruit jelly or used in preparation of beverages, sherbets, ice creams and syrups. A number of medicinal properties are attributed to leaves and juice.

## Karonda

*Carissa carandus* L.; **Common Name**- Karonda; **Family**- Apocynaceae

### *Habitat and Distribution*

Native of India spread in tropical Asia, Sri Lanka where it is sparingly cultivated. In Andamans, it is introduced by settlers and now spread to many Islands.

### *Botanical Description*

A shrub 1.25 -2.5m high with sharp spines, straight on stem and curved on branches. Leaves 3-6cm long, ovate, blunt with a micro, glabrous with rather indistinct lateral nerves, petioles less than one cm long. Flowers pink to white, fragrant, cory biform cyme, corolla tube nearly 1.5cm long. Berry purple 3cm long. Fruits available in June.

### *Propagation*

By seeds and vegetatively by layering.

### *Uses*

Fruits are eaten raw, pickled and preserved, As it is a rich source of iron and carbohydrates, the raw and ripe fruits are used to prepare preserved products like jam, jelly, squash and pickle. (Salvi *et al,* 2006).

## Carambola

*Averrhoa carambola* Linn; **Common Name**- Kamarakali; **Family**- Oxalidaceae.

### *Habitat and Distribution*

Occurs well in Indonesia and is now widely spread throughout tropics. In India, it is distributed and cultivated in Assam, West Bengal, foothills of

Tamil Nadu and other tropical areas. In Andaman and Nicobar Islands, it was introduced and planted in the settlement areas. It adapted exceedingly well in the Islands. Both forms with sweet and acid fruits occur.

### *Botanical Description*

The tree is pyramidal in shape when young but mature trees have generally rounded top with an open canopy growing up to a height of 10m. Leaves are imparipinnate, alternate, exstipulate, leaflets 5-11 and irritable to touch. Flowers are borne in short cymes chiefly axillary. Carambola fruits are technically a fleshy berry, which is five angled and star shaped in cross section. Fruits are yellow when ripe, tip grooved, seed ovoid, compressed, light brown and one cm long.

### *Propagation*

By seeds or propagated vegetatively.

### *Uses*

Fruits are used in salads and for making tarts, preserves and drinks. They may also be used for cleaning of brasswares. Dried fruits have cooling and anti scorbic effect. Dried fruit is employed against fever. Ripe fruits are used in dysentery. Ripe fruit is a remedy for bleeding piles and relieves thirst. It contains 0.7% acidity, 9.99% total sugar, 12. mg/100g ascorbic acid and 0.68% proteins.

## Bilimbi

*Averrhoa bilimbi* L.; **Common Name**- Bilimbi; **Family**- Oxalidaceae.

### *Habitat and Distribution*

Bilimbi is native to monsoon regions of Malaysia. It is now widely introduced and distributed in the tropics. Cultivated in gardens throughout plains of India and occurs as escape in tropical parts of India. In Andaman and Nicobar Islands, it was introduced and planted in settlement areas.

### *Botanical Description*

Bilimbi is a small tree up to 15m in height, leaves imparpinnate; leaflets 10-12 pairs, oblong, acute, 5-10 cm long; flowers dark red, 1-2cm long, from trunk and older branches; fruits resembling small cucumbers, cylindrical, green; seeds small, embedded in pulp, about 8 mm long.

### Propagation

By seeds or propagated vegetatively.

### Uses

Fruits astringent, stomachic, refrigerant. The very acidic fruits are used for making pickles, curries and preserves. It is used in the forms of curry and also useful in piles and scurvy.

## Wild Mango

*Mangifera andamanica* Kind;**Common Name**-Wild Mango; **Family**-Anacardiaceae.

### Habitant and Distribution

Endemic and rare, distributed in Andaman forests. Occurs wild in South Andaman, Chouldhari and Chidyatappu.

### Botanical Description

The tree is of upright growth having 30-40m. height. Leaves are 9-15cm long, obovate to broadly oblanceolate or eliptic, lateral nerves 10-12 pairs, curving upwards; petioles 1.5-3 cm long, channelled, flowers one cm in diameter; petals twice as long as the lanceolate sepals and drupes 3-4.5cm long. Fruits small, yellowish orange in colour and ripen in April. Fruits are free from insects, pests and diseases.

### Propagation

By seeds (stones) and by grafting on self-grown seedlings of *M.indica*.

### Uses

Edible fruits are good sources of vitamins A and C, which are consumed locally. Pickles and other preserves are being prepared from fruits.It is a good genetic stock for improvement of mango varieties.

## Wild Mango

*Mangifera griffithi*. **Common Name**- Jangli Aam; **Family-** Anacardiaceae

### Habitat and Distribution

Found growing wild in Mount Harriet (South Andaman). It is a wild species not recorded earlier from Indian sub-continent.

### *Botanical Description*

It is a small tree, 6-8m height, with greyish brown dark and dense green canopy. It is allied to *M.lagenifera* Griff, having a thin leaf stalk (± one mm) and prominently secondary veins along lower surface, besides being long. Red crimson coloured fruit is about 16g. in weight. Skin of fruit is thin.

### *Propagation*

By stones grasting and grafting on self seedlings and that of *M.indica.*

### *Uses*

The fruit is pleasant in taste and is quite rich in Vitamins C and A. Occurrence of this species having edible fruit has considerable botanical and horticultural significance.

## Wild Mango

*Mangifera camptosperma* Pier; **Common Name**- Jangli Aam; Family-Anacardiaceae.

### *Habitat and Distribution*

Not common, found in middle Andaman, long Island, Baratang, South Andaman and Rutland.

### *Botanical Description*

A large tree (50-100 x 2-4m) with greyish bark cut streaked dark reddish brown, with a little milky juice. Leaves 15-30cm long, lanceolate to oblong lanceolate, shortly acuminate, lateral nerves 16-24 pairs. Petiole is 3-9cm long, slender and thickened at base. Flowers white, in ample terminal panicle. Drupe as large as mango, much compressed and yellow when ripe.

### *Propagation*

By stones grasting and grafting on self seedlings and that of *M.indica.*

### *Uses*

The fruit is locally consumed. The fruit has average nutritional value. Wood is used for making houses.

## Wild Amra

*Dracontomelum dao* ; **Common Name**- Wild Amra; **Family-** Anacardiaceae

### Habitat and Distribution

Uncommon, originated from Indo-Malayan region. In Andaman and Nicobar, it is found in damp places along streams of South Andamans.

### Botanical Description

A tall handsome and conspicuous tree (50 – 100 x 5-10m) with a brownish grey bark cut reddish. Leaves usually imparipinnate 3-6cm long, leaflets 12-20, alternate to sub-opposite, the lowest 3-9cm long. The upper ones 21-24cm long, ovate oblong, acuminate oblique with tufts of hairs in the nerve axils beneath. Flowers greenish white 1.5cm in diameter in large axillary paniele, stamens 10, Drupe 3-4.5cm in diameter, depressed globose, yellow when ripe. Fruit ripens in June-July.

### Propagation

By seeds and cuttings.

### Uses

Acidic fruit is edible, generally used for making marmalade and chutney. Also used as vegetable and is a good source of Vitamin C.

## Amra

*Spondias cytherea* Sonner; **Common Name**- Amra/ Hog plum; **Family**-Anacardiaceae.

### Habitat and Distribution

Native of Polynesia. Distributed in tropical Asia; South Sea Islands, Fiji, Macarena and Pacific region. In India, it was introduced to Indian Botanical Garden; Calcutta .It is not known to occur elsewhere in India, either cultivated or wild. Occurrence of species *S.cytherea* in Andaman and Nicobar Islands was recorded. It appears possible that this taxon might have reached these Islands from adjoinng countries *viz.*, Malaysia and Jawa. In Andaman and Nicobar Islands, it is usually cultivated and also occurs in wild.

### Botanical Description

Erect spreading tree 18m in height. Leaves are imparipinnate, 20-30cm long; leaflets 13 or 15, elliptic or ovate, oblong. Panicles 20-30cm long; flowers polygamous, white, drupe ovoid, 8cm long; endocarp stony, fibrous with 5 main large spines, curving downwards and numerous spinose processes arising from fine lobes, seeds usually one, rarely two, remaining abortive, Flowers in March and fruiting is in Aug. – Sep.

### *Propagation*

By seeds and cuttings

### *Uses*

The edible fruit possesses a pineapple or mango like flavour. Unripe fruits are used for making pickles, jellies and marmalades and also as flavouring agent in fish and vegetable curries.

## Laseda

*Cordia oblique* Wild; **Common Name**- Laseda; **Family**- Boraginaceae.

### *Habitat and Distribution*

Native of tropical Asia and Australia, common throughout India, wild and often planted. In Andaman and Nicobar Islands, fairly common along sea shores, roadsides and usually passed unnoticed.

### *Botanical Description*

A small medium sized deciduous tree (15-30 x 2-3m) with a cracked bark, cut plate pinkish, streaked. Leaves up to 25cm long and 15cm broad, broadly ovate or apiculate, base rounded or sub cordate, lateral nerves 4-7 pairs, prominent, the lowest somewhat basal; petiole up to 10cm long, flowers handsome and conspicuous, 4cm diameter, funnel shaped with crumpled lobes, bright orange, stamens 6. Fruit 3cm diameter, hard, globose. Flowers and fruits available from March to July. Fruit round, 1-2 cm long, persistent calyx and style, light brown, surface smooth, one seeded, fruits have sweet smell.

### *Propagation*

By seeds and cuttings.

### *Uses*

Fruit astringent, anthelmintic, diuretic, demulcent, expectorant, and useful in infection in urinary passages, diseases of lungs and spleen. Kernel used as remedy for ringworms.

## Badam

*Terminalla catappa* L.; **Common Name**- Bengal Almond; **Family**- Combretaceae.

### Habitat and Distribution

Frequent along seashores of all the islands and never found naturally growing in the interior. Often planted and common in Ross Island.

### Botanical Description

A large tree (60 – 80 x 6-10m) some what resembling *Terminalia procerai.* Leaves 25-40cm long, 15-25cm broad, obovate, tapering to the small cordate or auricled base, apex rounded or apiculate. Flowers white star shaped. Fruits 3-6cm long, compressed and keeled all around. Flowering profusely in March - April.

### Propagation

By seeds.

### Uses

Kernels of the seeds are edible. The bark gives tannins.

## Chalta or Elephant Apple

*Dillenia indica;* **Common Name-** Chalta; **Family-** Dilleniaceae.

### Habitat and Distribution

Locally called as Chalta, is indigenous to an area stretching from India to China. It is distributed in the Indo-Malayan region, Indo-China, Nepal and North Eastern parts of India. In Andaman, it is found semi wild in South Andaman.

### Botanical Description

Chalta is a deciduous tree of moderate size with a short but straight trunk and spreading branches forming a dense head. The narrow and pointed leaves are set on short stalks near the end of branches. The large scented flowers grow strongly among the leaves. The thick rounded sepals form an outer covering for the large, spherical fruit.

### Propagation

By seeds.

### Uses

The thickened sepals of fruits are used for culinary purpose and made into jam and jelly, the juice sweetened with sugar is a cooling drink. Fruits are consumed as tonic and taxative. Juice of fruits mixed with sugar and used as a cooling beverage against fevers and as a cough mixture.

## Sapida

*Baccaurea ramiflora* Lour; **Common Name-** Sapida, Khattaphall; **Family-** Euphorbiaceae.

### *Habitat and Distribution*

This fruit is cultivated in South East Asia. It is found wild in sub-tropical Himalayan tracts, Assam, Myanmar and Andaman and Nicobar Islands. In Andaman and Nicobar islands, it is fairly frequent in evergreen forests, on hills along streams and in other localities.

### *Botanical Description*

A medium sized evergreen tree, locally known as Khatta Phall, grows 15m tall, flowers during March – April, while fruits are harvested during July- Aug. The tree adapts a wide range of soil and climatic conditions. Leaves are elliptic – oblong, lanceolate, acuminate, membranous and glabrous. Flowers are dioecious, apetalous, with shorter pedicles in densely fasciled racemes arising from old wood. Fruits are globose, yellowish brown, and endocarp non-separable. Seeds are orbicular in rose colour pulp (arilus).

### *Propagation*

By seeds.

### *Uses*

Fruit is highly nutritive and a very rich source of vitamin C and protein. Iron contents are also very high. A delicious juice is prepared from the fruit. Seeds are used against vomiting and asthmatic trouble. Every part of plant is equally useful in the antidotal treatment of snakebite or a scorpion sting.

## Mangosteen

***Garcinia mangostana* ; Common name-** Mangosteen**; Family-**Guttiferae.

### *Habitat and Distribution*

It is a native of Malaysia and has seldom been successfully established out side this area. In India, it is being grown in Kerala, foothills of Tamil Nadu, Karnataka and Assam. In Andamans, it is being introduced long back by Britishers and also by early settlers. Now it grows semi wild in Haddo and also in backyard gardens.

### *Botanical Description*

A slow growing glabrous, evergreen tree, 15m in height, with yellow latex in all its parts; leaves dark green, leathery, opposite, short-stalked, elliptic oblong with acuminate tips, 15-25 x 6-10 cm; flowers unisexual dioecious. Flowers borne terminally on branchlets, 5 - 6cm in dia; sepals 4 in 2 pairs, inner pair reddish; petals 4, yellowish, red edged, falling early; ovary 4-8 celled, stigma sessile with as many lobes as cells of ovary; fruits subglobose berry, usually parthenocarpic, 4 - 7cm in dia, with persistent calyx and stigma lobes; pericarp purple, tough, thick, with yellowish resin; 4 - 8 translucent white fleshy, segments with very delicate flavour; seeds 0 - 3, formed from nucellar tissue, about 2 cm long.

### *Propagation*

Mangosteen is difficult to establish as seeds have a low and short viability and must be planted within a few days after removal from fruits. Vegetative propagation is seldom successful.

### *Uses*

They are eaten fresh and are the most delicious of all tropical fruits.

## West Indian Cherry

*Malphigia glabra* L.; **Common Name**- West Indian Cherry; **Family**- Malphigiaceae.

### *Habitat and Distribution*

Native of West Indies and Northern South America to Southern Texas. It was further introduced to tropical and sub tropical areas like India. In Andaman and Nicobar islands, it was introduced recently from Kerala and Tamil Nadu.

### *Botanical Description*

A dense vigorously spreading evergreen shrub, 2-5 m in height is also valuable as ornamental bush. Glabrous; leaves opposite, shiny, dark green above, ovate to elliptic, 2.5 - 7.5cm. long, shortly stalked; flowers 1-2 cm in dia, in axillary cymes, hermaphrodite, calyx with 6-10 large sessile glands; petals 5, fringed, pink or red; stamens 10; fruit bright red, juicy drupe, depressed, ovoid, obscurely 3 lobed, 1-3 cm in dia. Borne 1-3 in axils on short pedicle; seeds 3, triangular ridged.

### *Propagation*

By seeds and vegetatively by cuttings and grafting.

### *Uses*

Fruit is eaten raw and is one of the richest sources of ascorbic acid (6 – 8 times more than the recommended dose). The acidic fruits are made into preserves like jam, jellies and candies. Juice of the fruit is used commercially to enrich other fruits, which are low in vitamin C. Since unripe fruits are equally rich in Vitamin C, they can be used for making other products like pickles.

## Jack Fruit

*Artocarpus heterophyllus* Lam.; **Common Name**- Kathal; **Family**- Moraceae.

### *Habitat and Distribution*

Jackfruit is of very ancient cultivation in India. It was taken early by the Arabs to the east African coast. It has now spread throughout the tropics including Andaman and Nicobar Islands.

### *Botanical Description*

An evergreen monoecious tree, 15-20m tall. The delicate tap-root of the seedling is easily damaged so they are difficult to be transplanted. White resinous latex is found in all parts. Twigs glabrous, stipules ovate-triangular, 2-7 cm long, leaves spirally arranged, phyllotaxis 2/3, usually glabrous, sapling leaves have 1-2 pair lobes, adult leaves entire, coriaceous, dark green and shiny above, pale green beneath; petiole 2-4 cm long; lamina 5-20 x 3-12cm, elliptic to obovate, bluntly tipped, base cuneate, pinntinerved with 5-8 pairs veins. Male inflorescence axillary on leaf twigs or from trunk and branches, but above female inflorescence, drooping, oblong, ellipsoldal or clavate, 5-10 x 2-3 cm, green; flowers numerous, densely crowded, minute, with 2 lobed calyx and single stamen. Female inflorescence borne on short stout twigs on the main trunk and branches; peduncle stout with green fleshy ring at apex, large than male inflorescence; flowers small, calyx tubular, ovary oblong and compressed, style terminal and oblique, stigma elevate. Fruit a gigantic syncarp, the biggest of all cultivated fruits, 30-90 x 20-50cm; barrel or pear shaped, glabrous, with short sharp hexagonal fleshy spines, rind yellow, flesh yellow or pear shaped, glabrous, with short sharp hexagonal fleshy spines, rind yellow, flesh yellow and waxy. Seeds with thick gelatinous yellow covering, large, oblong and about 3 x 2cm.The ripe male inflorescences have sweet scent of honey and burnt sugar attracts small flies and beetles which bring about pollination.

### *Propagation*

By seed. As the seedlings are very variable, it is desirable that selected clones should be propagated vegetatively, which can be done by inarching or grafting.

### *Uses*

Pulp of the fruit is eaten fresh or preserved in syrup. It is also dried. The juicy flesh round the seeds has a pungent odour and is sweet or acid and is eaten sometimes in salads. The large seeds are boiled or roasted and haves a chestnut flavour and are also ingredients in many culinary preparations in India. The unripe fruits are used as a vegetable or in soups and are made into pickles. The rind and leaves are fed to livestock. The latex is used to mend earthenware and other utensils. Yellow hard wood is a valuable timber and also yields a yellow dye.

## Monkey Jack/Barhal

*Artocarpus lakoocha* Roxb.; **Common Name**- Barhal; **Family**- Moraceae.

### *Habitat and Distribution*

In Andaman and Nicobar Islands, it grows wild in many localities especially Nicobar Islands.

### *Botanical Description*

A moderate sized deciduous tree (30 - 60 x 3-6m) with a dark coloured bark, leaves 25-40cm long and 12-24cm broad, broadly ovate or oblong, apiculate, base rounded and unequal, coriaceous with 14-16 pairs, having strong lateral nerves, the lower surface clothed with soft velvety green or rusty pubescence; petiole strong, 2-4cm long, globular, greenish 6-12cm long and velvety when ripe. Flowering in November. - February, and fruiting May - July. Its large pubescent leaves have a tendency to fold inwards.

### *Propagation*

By seeds and grafting and budding on seedlings.

### *Uses*

The fruit and pulp are eaten as such.

### Anjeer/Fig

*Ficus carica* L.; **Common Name**- Anjeer/ Gular; **Family**- Moraceae

#### *Habitat and Distribution*

A native of Asia Minor, it spread easily to Mediterranean region. It did not do well in low wet tropics but can be grown at higher elevations and areas of low rainfall. Early settlers and people coming from diverse geographical zones introduced it to Andamans.

#### *Botanical Description*

It is a large shrub or low growing deciduous tree growing to 10m height with large leaves, deeply 3-5 lobed, conspicuously palmately veined, and 10-20cm long. Pear shaped fruit is variable in size and colour. No staminate flowers are present and seedless fruits develop parthenocarpicaly.

#### *Propagation*

By cuttings.

### Wild Jamun

*Syzygium claviflorum* Roxb.; **Common Name**- Jangli jamun; **Family**- Myrtaceae.

#### *Habitat and Distribution*

Uncommon, found in evergreen forests and often bordering mangrove swamps. Found in Baratang, South Andaman and Navy Bay.

#### *Botanical Description*

A moderate sized tree (30-50 x 2-6m) with a whitish grey bark. Leaves are 12-18cm long, 4-7cm broad, elliptic lanceolate, acuminate. Flowers handsome, white, 1.5-2.0 cm dia. Calyx about 2.5 cm long, slender trumpets shaped. Fruit 2cm long, barrel shaped, white and turning scarlet to purple. Flowers in Jan – Feb. and fruiting in March. Visited by honeybees when in blossom.

#### *Propagation*

By seeds and cuttings.

#### *Uses*

Fruits are edible, good source of vitamins 'A' and 'C'. Seeds are used as medicines and contain elegiac acid. Fruits are used as astringent in bilious diarrhea and in diabetes. Fermented fruit juice is stomachic, carminative and diuretic. Nicobarese women consume fruits during pregnancy to improve the haemoglobin percentage.

## Jambose

*Syzygium jambolana* Lam.; **Common Name-** Jambose; **Family**- Myrtaceae.

### *Habitat and Distribution*

Usually found in South and Middle Andaman; Barren Island.

### *Botanical Description*

A deciduous tree (40-80 x 5-8m) with pale grey bark, reddish when cut. Leaves 10-15cm long, elliptic to oblong, bluntish acuminate, marginal vein distinct; petiole 2cm long, slender. Flowers 2 to 3 cm diameter, creamy yellow, in small panicles from base of previous year's fallen leaves. Petals 4, orbicular. Fruit 1.5-4cm long, ovoid-oblong, black or purplish and juicy when ripe. Size of seed is smaller than seeds of *Syzygium cuminii*. Fruits ripen in rainy season.

### *Propagation*

By seeds and cuttings.

### *Uses*

Fruit is edible, juicy. The ripe juicy fruit is stomachic, carminative and diuretic. Fruits astringent and used in bilious diarrhaea and in diabetes.

## Wild Rose Apple

*Syzygium samarangense* (BI) Merr & Perry; **Common name**- Wild Rose Apple/ Jangli Jamun; **Family** -Myrtaceae.

### *Habitat and Distribution*

Frequent along the seashores and often found growing along streams. Found in North, Middle and South Andaman, mostly in Rutland.

### *Botanical Description*

A moderate sized shady tree (25-35 x 4-6m) with a dark brown bark, often with white patches on it, cut dark brown. Leaves 18-30cm long, 9-12cm broad, ovate to oblong-lanceolate, blunt acuminate, base rounded to almost cordate, lateral nerves 9-12 pairs, not very looping intermarginally; petiole short. Flowers white, 3-5cm in twos or threes in short axillary or terminal corymbose racemose. Fruit 3-5cm dia, depressed turbinate, white and glossy.

### *Propagation*

By seeds and by cuttings.

### *Uses*

Fruit is edible used in liver complaints.

## Rose Apple

*Syzygium aquem* (Burm.f.) Merr. and Perry.; **Common Name**- Watery Rose Apple; **Family**- Myrtaceae.

### *Habitat and Distribution*

Native of India. In Andaman and Nicobar Islands being cultivated in backyards and also found semi wild or scattered in forests of South Andaman.

### *Botanical Description*

A medium sized evergreen tree. All the parts are glabrous. Leaves opposite, flowers usually clustered in leaf axils; calyx-cup shaped, usually with 4-5 lobes, petals 4-5, often fall as a cup on opening of flowers, usually with fluffy appearance, ovary inferior, 2-3 celled; fruit a berry crowned by persistent calyx cup, usually with 1 seed with two large cotyledons. Fruit is star shaped, yellow, white or red in colour.

### *Propagation*

By seeds.

### *Uses*

Fruit is edible.

## Kewra

*Pandanus andamananansium; Pandanus lerum; Pandanus tectorius* **Common Name**-Kewra,; **Family-** Pandanaceae.

### *Habitat and Distribution*

Endemic to Andaman and Nicobar Islands and distributed in all localities of swampy areas of Andaman and Nicobar Islands.

### *Botanical Description*

Endemic tree commonly known as elephant tree, 9-15m high, purple stemmed or branched. Stem is smooth with short, sharp prickles, supported at the base by stilt like roots, leaves 5-6m long, 10-15cm broad, sword shaped and spinulous. Fruit scarlet when ripe.

### *Propagation*

By seeds and side suckers.

### *Uses*

Fruit is boiled and paste baked into thin flat cakes. This is powdered for storage and eaten as it is or used in preparing other dishes. Fruits are edible and widely used in folk medicines. Leaves are used for thatching and for making mats, sacks, baskets, umbrella and other articles. The plant is useful for binding soils and sometimes used for fencing. The terminal bud and tender floral bracts are also eaten.It is also used in folk medicines.

## Passion Fruit

*Passiflora edulis* Sims; Common Name- Passion Fruit/ Sharbat phall; Family- Passifloraceae.

### *Habitat and Distribution*

Is a native of Brazil, commonly cultivated in warm tropical regions. In Andaman and Nicobar, it was introduced from Kerala and foothills of Tamil Nadu, at present popular among local farmers as sherbet phall.

### *Botanical Description*

Yellow type passion fruit is a woody perennial climber, 15m long, stem green grooved, glabrous: tendrils axillary, robust, green, and longer than the leaves, spirally coiled. The fruit is as big as a tennis ball, about 6.0cm long. In Andamans, it flowers throughout the year, however main harvesting seasons are Sep.-Oct. and Jan-Feb. producing about 75 fruits/ vine/ year.

### *Propagation*

By cuttings and seeds.

### *Uses*

It can be eaten raw and its juice is a good source of vitamins A and C. The juice is used in flavouring ice-cream and making refreshing drinks and jams.

## Wild Ber

*Zizyphus glabra;* **Common Name-** Wild Ber; **Family-** Rhamnaceae.

### *Habitat and Distribution*

Frequently found in middle and South Andaman.

### *Botanical Description*

A scandant shrub with sharp recurved prickles. Leaves 7-15cm long and 3-6cm broad, ovate or ovate oblong, bluntly acuminate base one sided, crenate-serrulate, perfectly glabrous and glossy with 3 strong unbranched nerved. Flowers slender, pedicelled, greenish white in axillary cymes. Fruit bluntly ellipsoid, 5-20 cm long, yellow when ripe. Flowers in Jan – Feb and fruiting is in April.

### *Propagation*

By seeds and budding.

### *Uses*

Rich in vitamin 'C', purifies blood and aids digestion. Dried fruits are laxative and expectorant. Seeds are taken with buttermilk, useful in bilious disorder. Seeds mixed with oil are useful against rheumatism.

## Bael

*Aegle marmelos* (L.) Corr.; **Common Name-** Indian Bael; **Family-** Rutaceae.

### *Habitat and Distribution*

Is native to India, cultivated throughout South East Asia and East Asia and East Indian Archipelago and introduced in a few tropical countries. In Andaman and Nicobar Islands, it is found in semi cultivated/ semi wild condition along roads and near streams.

### *Botanical Description*

A thorny deciduous tree with dimorphic branches, trifoliate leaves and leaflets erenate, globose fruits, 5.0 – 17.5cm in dia., softer flesh, which becomes reddish and hard on drying.

### *Propagation*

By seeds and vegetatively by cuttings and air layering.

### Uses

Sweet yellow pulp is eaten fresh or dried and sherbet is made from it. Stomachic, used for controlling dysentery.

## Rambutan

*Nephelium lappaceum* L.; **Common Name**- Rambutan; **Family**-Sapindaceae.

### Habitat and Distribution

Tropical fruit, cultivated extensively in its native area i.e., Indonesia and Malaysia. In India, it is grown in Ooty and Kodaikanal hills, W.B., Assam and Kerala. In Andaman and Nicobar Islands, its introduction is in last 25 years, still not a popular fruit and limited to backyards only.

### Botanical Description

An evergreen, bushy, dioecious tree, 15-20m height. Leaves pinnate; leaflets 2-4 pairs, elliptic, flowers in axillary panicles, which superficially appear terminal, small greenish white, about 4 mm in dia; fruits pendent in clusters, red or occasionally yellow, 4-6 x 2-4 cm, seeds covered with whitish aril.

### Propagation

By seeds. Seedlings fruit in 5-6 years, but wherever possible, it should be propagated vegetatively by budding.

### Uses

Fruit is edible; aril is plumpy, sweet and juicy and is highly esteemed.

## Bakul

*Mimusops elengi* Roxb; **Common Name**- Bakul or Khaya; **Family**-Sapotaceae.

### Habitat and Distribution

Not common, found in evergreen and semi evergreen forests.

### Botanical Description

A large tree with dark grey bark fissured length wise and transversely, cut bark with pale irregular streaks. Leaves scattered, 10-15cm long and half as broad, elliptic, shortly and bluntly acuminate, base somewhat rounded, dark green and glossy, petioles 2cm long. Flowers about 2.5cm in diameter, fragrant, axillary or a few cm in cluster, stalks1.5-2.0cm long. Berry 3 cm long, ovoid and yellow when ripe. Flowers Feb. - Sept. and fruits ripe in Sep. - June.

### *Propagation*

By seeds and cuttings.

### *Uses*

Most tasty and delicious fruit, which is fairly rich in vitamins A and C containing comparatively good % of TSS (Total Soluble Solids).

### *Benefits in underexploited and underutilized Plant species through Technology Transfer*

The International Plant Genetic Resources Institute (IPGRI) is leading a major initiative in support of the use enhancement of these species in partnership with other stakeholders around the world, including local communities. Conservation of land races, value addition and transfer of technologies and human resources development (HRD) are essential elements of these efforts (Padulosi *et al,* 2005). As per one estimate there are more than 3000 edible fruit and nut species. Many have unique taste and are good source of minerals and vitamins. Several other worlds' fruits hold great promise and are yet to be recognized and exploited for their potential importance. (Jalikop, 2006). Very little has been done to exploit genetic resources of these unexploited fruits of these islands, essential in the face of severe genetic erosion and disappearance of several species as a result of activities related to urbanization and intensive agriculture. Some fruits are being raised on limited scale in home gardens, small farm or as backyard trees, and several fruits also come up naturally in forests from where the local people gather the fruits for consumption. Realizing the importance of these untapped fruits new technologies were transferred and in this way, with the adaptation of technologies the area under utilization with these fruits may be expanded. Technologies were transferred for the potential fruits, which are rich in vitamins and minerals, and have ample scope for commercial exploitation due to their therapeutic value and also for market preparations like Jam pickle, murabba etc. Soursop (*Anona muricata*); West Indian Cherry (*Malpighia glabra*); Chalta (*Dillenia indica*); Carambola (*Averrhoa carambola*); Bilimbi (*Averrhoa bilimbi) ;*Custard apple (*Anona squamosa*)( Singh *et al.*, 2004) and Wild mangoes (*Mangifera griffithi, M.camptosperma, M.andamanica*) (Medhi *et al*,2004,2005) Fruits like Alligator apple (*Annona glabra*) and *Morinda citrifolia* were used in

Tsunami affected land due to their saline tolerant capacity and also *Morinda citrifolia* popularly known as Noni has healing properties as well as adaptability to any type of soil. (Nelson, (2001)

### *Adaptability of potential underutilized fruitcrops*

A good market for several less known fruits, in fresh or processed form exists and a few exhibit potential for direct improvement. Underutilized fruits should also help in the utilization of the land and improving the degraded habitats, besides adding to income of resource poor farmers. Adaptive plasticity of plants and diversity in people's concept of taste provide unique opportunities for systematic exploration of these underutilized fruits. These fruits need widespread adaptive trials. Since 2000, more than 40-50 rare and new fruit species were collected from diverse indigenous and also exotic agro-climatic regions at CARI,Portblair. Different propagation techniques were used for different underutilized fruit crops. In general, all the underutilized fruit crops are propagated by seed. However, efforts are made for propagation by grafting as well as by cuttings and 50-80% success was found in most of crops. In some of the prioritized potential fruits like *Morinda citrifolia,* soft wood and hard wood cuttings were done by using different concentration of growth regulators and 100% success was found. (Singh, 2005). The Vit. C supplement , West Indian Cherry(*Malpighia glabra*) was successfully grown with fruiting upto 75% after pruning during pre monsoon (Singh 2004). Alligator apple (*Anona glabra)* was tested in laboratory of CARI for fatty acids like Lauric acid, Capric acid, Caprylic acid, Linolenic acid, Arachidic acid etc. It can also be used as a salt resistant root stock for grafting with other Anona species like *Anona squamosa* and *A. muricata.* (Singh, 2004) and also the powder made out from different parts like seed, bark, leaf etc. can be tried in keeping away the African Snail (*Achatina fulica*). (Shyam Prasad *et al* 2004).Bilimbi (*Averrhoa bilimbi*) is not an important product on world market as a fresh fruit, although it is processed into jellies, sauces, pickles and juices. The juice can be effectively used as an eco-friendly cleaner. Since, bilmbi fruits contain high content of oxalic acid (6%) it is not advisable for people having stone problem. ( Singh *et al*,2004).It was estimated that the oxalic acid content was higher in tender fruits (0.015%) when compared to seed (0.0025%). There was no significant difference in oxalic acid content in mature, ripe and ripe (boiled) fruits (0.0112%). The powder form of it contains about 0.03% of oxalic acid.

### *Underutilized Potential fruit crops of the Andamans*

The Islands of Andamans have their own unique combination of living species, habitants and beautiful ecosystem which in combination makeup rich resources diversity. In the present scenario of changing food habits, job profile and health awareness, new and improved processed products are in demand world over. The concept of food is shifting from survival to the value added products. (Tandon, 2006). The underutilized fruits play an important role by providing, exotically flavored processed products. The emphasis in the past was only to the processing of major fruit crops of the country, and wealth of the indigenous fruit has been bought to limelight. Marketing of a number of processed products for therapeutic and food supplement due to high nutritive value (Aonla, *Morinda citrifolia,* Bael etc) brings hope for commercialization. Processed products like marmalade, pickle, squash etc were developed and transferred for fruits like West Indian cherry (*Malpighia glabra*) (Singh,*et al* 2004), Soursop (*Anona muricata*), Passion fruit (*Passiflora edulis*), Chalta(*Dillenia indica),* Carambola(*Averrhoa carambola*), Bilimbi (*Averrhoa bilimbi*) Alligator apple (*Anona glabra*), *Pandanus sp., Morinda citrifolia.* A few of them are fast growing in market popularity and acceptability.

There is no established orchard for these fruits and are mostly grown in homestead as a subsidiary crop in other fruit orchards. These underutilized fruits are hardy and hence they can be grown in various types of wastelands like acidic soil, undulating uplands and marshy lands. Increased production of these fruits will not only reduce dependence of income generation and nutritional requirements on a few fruit species but also help in diversification of agriculture besides catering multifarious human needs (Patel *et al,* 2006). Emphasis should be laid to utilize these indigenous fruit crops and market the processed products world over. It will not only bring significant economic benefits to the country but will also provide nutritious products without much competition in the global trade. (Tandon *et al,* 2006)

Physico-chemical analysis of some important underutilized fruits revealed that weight of fruit was 700g in Soursop and Chalta weighed 465g/ fruit whereas, Alligator apple weighed 235-300g/ fruit, and in *Morinda citrifolia,* the fruit weight was 147.93g/ fruit and the low fruit weight. was 5g in West Indian Cherry followed by 2g in Khattaphal and wild Amra. Further characterstics like juice(%) was recorded higher in *Morinda citrifolia* 54.6 followed by 50 in Carambola,and minimum 4.59 was recorded in *Mangifera camptosperma*. Acidity was the highest in passion fruit 5.09% and the lowest in kadam (0.1%). Acidity in *Morinda citrifolia* was 0.12% and higher T.S.S. was recorded in Bakul and Stargooseberry (18.5°Brix). In Khattaphal T.S.S. was (14.8 °Brix) followed by Alligator apple (6.10 °Brix) and *Morinda citrifolia* (8.45 °Brix). Ascorbic acid (mg/100g) was recorded in West Indian Cherry with maximum of 451.4 followed by 146.5 in *M.camptosperma* while *Morinda citrifolia* contained 139.09 and Khattaphal (42.64) and the lowest content of

A few promising fruit species of Andamans

| Scientific name | Common and Vernacular name | Family | Location | Technology developed and transferred |
|---|---|---|---|---|
| *Anona glabra* L. | Alligator apple/Jungli sitaphal | Anonaceae | South Andaman | Saline resistant root stock. |
| *Morinda citrifolia* | Noni /Indian Mulberry | Rubiaceae | In all the coastal belts, wastelands, forests of Andamans | Saline tolerant, highly medicinal, and highly adaptable to any kind of soil. |
| *Anona muricata* L. | Soursop | Anonaceae | South Andaman and Car Nicobar | Preserved as squash, marmalade |
| *Pandanus sp.* | Kewra | Pandanaceae | Nicobar group of Islands | Besides serving as food, scented water the fibrous helps clean the teeth, the tips of the aerial roots are used in traditional medicine. |
| *Dillenia indica* L. | Chalta | Dilleniaceae | South Andaman | For making chutney, pickle. |
| *Malphigia glabra* | West Indian cherry | Malpigiaceae | South Andaman | The fruit is a substitute for Vit. c, and the fruit can be blended to make preserves like pickle,marmalade etc. |
| *Baccaurea sapida* Roxb. | Sapida | Euphorbiaceae | All the Nicobar Group of Islands | - |
| *Baccuarea ramiflora* Lour. | Khattaphal | Euphorbiaceae | All the Nicobar group of Islands | - |
| *Averrhoa bilimbi* L. | Bilimbi | Oxalidaceae | A&N Islands | This fruit is good source of vit.C. ,it can also be used to prepare preserves like pickle, sauce, chutneys etc. |
| *Averrhoa carambola* L. | Carambola | Oxalidaceae | A&N Islands | Good quality of fruits can be used to make squash, jelly, preserves and candy. |
| *Passiflora edulis* Sims. | Passion fruit | Passifloraceae | Middle &South Andaman and South and North Nicobars | Rich in Vit. A and C and squash can be prepared. |
| *Mangifera andamanica* King. | Wild mango/Jangli Aam | Mangiferaceae | Andaman forest,South Andaman Chouldari, Chidyatapu | It is good genetic stock for mango varieties. Fruits are eaten locally as well as by birds , but the wood is of tremendous value. |

Vit.C (3.54) was in Alligator apple.Increased production of these fruits will not only reduce dependence of income generation and nutritional requirement of a few fruit species but also help in diversification of agriculture besides catering multifarious human needs.

Physicochemical properties of a few underutilized fruits

| S.No | Name of the species | Fruit wt.(g) | Juice % | TSS (°Brix) | Acidity (%) | Vit C(mg/100g) |
|---|---|---|---|---|---|---|
| 1 | *Anona glabra* L. | 235-300 | 25-30 | 5.90 | 0.14-0.16 | 3.14-3.54 |
| 2 | *Morinda citrifolia* | 147.93 | 54.60 | 8.45 | 0.12 | 139.09 |
| 3 | *Morinda trimera* | 24.6 | 41.60 | 5.0 | 0.08 | 125.9 |
| 4 | *Anona muricata* L. | 700 | - | 13.0 | 2.01 | 48.0 |
| 5 | *Dillenia indica* L. | 465 | - | 10.50 | 2.90 | 5.10 |
| 6 | *Malphigia glabra* | 5.00 | 36.11 | 5.44 | 1.18 | 45.14 |
| 7 | *Baccaurea sapida* Roxb. | 5.00 | 30.00 | 4.50 | 0.80 | 33.8 |
| 8 | *Baccuarea ramiflora* Lour. | 2.00 | 40.80 | 14.80 | 0.37 | 42.64 |
| 9 | *Averrhoa bilimbi* L. | 50.87 | 81.17 | 4.25 | 4.67 | 13.50 |
| 10 | *Averrhoa carambola* L. | 40.87-50.50 | 50.21 | 3.05 | 0.75 | 12.50 |
| 11 | *Passiflora edulis* Sims. | 69.76 | 22.93 | 16.00 | 5.09 | 26.80 |
| 12 | *Mangifera andamanica* King. | 18.30 | 42.73 | 8.6 | 0.37 | 98.50 |
| 13 | *Mangifera griffithi* | 12.0 | 3.06 | 11.0 | 0.128 | 78.50 |
| 14 | *Mangifera camptosperma* Pier | 80.82 | 4.59 | 6 | 1.25 | 146.5 |
| 15 | *Uvaria sp.* | 12.38 | 10.38 | 8.25 | 1.69 | 115.34 |
| 16 | *Emblica officinalis* | 3.58 | 41.89 | 18.50 | 1.50 | 13.00 |
| 17 | *Malphigia glabra* | 5.00 | 36.11 | 5.44 | 1.18 | 4514 |
| 18 | *Garcinia cowa* | 34.83 | 47.77 | 13.80 | 0.17 | 5.40 |
| 19 | *Anthcephalus chinensis* | 11.63 | 34.39 | 6.04 | 0.11 | 9.83 |
| 20 | *Mimusops elengi* | 7.58 | 33.95 | 18.50 | 1.50 | 5.20 |

## Significance of Underutilized Fruits for Tribals of Andamans

Wild fruits were important sources of food for mankind before dawn of civilization and domestication of present day fruits. Cavemen in forests also depended on theses fruits and passed on valuable information on utility and choice of wild species. Most of the tribals are located in the Islands of A&N. These islands have hot and humid tropical climatic conditions. Forests are the original habitat of tribal groups and are the repository of nutritious fruits which grow wild and are not marketed Tribals living as a part of nature, exploited natural resources to meet their food demands. They are semi nomadic people; their movements are regulated by the availability/ otherwise of fruits and game. There are five kinds of tribes or aboriginals in this Union Territory, Nicobarese (developing), Great Andamanese, Jarawa, Ongees and Shompens Various underutilized fruits grown in forests areas of Andamans are consumed by the tribals to meet their daily requirements. Wild varieties of plants, yielding edible fruits growing throughout the Islands, are eaten in plenty by local people, as they are commonly available in abundance in their habitats. A survey was conducted to find out edible fruits and their use in their medicines to cure a vast number of diseases. The list of such fruits is very large but the potential ones are *Pandanus* sp., Jack fruit, Tamarind, Bael, Khattaphal, Sapida, Bread fruit, Wild mangoes, Wild Amra, Hogplum,Chaur, Soursop, Bilimbi, Laseda, Modhuphal, Badam, Bridilia, Wild Fig, Monkey Jack, Rose apple, Wild Passion fruit, Wild ber etc. These fruits have their own history of consumption, tribals are well aware about their nutritional and medicinal properties. Value of underutilized fruits in traditional medicines is well known. These are major sources of raw materials for drugs since antiquity and have provided bulk of products used in traditional system of medicine. Fruits of Aonla, Harad, and Bahera are the most common entering into 219 patented drugs. Similarly Bael and Tamarind are used in 60 patented drugs. Aonla is the best of famous Ayurvedic medicine Chyavanprash. Ber in Joshanda, Jamun seeds in diabetes, *Morinda citrifolia* as food supplement and also in docking of AIDS virus on human cells are reported. (Abbott, *et al.,* 1985). Many fruits and their products help in treating diabetes and many obstetric diseases. (Hedwood, 1990).

Diversities of underutilized fruit crops in therapeutic use by the tribals

| S.No | Name of the species | Traditional uses by the aboriginals |
|---|---|---|
| 1. | *Pandanus andamanesium* | The lower portion of the fruits is fleshy and is consumed by all the tribals of the Bay Islands as staple food. The fruit is also largely used for making flour for bread by the local people, the leaves are used for thatching huts by the Shompens and the Nicobarese. The tender leaves pounded with coconut oil are rubbed on body by the Nicobarese young people before going to play games. |
| 2. | *Pandanus lerum* | The fruit is boiled and paste baked into thin flat cakes. This is powdered for storage and eaten in times of scarcity. The terminal bud and tender floral bracts are also eaten and used for flavoring. Leaves are used for thatching and for making mats, sacks, baskets, umbrella and other articles. The plant is useful for binding soils |

Contd...

| | | |
|---|---|---|
| | | and sometimes used for fencing. The spathe of the male inflorescence emits a fragrant smell which is of great use for making kewra water. It is also used in folk medicine. |
| 3. | *Morinda citrifolia* | Nicobarese call it Lorang, they use each and every part of the plant. They eat the fruit with salt or cook it, and also the leaves of the plant are used as vegetable as it contains protein. The bark of the plant is also used for making yellow dye. |
| 4. | *Morinda bracteata* | The leaves and fruits are eaten by the Nicobarese and Shompen tribels. The roots and leaves in combination with several other ingredients and coconut milk are made into a paste by the Nicobarese to treat bone fractures. |
| 5. | *Passiflora edulis* | The fruit is edible when ripe, but unripe is poisonous. The leaf juice mixed in equal amount of water is used by the Nicobarese as eye drops to treat conjunctivitis. The leaves are used as a dressing for wounds and in conjunctivitis by the Nicobarese. The fruits are used by the Nicobarese for stomach troubles |
| 6. | *Dracontmelum dao (Blanco) Merr.&Rolfe* | Most ripe fruit is stewed and used for jams, jellies and juices. Boiled and dried fruit can be kept for several months. Young steamed leaves are eaten as a vegetable. The wood is light brown and buoyant, and has no value for timber. There are diverse medicinal uses of fruits, leaves and bark in different parts of the world. The treatment of wounds, sores and burns is reported from several countries. |
| 7. | *Averrhoa bilimbi* | The fruit is occasionally eaten raw with salt or sliced thin and added to salad. It is used extensively as a souring agent for many native dishes. It may also be processed into candies or made into chutneys, relishes and pickles. Its raw juice is an effective remover of stains or spots on clothes, rust stains on brassware and stubborn food particles on enamelware. |
| 8. | *Anona squamosa L.* | Eaten as a dessert fruit, although pulp of fruit may be mixed with milk to form a drink or made into ice cream. The tree serves as host for lac-excreting insects. The crushed leaves are sniffed to overcome hysteria and fainting spells; they are also applied on ulcers and wounds and a leaf decoction is taken in case of dysentery. The crushed ripe fruit, mixed with salt, is applied on tumors. The bark and roots are both highly astringent. The bark decoction is given as a tonic and to halt |

*Contd...*

| | | |
|---|---|---|
| | | diarrhaea. The root, because of its strong purgative action, is administered as a drastic treatment for dysentery and other ailments. |
| 9. | *Averrhoa carambola Linn.* | Dried fruits are used in fever. Ripe fruits used in dysentery, remedy for bleedding piles. Relieves thirst. |
| 10. | *Baccaurea ramiflora Lour.* | The fruit is highly nutritious and very rich source of vitamin C and protein. Seeds are used against vomiting and aesthmatic trouble. Every part of the plant is equally useful in the antidotal treatment of snake bite or a scorpion sting.Jarawa tribes become aware of the rising tides with the ripening of the khattaphal |
| 11. | *Baccaurea sumatrana* | The shrub or small tree bearing leaves which are warmed in coconut oil and used as poultice by the Nicobarese. |
| 12. | *Dillenia indica* | The juice of the fruit is mixed with sugar and used as cooling beverage in fevers and as a coughing mixture. |
| 13. | *Syzygium claviflorum Roxb.* | Seeds used as medicine and contain ellagic acid. Fruits used as astringent in bilious diarrhoea and in diabetes. Fermented fruit juice stomachic, carminative and diuretic. Nicobarese women consume the fruits during pregnancy, to improve haemoglobin percentage. |
| 14. | *Ardisia solanacea* | *The fruits are consumed by the shompens of great nicobar. The roots are boiled in water and taken by the shompens for washing uterus after child birth.* |
| 15. | *Mangifera andamanica King* | Consumed locally as a good source of Vit.C. |
| 16. | *Sauropus androgynus* | The pounded leaves or fruits are boiled in coconut oil by the Nicobarese and rubbed on the body as a renovating agent |
| 17. | *Anona muricata L.* | The fibrous fruit has a juicy aromatic flesh and eaten by Nicobarese. The Nicobarese applies pounded leaves on skin to cure itch and with coconut oil, these are smeared to cure abscesses. |
| 18. | *Anona squamosa* | The leaves are ingredients of a formula of the Nicobarese to heal fractures. |
| 19. | *Syzygium aquem (Burm.f) Merr & Perry* | The leaves mixed with *Clerodendrum inerme* (Linn.) Gaertn. Leaves are pounded and tied on dislocated or fractured bones by the Nicobarese. |
| 20. | *Psidium guajava* | The leaves are ingredient of a formula of the Nicobarese for dysentery and Stomach irritation. |
| 21. | *Artocarpus lakoocha* | Ripe fruits are eaten or used as preserve for chutney. |

*Contd...*

| | | |
|---|---|---|
| 22. | *Salacia chinensis* | The ripe fruits are edible. The leaves with other ingredients are made into a paste by the Nicobarese and applied on abdomen for relieving labour pain in ladies. |
| 23. | *Passiflora foetida* | The fruits are edible when ripe, but unripe is poisonous. The leaf juice mixed in equal amount of water is used by the Nicobarese as eye drops to treat conjunctivitis. The leaves are used as a dressing for wounds and in conjunctivitis by the Nicobarese. The fruits are used by the Nicobarese for stomach troubles. |
| 24. | *Anthocephalus chinensis* | The fruit is edible and the Onges eat it. The bark is employed to make a vessel by the Shompens to cook *Pandanus*. This is also applied in joint pains by the Nicobarese. |

Plant diversity represents the primary source for food, shelter, medicines and many other products and means that make life on earth possible and enjoyable. Increased reliance on major food crops has been accompanied by a shrinking of the food basket which human kind has been relying upon for generations .This nutritional paradox has its roots in the agricultural "simplication", a process that favoured some crops instead of others on the basis of their comparative advantages for growing in a wider range of habitats.

The nutrients consist of carbohydrates, proteins, fats, vitamins and minerals. Present nutritional situation is alarming in the country because 36% of children affected with malnutrition in the world are Indians. Anaemia is prevalent in 60% of our children and 1/3$^{rd}$ children are born underweight which is maximum in the world. Moreover, the people are not aware of the nutritive value of the underutilized fruits. To combat malnutrition, the right strategy is to increase the production of local or non-traditional fruits, of which majority grows in wild, semi-wild condition as stray plants and in tribal areas and also it can generate income from disposal of the surplus fruits. Some of these fruits have already become popular in last a few years like *Morinda citrifolia*, Passion fruit, Alligator, Soursop etc. Since, the list of non-traditional fruits is very large and all fruits cannot attain the status of commercial fruits, some criteria may be adopted to determine the potential of a fruit for commercial exploitation. These criteria may be high productivity, market demand, freedom from the serious attack of pests and diseases, easier post harvest management, high nutritive value and availability of production technology. However, for some important species, information on their nutritional value are made available. Dry matter content was higher (88.10%) in Alligator apple followed by 75% in cultivated fig, and the least DM was found in Chalta (5.15%). Among the micronutrients, Calcium was higher in

Nutritional analysis of important under utilized fruit crops.

| Names of Minor fruits | Macronutrients | | | | | | Micronutrients | | | | | |
|---|---|---|---|---|---|---|---|---|---|---|---|---|
| | DM % | Ash % | Ca % | P % | EE % | CP % | K ppm | Zn ppm | Co ppm | Cu ppm | Mn ppm | Fe ppm |
| Jamun | 34.4 | 0.98 | 4.70 | 0.92 | - | 6.56 | 59.09 | 0.70 | 2.03 | 0.84 | 2.53 | 2.84 |
| *Morinda* | 38.52 | 3.34 | 3.49 | 1.56 | 3.41 | 6.25 | 48.62 | 0.06 | - | - | 0.48 | 3.30 |
| Wild Guava | 40.13 | 3.37 | 1.9 | 0.34 | 3.94 | 6.12 | 21.12 | 0.1 | 0.3 | 0.02 | - | 3.31 |
| Karonda | 70.74 | 10.0 | 10.0 | 1.06 | 9.53 | 9.38 | 33.11 | 0.09 | 0.86 | 0.003 | - | 5.0 |
| Wild Passion | - | - | 1.2 | 0.32 | 9.53 | - | 28.36 | 0.07 | - | 0.04 | - | 2.5 |
| Khattaphal | 12.28 | 5.72 | 4.48 | 1.36 | 5.44 | 9.63 | 44.95 | 0.05 | - | 0.08 | - | 3.98 |
| Velvet apple | 13.17 | 3.36 | 1.28 | 0.43 | 5.94 | 5.63 | 17.56 | - | - | 0.14 | - | 0.37 |
| Wild jack fruit | 17.35 | 14.46 | 5.34 | 0.05 | 1.88 | 10.88 | 33.5 | 0.21 | - | - | 1.17 | 24.74 |
| Attaphal | 37.83 | 3.3666 | 3.08 | - | 4.3 | 5.56 | 47.05 | 0.03 | - | 0.41 | - | - |
| Poon | 32.69 | 1.49 | 4.25 | - | 17.4 | 4.88 | 26.03 | 0.43 | - | 0.31 | 0.52 | 3.84 |
| Khariphal | 8.74 | 4.2 | 0.07 | 1.53 | 1.51 | - | 4.5 | - | - | - | - | - |
| Bael | 35.66 | 3.29 | 4.76 | - | 0.10 | - | - | - | - | - | - | - |
| Bethphal | 37.43 | 3.34 | 2.80 | - | 0.73 | 4.25 | 16.88 | 0.44 | - | 0.29 | - | 1.37 |
| Cultivated fig | 75 | 10.8 | 3.89 | - | 6.9 | 11.5 | 59.3 | 0.53 | 1.91 | 0.006 | 0.99 | 10.16 |
| Jambose | 13.82 | 24.34 | 3.82 | - | 1.1 | 4.3 | 36.52 | - | - | 0.16 | 0.33 | 8.01 |
| *M. trimera* | - | - | 0.0504 | - | - | - | 825 | Trace | Trace | 14.44 | Trace | 44.91 |
| Sapida | - | - | 0.0140 | - | - | - | 1200 | 11900 | 300 | 1700 | 4900 | 2800 |
| Khoonphal | - | - | 0.0140 | - | - | - | 1200 | Trace | Trace | 16.64 | 12.014 | 49.96 |
| Wild fig | 10.01 | 9.955 | 4.17 | - | 5.1 | 12.75 | 48.2 | 0.76 | - | - | 0.48 | 2.51 |
| Carambola | 12.38 | 5.55 | 0.50 | 0.072 | - | 5.25 | 41.7213 | 0.2736 | Trace | Trace | - | Trace |
| Bilimbi | 9.28 | 5.1 | 0.5 | 0.048 | 1.6 | 8.16 | 46.722 | 0.2579 | 0.035 | Trace | - | Trace |
| Alligator apple | 88.10 | 1.85 | 1.00 | 0.039 | - | 9.18 | 45.4503 | 0.1756 | Trace | Trace | Trace | 0.4069 |
| Pandanus sp. | 9.44 | 3.54 | 6.65 | 0.83 | 2.01 | 11.55 | 15.0248 | 0.0669 | Trace | Trace | 1.6466 | Trace |
| Chalta | 5.15 | 4.45 | 0.75 | 0.96 | 2.25 | 9.62 | - | - | - | - | - | - |
| West Indian cherry | 7.82 | 4.45 | 0.50 | 1.168 | 0.15 | 8.75 | - | - | - | | - | - |
| Soursop | - | - | 6.50 | 25.0 | - | 0.75 | - | - | - | - | | 0.5 |

Pandanus (6.65%) and 6.5% in Soursop and 5.34% in Wild Jack fruit, while in *Morinda citrifolia* calcium (%) was 3.49 and the lowest calcium was recorded from Sapida and Khoonphal (0.01%).Phosphorous (%) was maximum in Soursop (25) and minimum in Alligator apple (0.039).

Percentage of Crude Protein was higher in Wild Jack fruit (10.85), Khattaphal (9.63) and Chalta,whereas, 8.75 in West Indian cherry, while in *Morinda citrifolia* the percentage of crude protein was 6.25 and the lowest of 0.75 in Soursop. Higher content of Potassium (1200ppm) from Sapida and Khoonphal followed by 825ppm in *Morinda trimera* and 48.62ppm in *Morinda citrifolia.* Among the micronutrients, iron content was higher in Khoonphal (2800ppm) and (24.74ppm) in Wild Jack fruit. In *Morinda citrifolia,* it was 3.3ppm. (Singh, 2003, 2006).

## Therapeutic Value of Underutilized Fruits of Andaman

The tropical, coastal, humid and sub humid eco-zones of Indian peninsula like Andaman and Nicobar Islands, Tamil Nadu, Kerala, Assam, Orissa, West Bengal have vast variety and diversity in tropical fruit plants, many of which are of evolutionary climaxes in specific niche. Value of underutilized fruits in traditional medicines is well known. They have been major sources of raw materials for drugs since antiquity and have provided bulk of products used in traditional system of medicine. Herbal medicine is widely practiced throughout the world since ancient times. These medicines are safe and environmentally friendly. According to WHO about 80% of the world's population relies on traditional medicine for their primary health care. India, being one of the world's 12 mega biodiversity countries, enjoys export of herbal raw material worth U.S. $100-114 million per year approximately. These plants have close association with local beliefs and rituals and are used in health care needs in rich ethnic life of our people. Most of the trees have remained semi domesticated due to lack of research support for improvement as well as unorganized exploitation. While many species have become rare and endangered due to large scale urbanization, indiscriminate deforestation or due to extension of cultivation under high yielding areas. These underutilized fruits although having nutritional, commercial, medicinal and commercial importance are ignored and yet to be exploited to full potential. In general, fruits are rich sources of vitamins and minerals. Use of various kind of seasonal fruits helps in keeping human beings healthy by fulfilling their various requirements. (Singh, 2004).

| S.No | Name of the species | Potential value |
|---|---|---|
| 1. | *Morinda citrifolia* | Fruit is regarded as having a range of medicinal properties. The juice is high in vitamin C and there is a high demand for it as an alternative medicine for a host of illnesses. Arthritis, diabetes, high blood pressure, muscle aches and pains, menstrual difficulties, headaches both mild and severe, heart disease, AIDS, cancers, gastric ulcers, sprains, mental depression, senility, poor digestion, arteriosclerosis, blood vessel problems, drug addiction and more. The whole plant is known as a miracle plant due its curative properties. |
| 2. | *Aegle marmelos Correa* ex. Roxb. | Fruits are eaten fresh or prepared as sherbet, syrup or marmalade. From the fruit, household glue, dye, tanning agent and various medicines can be produced. Bark, leaves and roots have a number of medicinal uses and the bark can also be used as fish poison. The wood can be used for making small articles such as handles. The tree is very sacred in Hindu religion. |
| 3. | *Anthocephalus chinensis Rich.* | The fruit is said to be edible. The bark is employed to make a vessel to cook *pandanus.* This is also applied in joint pains. Young leaves as poultice and applied to stomach region; used together with *Ficus spp.* and *Bignonia spp.* |
| 4. | *Garcinia mangostana Linne.* | The highly praised fruit is mostly eaten fresh and only occasionally preserved. The fruit rind is used as a dyeing agent, and also, together with the bark, have several applications in traditional medicine. The wood is dark red, heavy and very strong and used in carpentry. |
| 5. | *Cordia oblique* Willd. | Kernel used as a remedy for ringworms. Fruit astringent, antihelmintic, diuretic, expectorant, useful in infection in urinary passages, diseases of lungs and spleen. |
| 6. | *Spondias cytherea* Sonnerat | Fruit is eaten raw or stewed and used for jams, jellies and juice. The green fruit is used in salads, curries or in pickles. The fruits can be fed to pigs and the leaves to cattle. Steamed young leaves are also eaten as a vegetable. The wood is not of much use but leaves, bark and fruits are used in traditional medicine. |
| 7. | *Pandanus andamanesium* | The lower portion of the fruits is fleshy and is consumed by all the tribals of the Bay Islands as staple food. The fruit is also largely used for making flour for bread by the local people. The leaves are used for thatching huts by the tribes. The tender leaves pounded with coconut |

*Contd...*

| | | |
|---|---|---|
| | | oil are rubbed on body by the tribal youths before going to play games. |
| 8. | *Pandanus lerum* Jones. | The fruit is boiled and paste baked into thin flat cakes. This is powdered for storage and eaten in times of scarcity. The terminal bud and tender floral bracts are also eaten and used for flavoring. Leaves are used for thatching and for making mats, sacks, baskets, umbrella and other articles. The plant is useful for binding soils and sometimes used for fencing. The spathe of the male inflorescence emits a fragrant smell, which is of great use for making kewra water. It is also used in folk medicine. |
| 9. | *Pandanus tectorias* Solander | The terminal bud and tender floral bracts are also eaten and used for flavouring. Leaves are used for thatching and for making mats, sacks, baskets, umbrella and other articles. The plant is useful for binding soils.The fruit is mixed with coconut milk or grated coconut and paste baked into thin flat cakes. |
| 10. | *Anona squamosa* L. | Eaten as a dessert fruit, although the pulp of the fruit may be mixed with milk to form a drink or made into icecream. The tree serves as host for lac-excreting insects. The crushed leaves are sniffed to overcome hysteria and fainting spells; they are also applied on ulcers and wounds and a leaf decoction is taken in cases of dysentery. The crushed ripe fruit, mixed with salt, is applied on tumors. The bark and roots are both highly astringent. The bark decoction is given as a tonic and to halt diarrhaea. The root, because of its strong purgative action, is administered as a drastic treatment for dysentery and other ailments. |
| 11. | *Anona muricata* L. | Soursop can be consumed as dessert. It is more often consumed after extraction of pulp. It can be made into fruit jelly or used in preparation of some beverages, sharbats, ice creams and syrups. |
| 12. | *Anona reticulata L.* | Leaves used internally against worms, and externally against abscesses; poultice from leaves for dyspepsia and when mixed with oil for scalp diseases; crushed leaves smell for fainting spells; fresh leaves as topicals to stomach due to indigestion. Unripe fruits and the bark used against diarrhoea and dysentery; root decoction as purgative; leaves, bark, pericarp used as astringent and anthelmintic. |
| 13. | *Emblica officinalis* | Aphrodisiac, diuretic, laxative, useful in anaemia, jaundice, diabetes, asthma, bronchitis, dyspepsia, haemorrhagic disorders, Bilionsness. |

*Contd...*

| | | |
|---|---|---|
| 14. | *Baccaurea sapida* Roxb. | A delicious juice is prepared from the fruit. Seeds are used against vomiting and asthmatic trouble. Every part of the plant is useful for treatment of snake bite and scorpion sting. |
| 15. | *Malpighia glabra* L. | Fruit is rich source of vitamin C and the fruits are made into preserves like jelly, jam and candies. |
| 16. | *Averrhoa carambola* Linn. | In various ways in traditional medicine (e.g. skin disorders, fevers).Crushed leaves or shoots for chicken fox, ringworm, and headache. Fruit is laxative, refrigerant, antiscourbic; stimulate appetite, febrifuge, antidysenteric. Infusion, decoction or tincture of crushed seeds is emmenagogue, lactagogue, abortifacient in large dose. Seeds as narcotic, emetic, powdered as anodyne in asthma, colic, jaundice. Counters ringworm. Leaves and flower consumed as antitusive and febrifuge and against cold. Fruit is ideal against hypertension, diabetes and as antiparalytic, haemostatic, antiemetic, and diuretic. |
| 17. | *Averrhoa bilimbi* L. | In various ways in traditional medicine (e.g. skin disorders, fevers). Leaf decoction heals inflammation of rectum. As paste, it is applied on the affected surface area for mumps, rheumatism and pimples. Infusion of flowers relieves coughs and thrush. Preserved fruit for beri -beri and disorders of the bile and liver. Leaves, flower and fruit are consumed for stomachache, wounds, stomatitis, whooping cough, bleeding gums, ache and hypertension; as well as antitussive and antidiabetic. Against boils. |
| 18. | *Dillenia indica* Linn. | The thickened sepals of fruits are used for culinary purpose and made into jam and jelly, the juice sweetened with sugar is a cooling drink. The fruits are consumed as a tonic and laxative. The juice of fruit is mixed with sugar and used as cooling beverage in fevers and is a laxative. The juice of the fruit is mixed with sugar and used as cooling beverage in fevers and as a coughing mixture. |
| 19. | *Anona glabra* L. | Dried fruits have cooling and antiscorbic effect. Dried fruits are also used to cure fever. Ripe fruits are used in dysentery. |
| 20. | *Dracontomelum dao (Blanco) Merr. & Rolfe* | Most ripe fruit is stewed and used for jams, jellies and juices. Boiled and dried fruit can be kept for several months. Young steamed leaves are eaten as a vegetable. The wood is light brown and buoyant, and has no value for timber. There are diverse medicinal uses of fruits, leaves and bark in different parts of the world. The |

Contd...

| | | |
|---|---|---|
| | | treatment of wounds, sores and burns is reported from several countries. |
| 21. | *Baccaurea ramiflora* Lour. | The fruit is highly nutritious and is a very rich source of vitamin C and protein. Iron contents are also very high. A delicious juice is also prepared from the fruit. Seeds are used against vomiting and asthmatic trouble. Every part of the plant is equally useful in the antidotal treatment of snake bite or a scorpion sting. |
| 22. | *Syzygium aquem* (Burm.f) Merr & Perry | All parts of java plum can be used medicinally and it has a long tradition in alternative medicine. The bark is used for anaemia; the bark and seed for diabetes, which reduce the blood sugar level quickly, the fruit for dysentery, leaves juice for gingivitis (bleeding gums). |
| 23. | *Passiflora edulis* Sims. | The fruit is edible when ripe. Leaves are used for dressing wounds and in conjunctivitis. The fruits are used for stomach troubles and it is rich source of Vitamin A. |
| 24. | *Ficus carica* L. | The fruits may be eaten fresh or stewed and may be dried or canned. A laxative is made from them. The fruits are rich in calcium and it is low in fat and low in cholestrol and is used against obesity. Laxative is made from them. |
| 25. | *Ficus racemosa* L. | Fruits are edible and variable in size and have high sugar content. Fruits laxative and seeds are cooling. |
| 26. | *Syzygium claviflorum* L. | Seeds used as medicine and contain ellagic acid. Fruits used as astringent in bilious diarrhoea and in diabetes. Fermented fruit juice stomachic, carminative and diuretic. Nicobarese women consume the fruits during pregnancy, to improve haemoglobin percentage |
| 27. | *Psidium guajava* L. | The fruit is eaten fresh or used for preserves, jam, jelly or juice. Leaves can be used for dyeing and tanning and in traditional medicine (against diarrhaea). |
| 28. | *Stercularia foetida* Linn. | Leaves employed as sedative, diuretic, expectorant, antitussive, diaphoretic, laxative and antiseptic. Use for hair washing. |
| 29. | *Carissa carandus* L. | Fruits are eaten, preserved and pickled. Its leaves are used in epilepsy. Its powder is used in curing of cracking skin of foot and hand. The powder is rubbed over the skin. |
| 30. | *Aegle marmelos L.* | Fruits possess aromatic pulp, eaten as such or as sherbets, squash, have medicinal value and cure chronic diarrhoea and dysentery pulp contains marmels which act as a |

Contd...

| | | |
|---|---|---|
| | | laxative and diuretic and in strong doses as cardiac depressant. |
| 31. | *Artocarpus lakoocha* Roxb. | Fruits either cooked when ripe or eaten when raw .Too much eating of the fruits may cause diarrhoea. |
| 32. | *Dracontomelon dao* | Bark is used in traditional medicine to provoke abortion. |
| 33. | *Callophyllum inophyllum* Linn. | Seeds used for illumination and medicine, mixed with coconut oil to give tongan oil for massage. Used as hair tonic/alopecia. It cures scable: kudis. |
| 34. | *Mimusops elengsis* | Seed as analgesic and emetic. Bark used as analgesic, emollient, diuretic, laxative, emetic, depurative and antispasmodic. |
| 35. | *Ananas comosus* Linn. | Fruit serves as digesting enzyme stimulant, vermifuge and laxative. |
| 36. | *Emblica officinalis* | Aphrodisiac, diuretic, laxative, useful in anaemia, jaundice, diabetes, asthma, bronchitis, dyspepsia, haemorrhagic disorders, Biliousness. |
| 37. | *Artocarpus heterophyllus* Lam. | Jackfruit pulp and seeds- as a cooling and nutritious tonic, useful in overcoming the effects of alcohol; in South-East Asia, seed starch used to relieve biliousness while the roasted are regarded as an aphrodisiac; heated leaves placed on wounds, and the ash of the leaves burned with maize and coconut shells used to heal ulcers; mixed with vinegar the latex promotes healing of abscesses, snakebite and glandular swellings; bark made into poultice; wood has sedative properties and its pith is said to induce abortion; root as a remedy against skin diseases and asthma, and extract for fever and diarrhoea. Charred and powdered leaves effective cicatrizant for wound due to surgical operation. Root decoction cures diarrhaea and lowers fever. Milky juice when mixed with vinegar is applied on glandular swelling for it promotes the formation or discharge of pus. Wood bark used for fever, diarrhaea; as expectorant and lactagogue. Against boils. |
| 38. | *Zizyphus glabra* Roxb. | The leaves of Ber are used for several eye ailments. Fruits are rich in minerals, it is for pickle preparation and also for gynaec problems. |
| 39. | *Garcinia cowa* Roxb | The fruits used medicinally for headache, stomach complaints and dysentery. |
| 40. | *Uvaria andamanica* King. | Fruits eaten for increasing haemoglobin (%). |

*Contd...*

| | | |
|---|---|---|
| 41. | *Mangifera andamanica King.* | Edible fruits are good source of vitamins A&C. Consumed locally, pickles and preserves are prepared from fruits. |
| 42. | *Mangifera griffithi* | Edible fruits are good source of vitamins A&C. Consumed locally. |
| 43. | *Mangifera camptosperma Pier.* | Edible fruits are good source of vitamins A&C. Consumed locally. |

Global and national assessment of number of plant species utilizable by man

| Author | Year | Number of species | Nature of use |
|---|---|---|---|
| Paroda and Mal | 1993 | 80,000 | Explored by man since dawn of civilization |
| Myers | 1983 | 75,000 | Edible |
| Kunkel | 1984 | 12,650 | Edible |
| Uphof | 1968 | 9,500 | Economic plants |
| Zeven and De Wet | 1982 | 2,489 | Cultivated species (excluding ornamentals, lower plants timber plants) |
| Vietmeyer | 1990 | 1,500 | |
| Arora and Pandey | 1996 | 1,000 (India) | Edible (wild) |
| Paroda and Mal | 1992 | 536 (India) | Economic plants |

Major research and development activities in underutilized fruit crops

| S.No | Problems | Outputs required | Relevant activities |
|---|---|---|---|
| 1 | Lack of genetic material | Improved availability of seed and other planting materials<br><br>Crop improvement programmes<br><br>Improved planting materials derived from traditional varieties | Set up local germplasm supply systems among rural communities. Initiate participatory and their improvement programmes to obtain clean planting materials and the improved varieties. |
| 2 | Loss of germplasm and traditional knowledge | Resource base of selected species secured through *ex situ* and on farm conservation documented and shared among stakeholders | Assess distribution of species and genetic erosion threats.<br>Sample germplasm for *ex situ* maintenance and use.<br>Implement on farm conservation through community-based actions.<br>Identify and collate traditional knowledge using participatory procedures based on informed consent (including e.g recipes on uses) |

*Contd...*

| | | | |
|---|---|---|---|
| 3 | Lack of knowledge on uses, constraints and opportunities | Enhanced information on production levels, use constraints and opportunities<br><br>Knowledge of gender and other socially significant factors identified | Participatory surveys on uses, constraints and opportunities with communities and<br>other levels of the "filieres"<br>Analysis of survey data for gender and other socially significant factors |
| 4 | Limited income generation | Strategies for adding value and increasing rural incomes using target crops<br>Enhanced competitiveness of selected crops | Development of value adding strategies (through processing, marketing, commercialization etc.)<br>Investigate and identify improved agronomic and production procedures |
| 5 | Market commercialization and demand limitations | Enhanced working alliances among stakeholders in "filieres"<br><br>Improved processing and marketing opportunities identified<br><br>Improved capacities of marketing associations and producer groups | Strengthen operational links in the "filieres" between seed supply system, processing and distribution stakeholders<br>Develop improved low cost processing techniques<br>Analyze and identify market opportunities |
| 6 | Lack of research and development activities and weak national capacities | Enhanced national capacities to work with neglected and underutilized crops.<br><br>Enhanced information and knowledge on the selected neglected and underutilized crops.<br><br>Method to improve nutritional values developed and documented | Carry out short training courses for researchers<br>Develop and undertake community -based participatory courses<br>Characterize crops for agronomic, nutritional and market related traits<br>Study formal and informal classification systems<br>Investigate methods of maintaining and enhancing nutritional value<br>Investigate new areas of crop growth |
| 7 | Lack of links across conservation and production to consumption "filieres" | "Filieres" established or strengthened<br>Participatory networking procedures established | Hold planning workshops for all stakeholders<br><br>Establish and strengthen operational links between stakeholders |
| 8 | Inappropriate (inadequate) policy and legal framework | Raised awareness among policy -makers of issues and options for improved policy and legal frameworks<br><br>Links to existing rural and economic development projects enhanced | Identify inappropriate policy/ legal elements<br>Undertake public awareness actions among policy makers<br>Establish close partnerships with extension workers and others involved in agricultural development |

## References

Abbott, I. and Shimazu, C., 1985. The Geographic Origin of the Plants Most Commonly Used for Medicine by Hawaiians, *Journal of Ethnopharmacology* 14: 213-222.

Alexander Dittmar, 1993. *Morinda citrifolia* L. – Use in Indigenous Samoan Medicine, *Journal of Herbs and Medicinal Plants*. Vol. 1 (3).

Awasthi, A.K. and Jacob John, 1987. A contribution to the forest resources of Great Nicobar. *J.And. Sci. Assoc.* 3:24-37.

Badal Kumar Basu 1990. The ASI, Andaman and Nicobar Island Tribe Series- The Ongee-Negrohunter- Gatherers of little Andaman.

Bordia P.C., 2004 Medicinal importance of underutilized fruits. *Exploitation of underutilized fruits*, Ch-4, pp-25-30

Elkins, Rita. ,1997. Hawaiian Noni (*Morinda citrifolia*). *Woodland Publishing; ISBN: 1885670672*

Dagar,H.S.1989. Endemic Plant species of Bay Islands. *J.And. Sci.Assoc.*, 5:161-166.

Hiremath S.M. and Halappannavar, A.B., 2006. Performance of custard apple seedling progenies for yield and its attributes. *National Symposium on Underutilized Horticultural crops*. Abstract of leading papers.IIHR, Hessarghatta, Bangalore. pp-49.

Jalikop S.H., 2006. Taxonomy , Growth habit and adaptability of some underutilized fruits. *National Symposium on Underutilized Horticultural crops*. Abstract of leading papers.IIHR, Hessarghatta, Bangalore. pp. 4-5.

Medhi, R.P., Singh D.R. and Senthil Kumar, 2006. Physico- chemical characteristics of wild mangoes in A&N Islands *National Symposium on Underutilized Horticultural crops*. Abstract of leading papers .IIHR, Hessarghatta, Bangalore. pp-79.

Meenu Srivastava and Dolly Mogra , 2004. Industrial application of underutilsed fruits. *Exploitation of underutilized fruits* Chapter-5,pp-31-38.

Mondal, A.B., Chattopadhyay. D. and Tarun Coomar, 2000. Rare and endangered flowering plants of Bay Island with special reference to endemics and extra Indian Taxa. *Indian Forester*.pp.389-396.

Nelson, S.C. ,2001. Noni cultivation in Hawaii. Univ. of Hawaii *CTAHR-Cooperative Extension Service PD-19.*

Padulosi, S. Noun, J., Giuliani, A., Shuman, F., Rojas W. and Ravi B., 2005. Realizing the benefits in Neglected and Underutilised Plant species through Technology Transfer and Human Resources Development (IPGRI*), Norway/UN Conference on Technology Transfer and capacity building* .pp-117-125.

Padulosi, S. , Hodgkin T., , Williams J.J. and Haq N., 2004 . Underutilised crops : Trends, challenges and opportunities in the 21st century. (IPGRI), Itlay, (ICUC),U.K.

Peter, K.V., Indira.P and Sarah George, 2006. Underutilized fruits and vegetables- A perpestive. *National Symposium on Underutilized Horticultural crops.* Abstract of leading papers.IIHR, Hessarghatta, Bangalore .pp-3.

Patel, R.K., Singh, Akath, Babu K.D and De L.C., 2006.Underutilised fruits- Potential fruit crops of the future for the North Eatern Region. *National Symposium on Underutilized Horticultural crops.* Abstract of leading papers.IIHR, Hessarghatta, Bangalore. Pp-23.

Salvi, B.R., Malshe, K.V., Bendale, V.W.,.Kshirsagar. P.J. and Jalgaonkar V.N., 2006. Survey, Collection, Maintenance and Evaluation of germplasm of Jamun (*Syzygium cumini*), *National Symposium on Underutilized Horticultural crops.* Abstract of leading papers.IIHR, Hessarghatta, Bangalore . pp-31.

Singh. I.S., 2001. Minor fruits and their uses. *Proceeding National Semninar on New Horizons in Production and Post – Harvest Management of tropical and subtropical fruits.* (HSI) pp-178-182.

Singh, D.B. Attri, B .L., Medhi, R. P., Suryanarayana M.A. and Sharma T.V.R.S. ,2002. Underutilised fruits of Andaman and Nicobar Islands *Published by CARI, Port Blair*.pp. 9- 32.

Singh, D.R., Medhi, R.P.,. Senani. S. , Senthil kumar R.and Rai R.B., 2004. The Alligator apple, *Published by CARI, Port Blair.*

Singh, D.R., Medhi, R.P., Senani, S., Rai, R.B. and Senthil Kumar, R., 2004. Nutritional exploitation of Pandanus fruits. *Published by CARI, Port Blair.*

Singh, D.R. Medhi, R.P., Senani S. and Rai, R.B.2003. Nutritional exploitation of underutilized fruits of Andaman (1st issue). *Published by CARI, Port Blair.*

Singh, D.R., Medhi, R.P., Senthil Kumar. R., Senani . S.and Rai, R.B., 2004. Carambola in Andamans. *Published by CARI, Port Blair.*

Singh, D.R, Medhi, R.P. , Dmaodaran. T.and Rai, R.B., 2004. Bilimbi in Andamans. *Published by CARI, Port Blair.*

Singh, D.R., Medhi, R.P. , Senani, S., Senthil Kumar R. and Pandey V.B. , 2004.Production technology of Custard apple for Andaman. *Published by CARI, Port Blair.*

Singh, D.R. and Medhi R.P., 2004. West Indian Cherry Cultivation in Andamans. Research bulletin-15. *Published by CARI, Port Blair.*

Singh, D.B., Attri B.L. and Sharma T.V.R.S., 1998. Nutritional value of indigenous fruits of Andaman. *Indian Food Packer* 52 (2): 30-31.

Sreekumar, P.V., Ray L.N. and Padhya ,P.M. , 1996. Economically important vascular plants occurring wild in Bay Islands.*J.Econ. Tax.Bot.,* 20(2): 407-421.

Singh, D.B., Attri, B .L. , Medhi, R. P. , Suryanarayana M.A. and Sharma T.V.R.S., 2002. Underutilised fruits of Andaman and Nicobar Islands. *Published by CARI, Port Blair*.pp. 9- 32.

Singh, D.R., Medhi, R.P., Senani , S., Senthil Kumar R. and Rai R.B., 2004. The Alligator apple, *Published by CARI, Port Blair.*

Singh, D.R., Medhi, R.P., Senani, S., Rai, R.B.and Senthil Kumar R. , 2004. Nutritional exploitation of Pandanus fruits. *Published by CARI, Port Blair.*

Singh, D.R. Medhi, R.P., Senani, S. and Rai R.B., 2003. Nutritional exploitation of underutilized fruits of Andaman (1st issue). *Published by CARI, Port Blair.*

Singh, D.R.,.Medhi, R.P., Senthil Kumar, R., Senani, S. and Rai R.B., 2004. Carambola in Andamans. *Published by CARI, Port Blair.*

Singh, D.R, Medhi, R.P., Damodaran T. and Rai R.B., 2004. Bilimbi in Andamans. *Published by CARI, Port Blair.*

Singh, D.R., Medhi, R.P., Senani, S., Senthil Kumar. R., and Pandey V.B., 2004.Production technology of Custard apple for Andaman. *Published by CARI, Port Blair.*

Singh, D.B., .Attri , B.L and Sharma ,T.V.R.S. , 1998. Nutritional value of indigenous fruits of Andaman. *Indian Food Packer* 52 (2): 30-31.

SreeKumar, P.V., Ray L.N. and Padhya ,P.M. , 1996. Economically important vascular plants occurring wild in Bay Islands. *J.Econ. Tax.Bot.*, 20(2): 407-421.

Tandon D.K. and Sanjay Kumar, 2006. Potentiality of underutilized fruits for value addition.. *National Symposium on Underutilized Horticultural crops.* Abstract of leading papers.IIHR, Hessarghatta, Bangalore .pp-11.

Yaganarsimhan, S.N., Keshavamurthy, K.R., Chelladur V. and Togunashi, V.S., 1983. Medicobotany of Andaman and Nicobar Islands. Physiography vegetation and enumeration .*Tax.J.Econ.Taxon.Bot.*, 4: 485-697.

**Underutilized and Underexploited Horticultural Crops, Vol. 2, 2007**
*K.V. Peter (ed.), pp. 55-60*
New India Publishing Agency, New Delhi (India)
E-mail : newindiapublishingagency@gmail.com
Web: www.bookfactoryindia.com

# 3 Giant Granadilla

**PRADEEPKUMAR. T**
College of Agriculture, Padannakkad, Kerala, India

Giant granadilla, *Passiflora quadrangularis* L. (syn. *P. macrocarpa* M.T. Mast.) is the largest fruit in the genus Passiflora. With more than 450 species, *Passiflora* is the most important genus of family *Passifloraceae*. It is further divided into 24 subgenera (Killip, 1938). Most of the species producing an edible fruit belong to subgenera *Passiflora* and *Tacsonia*. Many species have ornamental interest because of unique and spectacular flowers, or are used for their sedative, antispasmodic, antibacterial, and insect-resistant properties.

Taxonomy

| | |
|---|---|
| Kingdom | *Plantae* |
| Subkingdom | *Tracheobionta* |
| Superdivision | *Spermatophyta* |
| Division | *Magnoliophyta* |
| Class | *Magnoliopsida* |
| Subclass | *Dilleniidae* |
| Order | *Violales* |
| Family | *Passifloraceae* |
| Genus | *Passiflora* L. |
| Species | *Passiflora quadrangularis* L. |

## Origin

Crop is a native of South America and due to rapid degradation of their habitats, particularly in Andean valleys, wild forms are under extinction

(Morton, 1987). They have become naturalized in many tropical countries. Even though giant granadilla was brought under cultivation as early as $18^{th}$ century, popularization of crop remains as a mystery. From tropical South America, it was introduced first to Caribbean Islands and then to India, Africa and Australia. Out of edible 60 species of genus *Passiflora* only a few are known outside its limited geographical occurrence. Its use as juice and beverage contributed immensely to industrial development in America.

## Botany

Giant granadilla is a strong glabrous climber with fleshy tuberous roots. Stem is quadrangular on which simple leaves are produced alternately. Stem is prominently winged on angles, and produces axillary tendrils up to 30 cm long, flanked by leaflike, ovate or ovate-lanceolate stipules (2-3.5 cm) long. Leaves are broad-ovate or oblong-ovate, 8.25-15 cm wide, 10-20 cm long; rounded or cordate at base, abruptly pointed at apex, sometimes toothed near base; thin, with conspicuous veins sunken on upper surface and prominent beneath. Leaf lamina is characterized by shiny upper green colour and pale shadow below.

## Floral Biology

Flowers are bisexual in nature, produced singly on leaf axils with characteristic aroma.The solitary, fragrant flowers, up to 12-12.5 cm wide, have bell-shaped calyx. Sepals 5, greenish or reddish-green on outside, white, pink or purple inside; Petals 5, 4.5 cm long, white-and-pink; the corona filaments 2-ranked, 6 cm long, purple-and-white below, blue in middle, and pinkish-blue above. Ovary and five stamens are borne on an androgynophore. The unilocular and tricarpelar ovary is surmounted by three styles. Fruits are usually globular or oval berries. Flower opens only once, in early morning and main flowering period is from May- October. Outside the area of origin, poor fruitset is often reported and is due to protandrous nature of flowers and absence of pollinating agents. Hand pollination is recommended to obtain a reasonable crop. It should be done in late morning, no later than 4 to 6 hours after flowers open. Fruit is fleshy ovate - oblong berry with an approximate size of 25X15cm. Fruit color ranges from green initially to pale yellow or yellowish green at maturity. Flesh is whitish to which are embodied numerous obovoid seeds which are flat in shape and dark brown in color.

## Genetics

Cytologically, the species of *Passiflora* for which chromosome numbers are known can be divided into four karyological groups, represented by $x = 6$,

$x = 9$, $x = 10$ and $x = 12$. Different chromosome base numbers ($x = 3, 6, 9$) are proposed for the genus, without a clear understanding of this variation and of the phylogenetic relationships among species (Morawetz, 1986; Snow and MacDougal, 1993). Recently, Melo *et al.* (2001), revising cytotaxonomy of the group, considered $x = 6$ as the most probable base number for the genus, whereas $x = 9$, $x = 10$ and $x = 12$ were considered secondary base numbers. The second-most probable base number, $x = 12$, played an important role in evolution of group because it is better represented in other genera of the family. Granadilla is a diploid with $2n = 18$ (Beal, 1973). Secondary constrictions were recorded in the karyotype and Beal (1973) recorded up to 7 satellite pairs in *Passiflora quadrangularis.*

## Varieties

There are various strains producing fruits of different sizes and quality. One strain with especially large fruits and good flavor was formerly considered a separate species (*P. macrocarpa*), but it hybridizes readily with smaller strains and there are intermediate types. An ornamental form, 'Variegata', has leaves splashed with yellow.

## Climate

Being a crop of tropics, it relishes hot moist climate for its potential growth and it is a crop for plains. Ideal climate for giant granadilla is one that is truly tropical, warm both day and night, with little fluctuation, and with high humidity. It is grown between 213 and 457 m elevations in Jamaica and Hawaii, and up to 914 m in India (Pradeepkumar *et al*, 2005) to 1,800 or 2,200 m in Ecuador.

## Soil

For maximum growth and productivity, vine requires deep, fertile, moist but well-drained soil. Australians observed good growth on volcanic, alluvial, and sandy soil, and even decomposed granite. Vines planted in highly alkaline situations in Israel have perished showing symptoms of acute chlorosis. Crop loves forest soil and profusely grows in soils rich in organic matter.

## Propagation

Giant granadilla grows readily from seeds, which germinate in 2 to 3 weeks and seedlings can be set out when 15-30 cm high. There will be sufficient vegetative growth and root development to permit transplanting in 30 days. Even though seed propagation is possible, rooted cuttings are the most adopted planting material. Air- or ground-layers are also satisfactory. Shoot cuttings of this plant can be successfully rooted in a sand-soil mixture under 60% shade. Rooted cuttings are generally planted at 3 - 3.5m apart. They are usually trained on open overhead trellis, which permit fruits to hang.

## Aftercare

Crop loves heavy doses of organic manure and only deficiency of Nitrogen and Phosphorus affect normal growth of plant. Too much nitrogen will render plant in vegetative phase. Crop requires regular watering. It grows luxuriously during monsoon season. Crop is sensitive to water logging. Pruning should be done in late winter by removing weak shoots and surplus growth and cutting back side shoots to 15 cm. Fruiting begins usually in second year and continues yielding for next a few more years. In Indonesia and Queensland, a productive vine fruits more or less continually all year and annual yield ranges from 25 to 35 fruits in larger types to 70 to 120 fruits in medium to small types. Yield of 2 to 3year old vines varies from 16 to 50 fruits. Fruits are ready for harvesting when skin becomes translucent and glossy and begins to turn yellowish at apex. It is clipped from vine. Very careful handling and packing are essential.

## Pests and Diseases

Reports of pest and disease occurrence are minimum in this crop making it an ideal crop for homesteads. Young plants in nurseries may be severely defoliated by *Disonycha glabrata* in Venezuela. In Queensland, principal pest of giant granadilla is green vegetable bug, *Nezara viridula,* which punctures young fruits and sucks out juice, causing them to wither and fall; or hard lumps will form in flesh. To avoid damage by fruit flies, fruits are sometimes bagged.

Leaf spot, from fungal infection, occurs occasionally in Queensland but it is of little importance. Stem-end rot in East Africa is attributed to fungus, *Botryodiplodia theobromae.*

## Commercial use

Edible portions are mesocarp and endocarp of fruits, which can be eaten as melons are eaten and as fruit salad. Flesh of ripe fruit, with inner skin removed, is cut up and added to papaya, pineapple and banana slices in fruit salads, seasoned with lemon or lime juice. It is cooked with sugar and eaten as dessert, sometimes candied; but it is so bland that it needs flavoring. Jelly can be made from unpeeled flesh boiled for 2 hours and pulp simmered separately. Juice strained from both is combined and, with added sugar and lemon juice, is boiled until it jells. Its use as a flavoring agent in ice creams and sherbets is noteworthy. A nector prepared out of pulp of giant granadilla tastes much like that of pear. Unripe fruits can be boiled and eaten as vegetable. In Jamaica, fleshy tuberous roots are eaten as a substitute for yams. Leaves, skin and immature seeds contain a cyanogenic glycoside. Pulp contains passiflorine and, if indulged in excessively, causes lethargy and somnolence. Raw root is emetic, narcotic and poisonous.

Nutrient value Per 100 g of Edible Portion

| Nutrients | Thick Flesh | Arils and Seeds |
|---|---|---|
| Moisture | 94.4 g | 78.4 g |
| Protein | 0.112 g | 0.299 g |
| Fat | 0.15 g | 1.29 g |
| Crude Fiber | 0.7 g | 3.6 g |
| Ash | 0.41 g | 0.80 g |
| Calcium | 13.8 mg | 9.2 g |
| Phosphorus | 17.1 mg | 39.3 mg |
| Iron | 0.80 mg | 2.93 mg |
| Carotene | 0.004 mg | 0.019 mg |
| Thiamine | ——— | 0.003 mg |
| Riboflavin | 0.033 mg | 0.120 mg |
| Niacin | 0.378 mg | 15.3 mg |
| Ascorbic Acid | 14.3 mg | |

## Medicinal Uses

Fruit is valued in tropics as antiscorbutic and stomachic. In Brazil, flesh is prescribed as a sedative to relieve nervous headache, asthma, diarrhea, dysentery, neurasthenia and insomnia. Seeds containing a cardiotonic principle, are sedative, and, in large doses, narcotic. Leaf decoction is a vermifuge and is used for bathing skin afflictions. Leaf poultices are applied in liver complaints. Root is employed as an emetic, diuretic and vermifuge. Powdered and mixed with oil, it is applied as a soothing poultice.

## References

Beal P.R. 1973. Cytology of the native Australian and several exotic *Passiflora* species. 3. Morphology of satellited chromosomes. *Queensland Journal of Agricultural and Animal Sciences* 30: 19–24.

Killip E .P. 1938. The American species of Passifloraceae. *Publications of the Field Museum of Natural History, Botanical Series* 19: 1–613.

Morawetz W. 1986. Remarks on karyological differentiation patterns in tropical woody plants. *Plant Systematics and Evolution* 152: 49–100.

Morton, J. 1987. Giant Granadilla. p. 328–330. In: Fruits of warm climates. Julia F. Morton, Miami, FL.

Melo NF de, Cervi AC, Guerra M. 2001. Karyology and cytotaxonomy of the genus *Passiflora* L. (Passifloraceae). *Plant Systematics and Evolution* 226: 69–84.

Pradeepkumar. T, Peter, K.V. and Sajithbabu, D. 2005. Giant Granadilla-a crop for the future. *Indian Horiculture* 50: 35-36.

Snow N, MacDougal JM. 1993. New chromosome reports in *Passiflora* (Passifloraceae). *Systematic Botany* 18: 261–273.

**Underutilized and Underexploited Horticultural Crops, Vol. 2, 2007**
*K.V. Peter (ed.), pp. 61-87*
New India Publishing Agency, New Delhi (India)
E-mail : newindiapublishingagency@gmail.com
Web: www.bookfactoryindia.com

# 4 Apricot

***DEEPA HANSRAJ DWIVEDI AND **SANJAY K DWIVEDI**
* Department of Applied Plant Science
B B Ambedkar University, Lucknow 225025
** Field Research Laboratory, DRDO, 56 APO

Apricot (*Prunus armeniaca* L) is the most important fruit crop of cold arid regions of India. It is one of the most delicious of temperate fruits having pleasing flavour and truly one of the most delectable of the stone fruits. Exceptional fruit quality consists of blending of sugar and acidity and strong apricot aroma. It has been appreciated and grown for thousands of years on slopes of mountains in Asia and at least two thousand years in Europe. There are many uses of apricot. It is enjoyed as fresh fruit but a large portion of the production worldwide is preserved primarily by drying. In China, apricot was not only dried but also preserved by salting and smoking. In Japan, apricot (mume) flowers are revered for their beauty and are used since 7–8th century for ornamental purposes.

## Importance of the Crop

Apricot is an exotic fruit consumed all across the world. Ripe apricot is an excellent dessert and is used for table purposes. Apricot dehydration, with or without stone, is common since ancient times and is still practiced the world over. It is processed for canning, as candy, bar, frozen, juice/beverage, nectar and puree. Apricot especially its kernel is an important ingredient of confectionaries. Ripe apricot seed (kernel) is enclosed in a stony endocarp called the pit. The pit represents about 15% of ripe fruit and kernel content is about 25-32% in the pit. Kernel is used as dry fruit and is as nutritive as almond. Bitter kernel is used for oil extraction. It contains 36-62% valuable oil, an important constituent in cosmetic industry. It is also used for body massage, illumination in lamps and even for cooking purposes (Dwivedi, 2004).

### Composition

Apricot fruit is a good source of vitamin 'A', potassium and carbohydrates. Ingredients in the ripe edible portion (per 100 g) of fresh apricot are recorded as moisture 85.3%, protein 0.9%, fibre 2.0 %, energy 141 calories, soluble solids 12.5 °B, organic acids (malic and citric acids) 2.6 mg, total carotenes 617 mg, thiamine 0.02 mg, riboflavin 0.03 mg, niacin 1.4 mg, potassium 350 mg, sodium 1.0mg, iron 0.3 mg, calcium 16 mg, magnesium 9.0mg and zinc 0.1 mg (Wills *et al.* 1983). Nineteen amino acids are also identified in apricot and important ones are glutamic acids, aspartic acid, hystidine, arginine, tryptosine, cystine, methionine and hydroxyproline (Faust *et al.*, 1998). Major constituent sugars include glucose, fructose, sucrose, sorbitol, mesoinositol and oligosaccharides. Malic and citric acids are common acids in fruits. Dried apricots are high in calcium and iron. Edible portion of fruit is about 86.0 % and stone 14.0 %. Various ingredients in ripe and dehydrated apricot fruits were studied by National Institute of Nutrition, Hyderabad.

## Origin and Distribution

Apricot goes back to at least 5000 years in the history in China and 2000 years in Europe. It originated in Western China (De Candolle, 1886) and was known in China as far back as 658 BC and in Europe 406-250 BC. Grafting of apricot began in China in about 600 AD. Sogdiana in China was identified as the native place of apricot.

Apricot is introduced in cold arid Ladakh via Baltistan though some experts opine that the crop was introduced directly from China *via* Tibet. It also has the capability to tolerate as low as minus (-) 35°C temperatures during winter, when it is dormant. This perhaps makes it an ideal fruit crop for cold arid regions.

## Area and Production

As per FAO, world apricot production was around 2.3 million tonnes in 2001 but this constitutes less than 1 % of total fruit production which is increasing at a slow rate of 3.0 % /annum. Nearly 50% of crop is produced in Europe followed by Asia. The ten major producers are Turkey, Spain, Italy, France, USA, Pakistan, Iran, Hungary, Tajakistan and China. China grows a large quantity of apricot in Liaoning, Shantang and Hopei Provinces ((Faust *et. al.*, 1998). It is also found wild in Lachow in Kansu province. In India, it is grown in temperate Himalayas especially in the cold arid zones of Leh and Kargil in the state of Jammu and Kashmir, Lahaul Spiti in Himachal Pradesh and Kumaon and Garhwal hills of Uttarakhand.

Apricot is the main fruit crop grown in the cold arid regions of India which cover 3200 ha with a total production of 5200 tonnes (Dwivedi and Attrey, 2002). The trees grow well in dry rocky hill sides, in sandy soils and tolerate various stresses *viz.* moisture, nutrient and temperatures as low as -35 °C when dormant.

## Soil and Climate

Apricot is a very hardy plant and can be grown in a variety of soils from sandy to sandy loam, rocky and poor soils having moisture and nutrient stresses. It is sensitive to high moisture conditions and heavy soils. The best performance of apricot is noticed in well drained, deep (3.0 –4.0m), sandy loam soils, rich in organic matter. Soils having high lime content do not affect its growth if proper drainage is available. Apricot seedlings are noticed even on dry rocky hill sides where soil depth is below 1.0 m where there is no source of water. It is reported from all over the globe that apricot does not tolerate excess soil moisture and humid climate.

Apricot has wide climatic adaptability. It is grown in cold winters of Siberia, subtropical conditions of North Africa, cold desert of Central Asia to humid Japan and Eastern China. The apricot cultivars have very specific climatic requirement and are not very adaptable to variations in environment. Annual precipitation of 100 cm is sufficient for its good growth. It does well even in region where annual precipitation is below 50 cm (Faust *et al.,* 1998). It prefers dry cold winter temperatures, not below –35 degree Celsius without much fluctuation, and in warm spring. It thrives well in cold arid conditions with low atmospheric humidity and pressure. Humid conditions increase attack of diseases.

## Taxonomy of Crop

Most of the cultivated apricots belong to *P. armeniaca*. Trees of *P. armeniaca* are medium to large (4-6 m) and prefer dry conditions for growth. Leaves are serrated, ovate or elliptic, very often cordate and long having a reddish petiole with 1-5 glands. Flowers are white or pink. Fruits are drupe, mostly free stone and sometimes cling stone. Fruits are round to ovate, golden yellow / orange in colour with red blush on the side exposed to sun. Weight of ripe fruit ranges from 2-25 g in seedlings and 25-60g / fruit in cultivated cultivars but occasionally as large as 100g and even 200 g fruits are also reported from

China. Fruit colour ranges from creamy white-greenish-pale yellow-orange yellow to dark orange. Apricot occupies a place between plums and peaches. For more than 17 centuries, apricot was considered a form of early peach. They are graftable onto peach and plum and can hybridize with both species.

Apricot belongs to family *Rosaceae,* subfamily *Prunoideae,* genus *Prunus* L., subgenus *Prunophora* (Neck) Focke (apricot and plum) and section *armeniaca* (Mill) Koch (Rehder, 1940). All apricot species are regular diploids with 8 pairs of chromosomes (2n=16). No difficulty is reported in the inter crossing among species. Number of species ranges from 3-10 according to different taxonomists. A brief description of major species is given below (Mehelendbacher *et al.,* 1996).

### *P. armeniaca*

Most cultivated apricots belong to one species *P. armeniaca* L. , believed to have originated in the mountains of Northern and North eastern China.

### *P armeniaca var Ansu* Maxim

The Ansu apricot is cultivated in humid areas of China, South Korea and Japan. It is resistant to many fungal diseases. It has sour, edible, free stone fruits, used for processing into syrup and jam.

### *Prunus armeniaca var Holosericea*

This is called Tibetan apricot.

### *Prunus sibirica L.*

This is the Siberian apricot which is drought tolerant. The trees and buds are remarkably hardy and tolerant to -50 degree Celsius during dormancy. However, it has shorter rest period than *P. armeniaca*. Fruits are small (1.2 - 2.5 cm dia), free stone and have thin and yellow flesh.

### *Prunus mume (Sieb)*

Sieb Et Zucc.

This species is known as Japanese apricot which grows in areas warmer and humid than those required for *P. armeniaca*. Trees are small and shrub like. The species is not cold hardy. They are tolerant to wet soils, humid climate and diseases except Xanthomonas. This species is widely used for ornamental purpose since trees bloom early and have attractive fragrant single/double flowers which range in colour from white to pink to red. Fruits are clingstone and are not edible since they are highly acidic. They are used for processing as pickles and liquor. Kernels are generally sweet.

### *P. mandshurica (Maxim) Koehne*

It is called Manchurian apricot and found in Eastern Manchuria in a small area. It is a very cold hardy species capable of tolerating temperatures as low as -40 to -45 degree Celsius. Fruits are small (2.5cm) sour and astringent.

### *P.* X *dasycarpa* Ehrh.

The species is called black or purple apricot due to its black purple coloured fruits. It is cultivated in central Asia, Afghanistan, Baluchistan and Iran. Plants are resistant to many fungal diseases and tolerant to low temperatures and bloom late. Fruits are small, clingstone with dark purple skin colour and sour yellow flesh.

## Status of Genetic Diversity

Apricot is grown in the cold arid regions of India since ages. While traveling in Ladakh during 1819-1825 Moorcraft and Treback were highly impressed with the apricot varieties and their quality and desired to introduce some of the strains to European countries (Moorcraft and Treback, 1841). This highlights the value of genetic diversity of apricot in the cold arid regions of India. Since ages, apricot in the region has developed considerable variability. Little attention was paid by researchers to evaluate and conserve existing genotypes of apricot in the cold arid region.

The prominent apricot genotypes growing in cold arid regions are Amba, Afghan, Australian, Brochuli, Badamchuli, Chulikarpo, Charmagz, Daychuli, Germoy, Tago, Ghoun Chuli, Kaisi, Halman, Jeembo Chuli, Margulam, Narmo, Nari, Koban, Khantey, Rakchey Karpo, Rogan, Shakanda, Suka, Tripa, Tokpopa, Tokpopa and Khantey. 84 important genotypes were collected and evaluated which showed considerable variation in physico-chemical properties of fruits (Dwivedi *et al.,* 2002b).

Cold arid zones are bestowed with enormous diversity in apricot which needs to be evaluated, conserved and exploited for rootstock development, tree canopy structure, improvement in yield, quality and tolerance to biotic and abiotic stresses.

## Varietal Wealth

Cultivars found in cold arid region differ drastically from those growing in other temperate regions. These are very area specific cultivars suitably adapted to survival at high altitudes of 8,000-12,000 m above MSL and the temperature, nutrient and environmental stresses that they are subjected to under the cold arid conditions. Prominent cultivars of apricot found growing in cold arid region were studied by Bhatia *et al.* (1977) and Dwivedi *et al,* (2002a).

### *Halman*

The name is derived from Tibetan literature meaning grafted. This is the most popular and abundantly grown variety of apricot in cold arid zones. This is perhaps a seedling selection of local origin. It covers maximum area of plantation in Leh and Kargil districts of Ladakh. It is primarily used for the purpose of drying since it is less juicy with high TSS. Kernel is sweet and used as almond. Fruit is medium, round, pedicel end compressed, structure distinct, smooth skin, orange red in colour, pulp orange red, thick, sweet and mild acidic with pleasant flavour. Halman has the highest TSS in cultivated varieties, stone free, medium, round oval, pointed towards apex, prominent ridges on both the sides, rough, shelling (%) 72 and kernel 28 %.

### *Rakchey Karpo*

This is the second most popular variety for table purpose due to its sugar/ acid blend, juicy pulp and pleasing aroma. Name Rakchey (stone) karpo (white) means fruit with white stone, an uncommon feature in most of the apricot cultivars. This is also indigenous to the Ladakh region probably developed as seedling selection. It is an early cultivar, ripens in end ofJuly to mid August. Fruits are medium to large, round with compressed pedicel end, fruit size medium to large, glossy skin, pale yellow with red rose colour blushes where exposed to sun. Pulp is light pale, juicy, sweet and mild acidic with pleasant flavour. Stone is medium, free, round to oval, with sharp ridges, white creamish in colour, shell 68% and kernel 32%. Kernel is sweet.

### *Tokpopa*

This is an indigenous variety of Leh district. It is used for table and for drying purposes. Fruits ripen late and are available in August. Fruits are medium, round, compressed with smooth skin, dull yellow, juicy and acidic to sweet in taste. Stone is free, medium, round, rough and sweet. Shell is 68% and kernel 32%.

### *Rogan*

This is the earliest cultivar of apricot and is a table variety which grows in limited areas of Leh district. Fruits are small, highly juicy, round, glossy, skin straw yellow, very soft, juicy and slightly acidic sweet pulp. It bears the smallest fruit among all cultivated varieties. Harvest period is mid July. Stone is small, free, oblong, ovate, prominent ridges on both sides. Shell is 62% and kernel is 38%. Kernel is sweet.

### *Safaida*

It is also an exotic variety well adapted to cold arid conditions. Fruits are large round, skin is glossy, smooth, light yellow, flesh soft maize yellow in colour, very sweet, less acidic with pleasing flavour. Stone is small, ovate, apex, pointed with one ridge on both side. Shelling 72% and kernel 28%. The kernel is sweet.

### *Nari*

This is an exotic cultivar and is well adapted to cold arid conditions of India. It is a late ripening variety available after mid August. Fruits are medium, oblong, elliptical with truncate base, skin greenish to light yellow with red blush towards the surface exposed to the sun. Pulp is light yellow in colour, sugar/acid ratio is good and pleasant flavour. Stone is free, small, round-oval, shell is 70% and kernel 30%. Kernel is sweet. The cultivar is suitable for table purpose as well as drying. It is dried even on tree, if not harvested.

### *Australian*

This is also a variety introduced to cold arid regions of India. It ripens in end July to late August. Fruit size is extra large among the cold arid cultivars. Fruits are round in shape with acute apex, structure distinct, mild flavour, medium sweet and acidic, not liked for table purpose and good for processing. Stone size is big, free, round and compressed. Shelling (%) is about 72 and kernel 28%. Kernel is sweet.

### *Shakarpara*

This is also an introduced cultivar from Afganistan. Fruits are sweet with pleasant aroma. It can be used for table purposes. Ripening period is late July - mid August. Fruits are medium in size, round, skin is glossy, creamish yellow with rosy blush, pulp soft, light yellow in colour, sweet and less acidic and pleasant flavour. Stone is small, ovate, apex pointed one ridge on both sides, shell 78% and kernel 22%. Kernel is sweet.

Some of the important cultivars of apricot growing in other parts of the country are described below (Chadha, 2001).

### *Shipleys Early*

Fruit is medium-sized, creamy-white. Flesh is sweet and juicy and kernel is bitter. The fruit ripens early in middle of May and is of dessert quality.

### *New Castle*

It is an early cultivar ripening in May. The fruit is round, medium sized and barium-yellow. The flesh is sweet and juicy. It is a popular dessert cultivar for warmer locations.

### *Charmagz*

It is a self-incompatible cultivar and needs a pollinizer like Turkey. The fruit is medium in size and roundish flat in shape. The skin is straw-yellow with light yellow flesh which is very sweet and highly flavoured. It is suitable for dessert and drying purposes. The kernel is sweet. It has high chilling requirement and is a shy bearer.

This cultivar is also found growing in cold arid regions.

### *Kaisha*

Fruit is roundish and medium sized. Skin is yellowish with a red blush. The flesh is orange yellow, sweet and delicious. It ripens in early June and bears fairly good. It has bitter kernel.

### *Ambroise*

The fruit is oval, large, orange yellow with tiny dots of pink. The flesh is deep yellow, firm, juicy and sweet. It is a late season cultivar, ripening in end of June to first week of July. It is not suitable for lower midhills.

### *Moorpark*

The fruit is round and large. The skin is light orange with a light red blush. Flesh is juicy, sweet and excellent in flavour. Fruit ripens in first week of June. It is a shy bearer.

## Propagation

### *Seed Propagation*

Apricot is generally propagated on seedling rootstock world over. Freshly harvested seeds are collected and soaked in water for 10-15 days in end of February so that its hard seed coat gets softened. Thereafter, seeds are stratified by putting them in alternate layer of seed and sand in a box/container which is kept away from the sun, in shade. Regular watering is done to keep it sufficiently moist. When, 25-30 days after stratification, seeds break dormancy and start sprouting. The seeds should be separated from the box and sown in well prepared nursery bed. Apricot seed should be sown at a distance of 45 x 20cm. First fortnight of April is ideal for sowing of stratified apricot seeds for raising rootstock. The nursery bed should be kept moist till germination of seedlings is completed. For this light irrigation should be given twice a week. After germination of seeds, normal cultural practices are adopted for healthy raising of rootstocks. About 80 % stratified seeds germinate and 70-75% seedlings attain graftable diameter of pencil thickness, in November of same year, in a properly managed nursery (Anon, 2003). These seedling rootstocks can be used in future for commercial propagation of apricot.

### *Vegetative Propagation*

Various methods of grafting and budding are developed world over. In cold arid zones of India, vegetative propagation is not practiced by locals which often result in long gestation period and wide variation in fruit quality. This poses serious problem in development of apricot nursery and availability of genuine planting material for commercial cultivation of apricot in such areas. There is lack of

a) standard clonal propagation methods,
b) available standard rootstocks and
c) scion materials from superior cultivars.

To meet demand of local orchardists, the State Department imports clonal planting material of apple plants from Kashmir valley and Himachal Pradesh. This is adversly affecting expansion of new plantation of apricot in cold arid zone. To overcome such problems, efforts are made to propagate apricot through asexual methods under cold arid conditions. Mir *et. al.,* (1992) recommended annular method of budding of apricot over ring budding during third week of June to first week of July. Chip budding (Dwivedi and Singh, 2001), 'T' budding (Dwivedi *et. al.,* 2000) and cleft grafting (Dwivedi *et al.,* 2000) are successful at different periods of year in cold arid zones of Ladakh. Chip budding was the most suitable method for cold arid zones (Dwivedi *et al.* 2002b).

### *Micro Propagation*

Development of protocols for micropropagation of apricot in the cold arid zones is the need of the hour, not only to provide elite planting material for commercial cultivation but also to conserve valuable genotypes of such areas. Various explants viz. immature embryos, shoot tips, excised shoot segment, etc. were used by various researchers using ½ MS, WPM and Lloyd and McCown nutrient medium with varying levels of plant bio regulators viz NAA, IBA, 2,4-D (0.05-0.5 ppm) and BA (1.0-2.5 ppm) (Lane and Cassio, 1986). Shoot proliferation protocols were standardized for 'Halman' cultivar of apricot under cold arid conditions of India using ½ MS medium and 2.0 ppm BA and 0.1 ppm IBA (Anon, 2005).

## Rootstock

Unlike other temperate fruit crops, little work was done on breeding and development of rootstock in apricot. Choice for rootstock is still limited in apricot. Apricot seedlings are the most commonly used rootstock for vegetative propagation of apricot. It is ideal for well drained and drought affected areas. The deep and extensive root system provides a strong anchor even in rocky soils having poor depth and nutrition stress. They perform well in ravine and affected soil. It is sensitive to high moisture and heavy soils. Apricot stem cuttings are difficult to root. Hard wood cuttings with varying levels of IBA (2500-5000ppm) and NAA (1500-5000 ppm) and their combinations yield poor rootability (below 30% success).

Apricot can also be grafted on seedling or plum rootstock. It is recommended for heavy soils and soil having high moisture conditions. Peach seedlings can also be used as rootstock for apricot scion. Apricot seedling does not have suckering habit whereas, plum and peach rootstocks have suckering habit. Apricot is compatible with apricot, peach and plum and no incompatibility problems are reported in such cases.

## Rejuvenation of Old Unproductive Orchard

Primary seedling trees of apricot are available abundantly in cold arid region of India. They have well established root system and have the ability to tolerate moisture and nutrient stress. They are unproductive since they are old and produce poor quality and small sized non edible fruits. These unproductive plants can be converted to productive ones successfully by grafting/budding with superior cultivars. Shield method of budding performed during June-early July was the most successful (90%) for cold arid conditions (Dwivedi and Singh, 1999). This technique can also be used for developing orchard of a tree and conservation of valuable germplasm.

## Agro Techniques

### *Planting System*

Apricot is generally planted in and around the cultivated fields as a mixed cropping system. No planting system is adopted even for commercial cultivation by farmers in cold arid regions. Square method having a spacing of 5-6 m is recommended for these zones. Rectangular system can also be adopted because more plants can be accommodated in a row (Westwood, 1978). A planting density of 2500-3333 plants / ha can be obtained by adopting closer planting distances. High density planting in apricot is still not popular due to absence of suitable dwarfing rootstock and scion cultivar. High density orchards can be managed by proper training and pruning systems (Rathore, 1991). Tree canopy in high density plantation can also be managed by using intermediate rootstock and application of plant bioregulators like paclobutrazol and adopting training systems like free spindle, tatura, open centre, dwarf pyramid and cordon.

## Irrigation Management

Water requirement of apricot depends on climatic conditions, type of soil, cultivars, tree age and method of irrigation. Though, apricot is tolerant to low moisture conditions, yet assured irrigation is essential for its commercial cultivation. Water stress at the time of fruit set and fruit growth affect the size, quality and yield of fruits adversely. Feeder roots of apricot are shallow and hence, require light and frequent irrigation under sandy soil conditions of cold arid conditions. Mulching of plant basins with dry grasses and black polythene reduce frequency of irrigation (Dwivedi and Attrey, 2002). Drip followed by sprinkler methods are advocated over furrow or ring methods of irrigation. Drought tolerant cultivars are also developed in Europe. Such studies need to be initiated for cold arid conditions.

## Integrated Nutrient Management

Nutrients play a major role in production of healthy apricot trees and good quality fruit. Soils of cold arid region are very poor in nutrients like N, P and micronutrients especially Zn, Fe, Cu and Bo. All regions have adequate levels of potassium. Since there are no planned orchards and farmers generally plant apricot trees bordering fields of wheat, barley and vegetables, nutrient management for apricot, individually, is lacking. Trees depend for their nutrition on fertilizers applied to field crops and these are thus, shared between the two. A study conducted by Narboo (1994) has clearly shown acute deficiency of phosphorous, calcium, magnesium, sulphur and most of the micronutrients in bearing and young trees. There is no recommended dose of nutrients and stage and method of application of nutrients for apricots in cold arid regions.

There is immense scope for organic production of apricots and thus, use of biofertilizers should be promoted on a large scale for standardisation of nutrient management practices.

## Canopy Management

### *Pruning*

In Apricot pruning depends upon soil, cultivar, rootstock, planting density, tree type, training system and harvesting method adopted. In fertile soils and vigorous cultivars, heavy pruning is recommended which allows exposure of plants to sun. Under poor soil conditions and short growing period of cold arid conditions, light to medium pruning is advisable. In initial four years, the basic aim of pruning is to provide desirable shape to the tree and to train it to a desirable form. Thereafter, thinning of overcrowded, criss-cross branches and removal of excess vegetative growth is done. The basic concept which should be kept in mind while pruning apricot is that its flowers are borne on lateral buds, one year old shoots and two year old spurs (Westwood, 1978). The apricot spurs have a short life of 2-3 years. Thus, pruning should induce new growth every year, essential for induction of new spurs. The best time of pruning is spring, March. Average annual growth of shoots in young trees should be 40-50 cm and 20-30 cm in old trees. The old trees can be pruned heavily in March to rejuvenate the trees. Pruning is done to remove dried and diseased branches and to maintain proper ratio between root and shoot growth. It is also essential to provide proper and strong shape to the tree to enable it to bear the load of quality fruit.

### *Training*

Training is one of the most important operations for quality production of apricot in cold arid zones since apricot fruits develop red blush on fruit surface which is exposed to sun. Orchardists do not follow any training practice in cold arid zones. Trees are allowed to grow freely which cause substantial damage of fruits during harvesting (Dwivedi and Dwivedi, 2004). Modified leader system is the most common method recommended for apricot. Open vase system can be adopted if mechanical harvesting facilities are available. The cold arid zones are not prone to damage from hail storms, which should be considered while deciding method of training. Modern training systems viz. palmette, vase palmette, cordon, tatura, kecement hedge and border hedge can also be adopted for commercial production of apricot.

## Weed Management

Weeding can be done manually or by use of chemicals. In manual weeding, weeds are removed manually or by frequent light hoeing. Chemical method of weeding is much cheaper than manual methods. Application of weedicides largely depends on nature of soil, weed and intercrop or cover crop grown in the orchard. Also the effectivity of herbicide depends upon stage of growth of intercrop/cover crop and intensity of weed flora in crop field. In general pre-emergence herbicides are applied only during initial land preparation for the annual or perennial fruit crops but post emergence herbicides are used frequently in the fruit tree orchards to control weeds and reduce cost of production due to weeding, to increase yield / unit area, to minimize labour problem, to make mechanized harvesting easier, to prevent disease and pest infestation (Hayes, 1970).

### *Flowering and Fruiting*

Usually three buds develop in axil of a leaf at each node on a shoot and spur in apricot, the central one being a vegetative bud while the two side ones are floral. Number of flower buds on a shoot decreases from base towards the tip. Duration and time of flowering largely depends on cultivar and prevailing weather conditions. Apricot flowers between first week of April to first week of May under cold arid conditions depending upon the altitude and cultivar. Gotso variety is the earliest to bloom in Ladakh.

Most of the commercial apricot cultivars are self-fruitful .Pollination studies revealed fruitset without pollinizers but open-pollination has always given better fruit set. All the apricot cultivars are self - compatible except Charmagz, Perfection and Riland which are self-incompatible. The mode of pollination in apricot is through insects, especially honey bees.

Fruit set in apricot is rather heavy which results in under sized fruits and increases the tendency of biennial bearing. There is need for fruit thinning which results in increased fruit size, promotes regular bearing, decreases limb break and maintains tree vigour. It should be done within forty days after full bloom. For fruit thinning both manual and chemical methods are employed. In manual thinning, flowers/fruits are removed to make 6-10 cm space in between fruits. But being a labour intensive and costly method, in general, chemical methods are preferred over manual thinning. Such studies need to be restandardised for cold arid conditions.

## Harvesting and Yield Maturity Indices

Fruits are mature when they attain their full size with proper completion of ripening process. Apricot fruits do not mature uniformly, so selective harvesting has to be done. Systematic studies need to be undertaken for cold arid regions.

In USA, under Californian conditions, at least three-fourths of outside surface area must have attained a yellowish green colour and flesh should be at least half yellow as described in a standard colour chart for harvesting of fruit. Number of days from full bloom and the accumulated heat units above 7 C are used as guides by growers in setting the harvest dates. The first six weeks after full bloom is critical in respect of temperature effects. High temperatures (39-39.5° C) late in growing season cause softening of flesh near the pit and subsequent browning.

Apricots must be tree ripe to attain a pleasing flavour characteristic of fruit. Neither fruit with greenish colour that are firm to hard should be selected nor soft ones. Fruits with full colour, firm ripe to ripe, should be selected but not those that are mushy and overripe. Soft spots on fruit surface indicate beginning of decay.

Maturity test of Titlon variety of apricot using Magness-Taylor Pressure Tester with 5-16 inch diameter plunger was as follows.

| | TSS (%) | Pressure test (lb) | Consumer acceptance (%) |
|---|---|---|---|
| Mature but not ripe | 11.3 | 8.5 | 85 |
| Av. commercial Maturity | 10.0 | 11.3 | 56 |

## Harvesting Practices and Yield

### *Harvesting*

Ripening of apricot generally sets in by mid July and lasts upto first week of October as it is affected by altitude and cultivar. Rogan is first to ripen while Dey Chulli is last to ripen under Ladakh conditions. Harvesting time of apricot depends mainly upon method of harvesting, cultivar and use of variety. For table purpose, fruits are harvested when they change colour from green to yellow but slightly more ripe fruits are harvested for freezing. Fruits which are less mature but still firm enough to be pitted and processed are good for canning while for drying fully ripened fruits are harvested. The best method to judge time of harvest can be determined by measuring fruit firmness, dry matter content and refractive index of the cultivar. Fruits should be harvested with pedicel attached else there are greater chances for damage by micro-organisms.

### *Yield*

Yield of apricot showed vast variation depending upon soil conditions, cultivar, orchard management practices and altitude of orchard. Vegetatively propagated apricot starts bearing at age of four years and attains full bearing after 10 years. Fruit yield varies from 20-100 kg / tree / year. An average yield of 60-80 kg fruits / tree was recorded under normal conditions. Productive life of tree is about 40-45 years.

## Post Harvest Technology and Post Harvest Handling

Lack of knowledge regarding maturity indices, harvesting, grading, packaging methods and poor storage facilities result in 40-60 % loss in apricot. Apricot is generally preserved by open sun drying which affects fruit quality. Solar driers viz. polyhouse, tent and tunnel driers are developed for hygienic dehydration of surplus fruits and vegetables in the region. Methods for value addition viz. osmotic dehydration, intermediate moisture, apricot bar, hurdle technology, squash, nectar, and jam are also developed.

Various products like apricot bar, apricot juices/beverages are developed by Osmotic Dehydration (OD), Intermediate moisture (IM) and Hurdle Technology (HT).

Another potential area is extraction and processing of apricot kernel oil which is of medicinal and cosmetic use and is found in abundance in the region.

## Storage

Apricot is a soft and delicate fruit which needs to be consumed/ processed within 3 days after harvesting. Fruit harvested at optimal maturity can be stored for 72-80 hrs at room temperature. Storage of ripe fruits can be extended to 7-8 days if stored at 5 degree C and 85% relative humidity. Under controlled atmospheric conditions, it can be stored for a period of 10 days at $3^{o}$ degree C.

## Processing

Most of the cold arid regions of the country are remotely located and this poses a problem for marketing of fresh fruits due to their perishable nature. Ripe fruits can be marketed in local markets within 3 days. During storage at ambient temperatures, apricot often develops abnormal flavour within 3 days of harvesting. To avoid further spoilage of fruits it should be processed or preserved in the form of pulp or other products for subsequent utilisation. Some of the products that can be prepared from apricot are listed below:

### *Apricot Pulp*

There is good demand for apricot pulp from fruit processing factories. lApricot pulp may be used as a base for preparing apricot nectar/squash, RTS-beverage, jam, mixed fruit-jam, apricot leather, sauce, baby-food and ice cream, . Bulk preservation of apricot pulp in its growing areas should receive priority attention. But before commencing for such production, a project tie-up should be made with big fruit processing houses/factories. Apricot pulp can be preserved by chemicals or by thermal processing and by the latest technology of "aseptic packaging".

### *Apricot Jam*

Apricot jam can be prepared either from fresh fruit or preserved pulp. It has good demand in the market and can also be incorporated in mixed fruit jam. Apricot jam is rich in carbohydrate, carotenoids and amino acids and is a good source of energy.

### *Apricot Compote*

Apricot compote is like a preserve (murabba) and is prepared by steeping or preserving fruit pieces (or halves) in syrup. In this product, fruit pieces are embedded in syrup and fruit identity is not lost unlike in jam.

### *Apricot Fruit Bar*

Those varieties of apricot like Kho Chuli, which do not fetch a good price as dried fruits can be utilized profitably for making apricot leather or fruit bar, a product similar to mango leather (Aam papar).

### *Apricot Squash/Nectar/RTS Beverages*

Apricot pulp is used extensively for production of various beverages and juices. It is processed alone or blended with other fruit juices/pulp. It has good market in India and abroad.

### *Apricot Chutney*

Delicious, appetizing fruit chutney can also be made from low grade apricots.

### *Apricot Baby Food*

Baby food prepared from apricot pulp is nutritious, palatable and a good source of phosphorous, iron and calcium. Apricot is a good source of carbohydrates and vitamin 'A' and contains sufficient amount of thiamine.

### *Apricot Fruit Based Products*

The term 'fruit based product' is used in dairy industry for use in dairy products like ice-creams, yoghurts, sour-cream, whey-based topping and spreads – cheese with fruits.

The product may be in the form of small fruit pieces or in puree form with sugar, stabilizers or other ingredients and is mostly in bulk packs. It is a tailor made product and has application in jam industry, dairy industry, cereal baby/health foods and in hotels and restaurants. In common terminology it is known as 4 + 1 pack and 3+2+1 pack meaning 4 parts fruit and one part sugar (for 4+1 pack).

### *Canned/Bottled Apricots*

Among various varieties of apricot grown in Ladakh region Halman, Shakarpara, Sufaida, Nari, Rakchey Karpo and Tokpopa (in order of preference) are suitable for canning or bottling in terms of texture, flavour, colour and overall quality of the product.

Apricots are dried as whole fruits or as halves after removing the internal pit.

### *Dried Apricots*

Brown discoloration is a major problem in drying of apricots. To get a good golden yellow coloured dried product, it is necessary to give the whole or half apricot pieces certain pre-treatments like exposing to sulphur dioxide fumes or dipping in potassium Meta bisulphite solution. Such pre-treatments help in checking brown discoloration of the product to some extent during drying and also during storage. Maximum permitted limit of $SO_2$ in dried products is 2000 ppm. Fully mature and ripe fruits should be selected for drying. Drying can be accomplished either by direct sun drying or it may be dehydrated in electrically operated dehydrators. Osmotic dehydration is another alternative. Yield of whole dried apricot is 30-35 %. In order to have good keeping quality and tender texture, apricots should be dried to a moisture content of about 16-18 %. Statutory maximum requirement of moisture content is 20 % in dehydrated apricots and 24 % in sun dried apricots under Ladakh conditions.

### *Hurdle Technology*

In recent times self stable high moisture fruit slices (50-60 % moisture) have been developed in many developed countries. This is based on a combination of various preservation techniques by creating several hurdles for micro organisms responsible for spoilage. It involves slight reduction in moisture, acidity (av 0.92-0.95), lowering of pH (below 4.5) and mild heat treatment (pasteurization at 85 °C).

### Osmotic Dehydrated Apricot

This is also one of the modern methods of fruit preservation. Besides apricot the fruits of aonla, apple, cherry, citrus, pear, peach, papaya and plum, can also be preserved through this method.

### Apricot Cake

It is reported that apricot cake after oil extraction, contains 0.06 % hydrocyanic acid and is thus, not recommended for animal feed as such. It can be used as animal feed only after removal of essential oil (1.2-1.8 %) by distillation process. The apricot cake is used for feeding animals in Ladakh, only after boiling.

### Apricot Kernel

Apricot seed/pit, which constitutes about 10-15 % of fruit, yields about 22-30 % of kernel. Wild apricot seedling yields bitter kernel which constitutes about 22 - 34% of the pit. These bitter kernels are utilized for oil extraction. The oil is traditionally extracted from mildly roasted kernels by crushing along with water in a stone mortar. Oil yield by this method is quite low to the tune of 30 %. The oil content in kernel varies from 36-62 %. The apricot oil is rich in ? carotenoids and Vitamin E. It contains 70-75% oleic acid and 16-22% linoleic acid as major fatty acids. The protein content in kernels ranges from 20.5-45.2% (Dwivedi, 2004). Both the bitter and sweet kernel may yield oil and show almost similar physico -chemical characteristics. Bitterness in apricot kernel is due to presence of "Amygdalin" which on hydrolysis yields hydrocyanic acid, harmful for human consumption. Thus oil cake must be detoxified before using it for food purposes.

Kernels may be sweet or bitter. Composition of kernels is reported to be:

| | | |
|---|---|---|
| Moisture : 4.3%, | Protein : 31.4 %, | Oil : 53.4% |
| Sugar : 8.1 %, | Fibre : 4.8 %, | Ash : 2.6 % |

## Apricot Oil

A sizable quantity of apricot kernels are available in Ladakh which are mostly used for the oil extraction. Two types of apricot seed are found –those with sweet kernels and others having bitter kernels. No morphological difference is recorded between these two.

Fruits of sweet kernel varieties are dried along with stones for marketing for consumption as dry fruits and thus little quantity of sweet kernels are available. At present sweet apricot kernel fetches about Rs. 100.00 / kg whereas the bitter kernels are unfit for consumption as such. These originated from semi wild species and are being utilized mostly for oil extraction purposes. There are only a few varieties which yield bitter kernels.

The oil is traditionally extracted from mildly roasted kernels using a stone mortar. The wild and semi wild trees are the main source of kernels used for oil extraction purposes. The physico-chemical studies of oil indicated no difference in the oil extracted from bitter and sweet kernels (Bhatia et. al, 1996.) and have found it nearly similar in composition and appearance to almond oil. Oil content in apricot kernels varies from 36-62 % with stage/ type of kernel used and method of extraction. Oil colour is generally light to deep yellow. Oleic acid (70-75 %) and linoleic acid (16-20%) are the major fatty acids.

The sweet kernel apricot oil is consumed by humans as well as animals whereas bitter kernel oil cakes are harmful for human consumption as it contains amygdalin which yields toxic hydrocyanic acid. The sweet kernel oil is used for cooking purposes and that from bitter kernels as hair oil and for illumination purposes. Apricot oil is the best for warmth effect when applied to body and prevents dryness of skin. It is also extensively used in joint pains and for body massage.

## Plant Protection

Cold arid regions do not host many diseases and pests of apricot because of climate advantages, i.e. ultra low temperature, low humidity and long winters. A few pests and diseases affecting yield and quality of apricot are discussed below:

## Insect Pest Management

### *Codling Moth (Cydia pomonella)*

Codling moth (*Cydia pomonella*) is a major pest of apple in cold arid zones. The pest causes severe damage in apricot also. Since the pest is prevalent only in Ladakh region, the government of Jammu and Kashmir has imposed strict restriction on movement of fresh fruits from Ladakh under SRO 397 dated 8$^{th}$ Septmeber, 1981 issued under plant disease act, 1973 to avoid entry of this pest in other temperate zones of the country (Dwivedi, *et. al.*, 2001). The female moth after mating lays eggs on fruits, leaves and twigs. After hatching, larvae come out and bore into fruits generally from calyx end. Adult emergence takes place in June. The infested fruits drop prematurely which can not be marketed or used (Waadi and Sethi, 1975). Use of light trap, sex pheromone, burlapping of trees trunk in autumn and removal of dried/dead bark which provide shelter to larvae during winter are bio control measures, besides release of *Trhichogramma embiophagum* and 2-3 sprays of suitable insecticides at 10-15 days interval from pea stage of fruit growth.

### *The Blossom Thrips (Thrips spp.)*

The blossom thrips attack flowers and buds which affect fruit set and fruit quality. They also infest delicate leaves and cause necrosis and silvery appearance.

### *Two spotted Spidermite*

This is a minor pest of apricot in cold arid zones. It attacks foliage and feeds on green matter causing loss of chlorophyll and dropping of leaves. Infestation is noticed only in lower altitudes. Spraying of wetable sulphur and dicofal are recommended to manage the pest.

### *Gypsy moth (Lymantaia obfuscata)*

Gypsy moth is a polyphagous pest and mostly attacks apricot especially in lower belts i.e. altitude below 3000m. The caterpillars are nocturnal and gregarious feeders and in severe cases they defoliate the plant completely and affect fruit formation. The eggs can be destroyed in early spring.

### *Aphids*

Aphids also attack apricot plantation especially at young stage or new and tender leaves in spring and suck the sap. In severe cases, they result in complete defoliation of tree and even premature fruit drop. Nymphs and adults suck the sap from tender leaves, petiole blossoms resulting in curling of leaves.

Other pests attacking apricot in the cod arid zones are white grub, pin hole borer, leaf minor and scale insects which are recorded in apricot growing areas. These pests are not reported to cause significant damage.

## Disease Management

### *Gummosis*

Exudation of gums from stem and branches is the major problem in apricot in cold arid conditions due to poor nutrition and soil conditions. Most of the cold arid cultivars are susceptible to this disorder. Gummosis loosens the bark which splits out causing secondary infection and leads to drying of branches/tree in severe cases.

### *Blight*

Blight also attacks twigs, leaves, blossoms and fruits of apricot in the cold arid regions. Brown lesions develop on plant parts and spread out. This results in formation of short holes on leaves. Application of carbendazim (0.05%) controls the disease.

Physico-chemical characteristics of genotypes of apricot from cold arid region of Ladakh

| Parameters | Range | | Mean |
|---|---|---|---|
| | Min | Max | |
| Moisture (%) | 70.4 | 82.5 | 75.0 |
| Fruit weight (g) | 8.7 | 35.5 | 19.04 |
| Fruit volume (ml) | 9.3 | 34.6 | 20.4 |
| Specific gravity | 1.02 | 1.14 | 1.10 |
| Fruit length (mm) | 26.5 | 44.0 | 34.00 |
| Fruit width (mm) | 24.9 | 41.8 | 33.20 |
| Cheek diam (mm) | 22.9 | 38.8 | 31.00 |
| T.S.S. (°B) | 12.0 | 33.4 | 19.35 |
| Pulp weight (g) | 2.5 | 31.2 | 16.97 |
| Stone weight (g) | 1.0 | 20.0 | 2.37 |
| Pulp:stone ratio | 1.5 | 17.86 | 9.42 |
| Kernel oil (%) | 27.0 | 45.0 | 35.0 |
| Av. kernel weight (g) | 0.3 | 0.5 | - |

## Future Research Needs

Development of techniques for post harvest handling of fruit crops is the immediate requirement for cold arid region. Lack of knowledge regarding maturity indices, harvesting, grading, packaging methods and poor storage facilities result in 40-60 % loss in apricot alone.

Since the fruit is highly perishable and the growing areas remote, fruit processing and preservation techniques need to be developed for complete utilization of quality produce. Quick transportation of fresh fruit to better and more lucrative markets by air or refrigerated vehicles needs to be explored.

Another potential area is extraction and processing of apricot kernel oil which is of medicinal and cosmetic use and is found in abundance in the region.

Composition of fresh and dehydrated Indian apricot (per 100g edible portion)

| Constituents | Fresh apricot | Dehydrated apricots |
|---|---|---|
| Moisture (%) | 85.3 | 19.4 |
| Protein (%) | 1.4 | 1.6 |
| Fat (%) | 0.39 | 0.7 |
| Mineral (%)s | 0.7 | 2.8 |
| Fibre (%) | 0.6 | 2.1 |
| Carbohydrate (%) | 11.12 | 73.61 |
| Sugar (%) | 1.57 | 11.85 |
| Energry (kcal/100g) | 48.0 | 306 |
| Calcium (mg/100g) | 20 | 110 |
| Iron (mg/100g) | 0.54 | 4.6 |
| Potassium (mg/100g) | 89 | – |
| Phosphorus (mg/100g) | 25 | 70 |
| Thiamine (mg/100g) | 0.04 | 0.22 |
| Riboflavin (mg/100g) | 0.13 | – |
| Niacin (mg/100g) | 0.6 | 2.3 |
| Vitamin C (mg/100g) | 10 | 2.0 |
| Vitamin A (mg/100g) | | |
| Carotene (IU/100g) | 2612 | 18 |

*Source:* Nutritive value of Indian foods, National Institute of Nutrition, Hyderabad

Besides the above, lack of quality planting material from registered nurseries is a constraint. Propagation of superior varieties through methods already standardised for the region needs to be given an impetus.

### *Future Thrust*

i) Standardization of fast methods of propagation for enhancing local availability of quality planting materials

ii) Development of modern agro techniques and canopy management suitable for cold arid regions

iii) Identification / development of improved cultivars for different purposes

iv) Collection, evaluation and conservation of valuable germplasm

Physico-chemical characteristics of common apricot cultivars of Ladakh

| **Cultivars** | **Length (cm)** | **Breadth (cm)** | **Chick diameter (cm)** | **Fruit weight (g)** | **Pulp weight (g)** | **Stone weight (g)** | **Pulp stone ratio** | **TSS (°B)** | **Titratable acidity (%)** |
|---|---|---|---|---|---|---|---|---|---|
| Afghani | 2.63 | 2.64 | 2.63 | 13.2 | 12.0 | 1.2 | 10.0 | 19.6 | - |
| Australian | 4.15 | 3.86 | 3.67 | 33.2 | 30.8 | 2.0 | 15.4 | 15.0 | 0.92 |
| Charmagz | 3.44 | 3.76 | 3.42 | 24.8 | 23.0 | 1.8 | 12.8 | 19.0 | - |
| Halman | 2.43 | 2.43 | 2.39 | 11.6 | 10.2 | 1.2 | 8.5 | 18.4 | 0.67 |
| Raktsey Karpo | 3.85 | 3.82 | 3.63 | 34.6 | 32.0 | 2.4 | 13.3 | 21.2 | 0.61 |
| Safeda | 3.35 | 3.17 | 3.42 | 19.5 | 18.1 | 1.4 | 12.9 | 18.0 | 0.64 |
| Shakanda | 3.05 | 2.89 | 2.78 | 11.5 | 10.0 | 1.5 | 6.6 | 19.3 | 0.95 |
| Shakarpara | 3.73 | 3.58 | 3.64 | 28.0 | 25.3 | 2.7 | 9.3 | 12.0 | 0.13 |
| Tokpopa | 3.30 | 3.34 | 3.39 | 14.5 | 12.8 | 1.7 | 7.5 | 12.2 | 1.00 |
| CD (0.05) | 1.51 | 1.41 | 1.15 | - | - | - | - | 3.93 | 0.09 |

Performance of different methods of vegetative propagation in apricot in cold arid region of Ladakh.

| **Method/ Techniques** | **Time** | **Sprouting (%)** | **Bud take (%)** | **Days taken in Sprouting** | **Saleable plant (%)** | **Linear growth (cm)** | **Radial growth (cm)** |
|---|---|---|---|---|---|---|---|
| Grafting – Tongue/Cleft | First week of February | 95.0 | 90.0 | 73.5 | 82.5 | 164.0 | 0.55 |
| Budding - Chip | Fourth week of February | 90.0 | 85.0 | 71.4 | 77.5 | 28.4 | 0.58 |
| Budding - Shield | Second fortnight of August | 88.3 | 81.6 | 29.5 | 76.5 | 17.0 | 0.45 |
| Top working - Shield | Second fortnight of July | 93.3 | 80.0 | 31.5 | - | 76.5 | 0.80 |

Physico-chemical composition of apricot kernel oil of Uttaranchal and Ladakh and comparison with almond oil and foreign apricot oil

| Particulars | Uttarakhand | | Ladakh | | Foreign apricot oil | Indian almond oil |
|---|---|---|---|---|---|---|
| | Sweet apricot oil | Wild apricot oil | Sweet apricot Oil | wild apricot oil | | |
| Oil content of Kernel (%) | 44.21 | 44.55 | 37.6 | 41.4 | --- | 52.31 |
| Colour | Pale yellow | Light pale yellow | Pale yellow | Yellow | --- | Colour less |
| Sp gravity at 20 | 0.9230 | 0.9154 | 0.9235 | 0.9118 | 0.912-0.916 | 0.9254 |
| Refractive Index | 1.4722 | 1.4718 | 1.4679 | 1.4679 | 1.462-1.465 | 1.4718 |
| Acid value | 27.14 | 20.10 | 0.41 | 0.81 | 0.2-40 | 1.49 |
| Saponification value | 191.3 | 193.8 | 294.5 | 293.0 | 188.2 | 200.00 |
| Iodine value | 102.90 | 99.58 | 85.0 | 104.0 | 97.109 | 97.09 |
| Palmitic | -- | -- | 10.68 | 3.37 | --- | -- |
| Oleic (%) | | 90-93 | 79.80 | 37.98 | 60-79 | 77.0 |
| Linoleic (%) | | | 9.62 | 28.25 | 18.32 | 17.0 |

v) Detailed scientific studies on post harvest and storage technologies

vi) Creation of basic infrastructure for harvesting, grading, packaging, storage, transportation, processing, marketing and export.

vii) Emphasis on production of organic apricot fruits from the region

viii) Standardization of modern scientific methods, like plant biotechnology and high density orcharding. for the region.

## References

Anonymous 1994. Report of the central team of experts on visit to Ladakh region. Deptt. of Ag. & Cooperation, Min. of Agriculture. Krishi Bhawan, Govt of India. New Delhi-Oct 1994.

Anonymous 2003. Annual Report-2002. Field Research Laboratory (DRDO), Leh.

Anonymous 2004. Annual Report-2003. Field Research Laboratory (DRDO), Leh.

Anonymous 2005. Annual Report-2004. Field Research Laboratory (DRDO), Leh.

Bhatia, A.K.; Bedi, K.L. and Sarain Y.K. 1996. Survey of availability of apricot kernel. Personnel communicated note.

Bhatia, A.K.; Singh, R.P. and Gaur, G.S. 1977. Important apricot varieties of Ladakh. *Prog. Hort.* 8(4). 19-25.

Chadha, T.R. 2001. Textbook of Temperate Fruits. Indian Council of Agricultural Research, New Delhi. pp.26-27.

De Candolle, A. 1886. Origin of cultivated plants. 2nd Ed. Reprinted in 1969. Hanfer, New York.

Dwivedi, Deepa H. 2004. Apricot- A potential oil yielding plant of Ladakh. In: First Indian Horticulture Congress. 6-9 Nov. 2004, New Delhi. Book of Abstract (poster paper). pp. 30-31.

Dwivedi SK and Dwivedi DH 2004 Cold arid horticulture-An overview. In: *Advances in arid horticulture* Vol. I : eds Saroj *et. al.* pp 71-90.

Dwivedi, Deepa H.; Kareem A.; Dwivedi, S.K. and Mir, A.A. 2002 a. Physico-chemical charactersitics of important apricot cultivars of the cold arid region of Ladakh. Ind. *J. Hort.* 59 (2): 118-12.

Dwivedi, S.K. and Attrey, D.P. 2001. Possibilities of apricot production in high altitude regions of Uttaranchal. In: National Seminar on Development of Horticulture in Uttaranchal. Directorate of Horticulture, Govt of Uttaranchal, Dehradun. 3-4 September, 2001. pp. 17-20.

Dwivedi, S.K. and B. Singh 1999. Topworking of apricot in Ladakh. *Prog. Hort.* 3 (1-2): 29-31.

Dwivedi, S.K. and Singh, B. 2001. Studies on chip budding in apricot in cold arid conditions in Ladakh. *Indian J. Hort.* 58 (3): 215-217.

Dwivedi, S.K.; Attrey D.P. and Paljor Eli 2000 a. Standardization of budding in apricot in cold arid conditions of Ladakh. *Prog. Hort.* 32: 117-120.

Dwivedi, S.K.; Singh, B. and Paljor Eli 2000 b. Studies on vegetative propagation of apricot through grafting in Ladakh. *Indian J. Hort.* 57 (1): 39-41.

Dwivedi, S.K; Attrey, D.P.; Paljor Eli and Singh B. 2002 b. Studies on conservation of genetic diversity in apricot in cold arid regions of India. (In) 26th International Horticultural Congress, Toronto (Canada). 11-17 August, 2002. pp. 224-225.

Dwivedi, S.K; Narboo Sonam, Dwivedi DH and Abdul Kareem 2003. Codling Moth-A major threat to the fruit industry of Ladakh. (In) National Symposium on Sustainable Pest Management. Organised at BHU-Varanasi, Jan 2-4, 2003. pp. 79-80.

Faust, M; Suranyi D and Nyujto F 1998. Origine and disseminaiton of Apricot. *Hort Rev.* 22; 225-266.

Hays W.B. 1970. Fruit growing in India. Kitab Sadan. Allahabad.

Jayaraman, K.S.; Vibhakaran, H.S. and Ramanuja, M.N. 1997. Preparation and Evaluation of Shelf stable fruit slices for ambient storage using combination preservation (Hurdle) technique. *Indian Food Packers.* May_June. 5-11.

Mehlendbacher, SA; Cociu, V and Hough, LF 1996. Apricot. *Acta Hort.* 290(1). 65-107

Mir, M.S.; Dhar, M.A. and Sofi, A.A. 1992. Study on vegetative propagation of apricots under cold arid conditions of Ladakh region Pb. *Hort. J.* 32 (1-4): 114-117.

Moorcraft, W. and Treback, G. 1841. *Trends in Hindustan from* 1819-1825. Vol. I. London.

Nurboo, S. 1994. Nutritional status of apricot plantation in some orchard areas of kargil district. M.Sc. (Ag.). Thesis. Pomology Division, SKUAST, Shalimar-Srinagar (J&K).

Rathore, D.S. 1991. Apricot. In: Temperate fruits. Eds. Mitra, S.K., Rathore, D.S. and Bore T.K., Noya Prokash. Calcutta. Pp. 279-308.

Rehdar, A. 1940. Manual of cultivated trees and shrubs hardy in North America. Macmillan, New York.

Sharma, T.R.; Sekhon, K.S. and Sainis 1992. Studies on canning of apricot. *J. Fd. Sci. Technol,* 29 (1): 22-25.

Wadi, S.R.A. and Sethi, G.R. 1975. Eradication of codling moth- A suggestion. *J. Nuclear Ag. Biology.* 4: 18-19.

Westwood, M.N. 1978. Temperate zone pomology. W.H. Freeman & Co., Sanfransisco.

Wills, RBH, Scriven, F.M. and Greenfield, H. 1983. Nutrient composition of stone fruit cultivars. *J. Sci Food and Agri.* 34: 1383-1389.

**Underutilized and Underexploited Horticultural Crops, Vol. 2, 2007**
*K.V. Peter (ed.), pp. 89-103*
New India Publishing Agency, New Delhi (India)
E-mail : newindiapublishingagency@gmail.com
Web: www.bookfactoryindia.com

# 5 Low Chilling Peaches

**AKATH SINGH, R. K. PATEL, K. D. BABU AND L.C. DE**
Division of Horticulture, ICAR Research Complex for NEH Region
Umiam - 793 103, Meghalaya

India produces all deciduous fruits including pome fruits (apple and pear) and stone fruits (peach, plum, apricot and cherry) in considerable quantity. These are mainly grown in the North-Western Indian States of Jammu and Kashmir (J&K), Himachal Pradesh (H.P.), Uttarakhand and in Uttar Pradesh. The North-Eastern Hills region, comprising of the States of Arunachal Pradesh, Nagaland, Meghalaya, Manipur and Sikkim also grow some of the deciduous fruits on a limited scale. The sub tropical region lies between $15^0$ and $30^0$ north of equator covering North Indian plains, the gangetic basin, large parts of North eastern state, western ghats, nilgiri regions etc. In this zone spring is generally frost free and chilling availability ranges from 50-500 hours. This region is not considered proper for temperate fruit growing but grows several low chilling varieties of peach, plum, pear etc. Deciduous fruits, such as peach and nectarine, require accumulated exposure to cool temperatures during winter dormancy for bud break and the resumption of normal growth the following spring. Minimum amount of accumulated cool temperature exposure needed for normal growth, which varies by cultivars, is referred to as the chilling requirement. Due to introduction and adaptation of low chilling cultivars of peach, plum and pear, they are also now being grown commercially in certain areas of the north Indian plains. Out of all the deciduous fruits, low chilling peaches is the most important in terms of adaptability and extent of area. Peaches mature early in subtropical areas and thus find an excellent market. Stone fruits, especially peaches are fast growing in sub tropical climate and catching the attention of both growers and consumers. The development of high quality peach and nectarine (*Prunus*) cultivars with low chilling needs of a mild climate has greatly expanded the potential for commercial peach and nectarine production in north and northeastern India.

## Botany and Taxonomy

Peach (*Prunus persica* (L) Batch.) belongs to family Rosaceae, sub family Prunoideae, genus *Prunus* and sub genus *Amygdalus*, with 8 basic and 16 somatic chromosome numbers. All commercial cultivars belong to *Prunus persica* (L) Batch. The nectarine is a botanical variety of *Prunus persica* and is classified as *P. persica* var. *nucipersica* (Hesse, 1975).

Low chilling peaches is mostly low headed widespread tree with glabrous twigs. The attractive pink flowers are produced in advance of leaf emergence. Flowers are solitary, sessile or short stalked, born on one-year-old shoots. The flowers are simple, perfect, complete, perigynous and lateral. Botanically fruit of peach is a drupe. It develops entirely from a superior ovary. The fleshy mesocarp may adhere to the stone (clingstone) or may separate from it (freestone).

## Composition and Uses

Peaches are recommended for low-cholesterol, low fat, low sodium, and weight reduction or diabetic diets. Peaches are a good source of sugar, vitamins, calcium, potassium etc. A medium size peach contains about 40 calories. Prunacin is the principal glycoside present in the pulp while amygdalin is present in the seeds of peach (Voldrick and Kyzlink, 1990). Principal acid in the pulp is mallic acid followed by citric acid.

Various processed products like canned peaches, dried, frozen preserve, jams, nectar juice and beverages are prepared from peaches. Peach kernal is a rich source of protein and oil.

## Origin and Distribution

Evidences proved that peach originated near the city of Xian, China and recorded to be grown as far back as 2000 BC (Childers, 1983). Peaches are commercially grown around the world between $25^0$ and $45^0$ latitude above and below the equator although it is now grown almost all over the world, extending from 10-$49^0$ N (Rana and Sharma, 2002) The total world production of peaches during 2002 are 13400 thousand metric tones (FAO, 2002). It is commercially cultivated in USA, France, Japan, Australia, Italy, Spain, China and India. Maximum production of peach in the world is in USA, Italy, Spain and the lowest in Australia and India .In India, it is being grown in the mid hills zone of Himalayas extending from Jammu and Kashmir to Khasi hills of Meghalaya at an altitude of 1000-2000m above mean sea level. Low chilling peaches are grown in submountaneos regions and Punjab, Haryana, Delhi and Western Uttar Pradesh. It is also grown to a limited scale in hills of South India and in northeastern region of the country.

| Country | Area (000ha) | Production (000 tonnes) | Productivity (tonnes/ha) |
|---|---|---|---|
| China | 1002.50 | 4026.00 | 4.01 |
| Italy | 92.95 | 1726.80 | 18.57 |
| USA | 78.00 | 1449.50 | 14.73 |
| Spain | 70.00 | 1187.80 | 16.96 |
| Iran | 24.02 | 270.00 | 11.23 |
| France | 23.80 | 417.50 | 17.54 |
| South Africa | 22.00 | 218.00 | 9.91 |
| Turkey | 23.90 | 400.00 | 16.73 |
| India | 18.50 | 114.00 | 6.16 |
| Ukrain | 15.00 | 42.00 | 2.80 |
| Japan | 10.70 | 174.60 | 16.30 |
| World | 1380.87 | 10026.20 | 134.94 |

(FAO, 2002)

Area and production in India

| State | Area (ha) | Production (tonnes) |
|---|---|---|
| Punjab | 3569 | 53635 |
| Uttar pradesh | 3015 | 27705 |
| Himachal pradesh | 2167 | 1087 |
| Jammu and Kashmir | - | - |
| Arunachal Pradesh | 440 | 981 |
| Nagaland | 162 | 206 |
| Bihar | 10 | 80 |
| Haryana | 75 | 125 |

*Source:* Horticultural Database NHB (1992-93)

## Soil and Climate

Although low chilling peaches grow well in a wide range of soil types, a deep soil ranging in texture from a sandy loam to a sandy clay loam is preferred. Peach trees are extensively sensitive to poorly drained soils. In areas of poor drainage, roots die, resulting in stunted growth and eventually death of tree. Fertile and heavy soils are hazardous as it makes heavy growth and hence results in winter injury. Peaches grow the best where soil pH is near neutral, below this level; it is necessary to incorporate lime before planting.

Low chilling peaches require warmer climate and is grown in low, mid hills and subtropical plains which require only 100-400 chilling hours during the years for proper flowering and fruiting. The limiting factors in low chilling cultivation are the minimum winter temperature, spring frost, hailstorm, high humidity and desiccating winds during summer. Generally valley areas are not the ideal site because cold air settles in these area and frost and freez injuries are very common. Plain land or land with gentle slope is ideal for low chilling peaches.

## Propagation

Commonly used rootstocks for low chilling peaches in India are:

Peach seedling

Almond seedling-*Prunus amygdalis* (dwarfing root stock)

Plum seedling- *Prunus saliciana* (produce medium to large size trees)

Apricot seedling- *Prunus armeniaca* (highly resistant to root knot nematode)

Most commonly used rootstock is peach seedling. Exceptionally propagated by seed. Seed should not be taken from early maturing varieties, as the germination percentage is low. Seeds require stratification for breaking dormancy. Removal of seed testa followed by GA3 treatment reduces stratification period. Kept seed under alternate layer of moist sand for 1-3 months at temperature 3-5 0C

### *Vegetative Propagation*

#### Tongue Grafting

Best time of tongue grafting is December-January when the stock and scion are dormant. Stock and scion of equal diameter are selected. A smooth slanting cut of 4-5 cm long is made on the rootstock at about 15-20 cm above the ground level. Another downward cut is given starting approximately 2/3$^{rd}$ from top of slanting cut and about one cm in length. This formed a tongue like structure on the stock. Similar cut was made on the lower side of the scion exactly matching cut given on rootstock. The scion having 2-3 buds is then fitted tightly with rootstock and wrapped with polythene strips. The union completed within 1-1$_{1/2}$ months and after that polythene strip is also removed to avoid girdling.

### Shield or T-budding

Shield or T- budding is conducted in April-May or August-September when the bark starts splitting and shows that cambium and tissue responsible for union are active. The bud stick from mother plant is collected and kept under polythene bags containing sawdust or moist gunny bags to avoid drying. The leaves on rootstock around the place at about 10-15 cm height from ground level are removed. Thickness of the stock should be of pencil like and should be healthy. A horizontal cut on the bark up to 1/3$^{rd}$ of the circumference around the stock is given at about 15-20 cm height from ground level. A vertical cut measuring 2-3 cm is given on middle of horizontal cut and bark is loosen with help of budding knife. A single bud from bud stick is removed in the shape of shield measuring about 1.25cm at lower end and 1.25 to 2.0 cm above with an appropriate slanting cut. The shield is pushed into two raised flaps of bark on stock from upper side in such a way that the shield fits properly in place and is covered well by the two flaps of the bark, but the bud should remain exposed. The bud on rootstock should be wrapped with polythene strips. When the bud sprouts and grows up to 10-15 cm, the stock above the union is cut off to encourage scion growth. Polythene strip is also removed to avoid girdling.

## *Varieties*

The most outstanding works on development of low chilling peaches were done by Prof. RH Sharp in Florida in 1952 and several peach and nectarine cultivars were released for commercial cultivation in sub tropical areas. The cultivars are grouped according to estimated chilling requirement, which is given in chill units (cu) (low chill = less than 300 cu; medium chill = 300 to 525 cu). One chill unit is equal to one hour of exposure to the optimum chilling temperature (about 7.2$^{0}$C).

Chilling requirements of peach varieties

| Cultivars | Chilling hours (below 7.2$^0$C) |
|---|---|
| Red Ceylon | 50 |
| Florda Grande | 100 |
| Florda Red | 100 |
| Okinawa | 100 |
| Florda Bell | 200 |
| Florda Won | 200 |
| MC Red | 200 |
| Early Grande | 250 |
| Florda Sun | 300 |
| Sun Red | 300 |
| Desert Gold, Early Amber, Sun Ripe | 350 |
| Florda Home, Florda King | 450 |
| Lapekar, Rio Grande | 400 |
| Rechan | 450 |
| Florda Queen, Sun Gold | |
| Ventura, June Gold, May Gold, Su | 500 |

*Source :* Uppal, Chopra, and Chanana (1993)

| Cultivars | Characters |
|---|---|
| August Pride | Large fruited, all purpose, yellow, freestone for mild winter climate. Sweet, aromatic, rich flavor. Chilling requirement less than 300 hours and self-fruitful. |
| Desert Gold | Very early ripening, yellow clingstone. Heavy bearing. Self-fruitful. |
| Florda King | High quality early season peach. Large, firm, semi freestone, very sweet when fully ripe. Self-fruitful. |
| Florda Prince | Yellow, semi freestone, Large, firm, very good quality fruit when tree ripe. Chilling requirement of 150hours. Self-fruitful. |
| May Pride | Very early ripening, semi free stone peach for warm winter climates. High quality, large fruited peach. Chilling requirement of 175-200 hours. Self-fruitful |
| Tropic Beauty | Excellent quality yellow flesh, firm, freestone peach ripens early May. Produce large fruit when thinned and girdled properly. Chilling requirement of 100-200 hours. Self-fruitful |
| Tropic Snow | Medium sized fruit, skin is white with red blushed. White sweet flesh with free stone. High quality early season peach. Chilling requirement of 175-200 hours. Self-fruitful |
| Vallegrande | This is an excellent quality, semi free stone, yellow fleshed peach. Flesh is firm and ripens to 60% blush at maturity. Flesh becomes red as fruit matures. Chilling requirement of 100-200 hours. Self-fruitful. |

Performance of low chilling peach cultivars under mid hill conditions of Meghalaya.

| Characters | Cultivars | | | |
|---|---|---|---|---|
| | Flordasun | TA-170 (Partap) | Shan-e-Punjab | Meghalaya local |
| Leaf/fruit ratio | 6.50 | 5.16 | 6.57 | 7.67 |
| Fruit/tree | 343.33 | 347.83 | 319.50 | 214.17 |
| Yield kg/tree) | 15.35 | 20.33 | 14.06 | 12.77 |
| Fruit weight (g) | 45.68 | 61.96 | 46.59 | 49.45 |
| Fruit size (length x dia). (cm) | 4.39 x 4.35 | 4.83 x 4.80 | 4.39 x 4.35 | 4.72 x 4.42 |
| Pulp: stone | 12.69 | 13.83 | 11.95 | 13.77 |
| T.S.S. (%) | 9.27 | 9.23 | 9.13 | 7.33 |
| Acidity (%) | 0.62 | 0.79 | 0.66 | 0.52 |
| Ascorbic acid (mg/100 g) | 53.86 | 41.23 | 49.93 | 13.71 |
| Reducing sugar (%) | 1.10 | 1.30 | 1.28 | 0.83 |
| Total sugar (%) | 5.51 | 5.59 | 6.05 | 2.79 |

Shan-e-Punjab

Prabhat

Flordasun

## Orchard Establishment

The planting area should be ploughed or dug and the arrangement of the drainage be made in advance of planting. In hilly areas, contour planting system should be adopted, if the slope is steep, contour lines are marked and half moon terraces (1 meter dia.) are prepared. Pits are dug at 4.5x4.5 m apart on half moon terraces in hills and 6.5 x 6.5 m in the plains. The size of the pit is kept 0.75 x 0.75 x 0.75 m and pit refilled with upper 30 cm soil along with 15-20 kg FYM, 100g Urea, 100g MOP, 300g SSP and 100g chloropyriphos dust or granule. The pits are filled about 10cm above ground level. The best time for peach planting is winter season but in northeastern hill region, June to August is recommended. If there is no rain after transplanting, light irrigation is given.

## Manures and Fertilizers

In low chilling peaches, balance nutrition is of utmost importance to maintain tree health and productivity. Depending upon soil and climatic condition, following manure and fertilizer schedule is given.

Fertilizer recommendation for full grown tree

| States | Age of full grown tree | FYM (kg/ tree) | N (g/tree) | $P_2O_5$ (g/tree) | $K_2O$ (g/tree) |
|---|---|---|---|---|---|
| Mid hills of Himachal Pradesh | 7 | 40 | 500 | 250 | 700 |
| Western U.P. | 7 | 35 | 500 | 200 | 300 |
| Uttaranchal | 10 | - | 350 | 500 | 350 |
| Punjab | 4 | 26 | 500 | 200 | 600 |
| Haryana | 6 | 35 | 540 | 180 | 540 |
| Arunachal Pradesh | 7 | 50 | 350 | 210 | 210 |
| Mid hills of Meghalaya | 5 | 25 | 530 | 250 | 500 |

*Source:* Rana and Sharma (2002)

The whole quantity of FYM along with phosphorus and potash are given during December- January. Half of the nitrogen should be given 15 days before bud break and the remaining half a month later if irrigation facilities are available.

Low chilling peaches are very susceptible to micronutrient deficiency like iron, zinc copper etc. which can be controlled by foliar spray of 0.5-1.0 % ferrous sulphate, 0.5% zinc sulphate (Sidhu et al, 1980) and 0.2 % $CuSO_4$.

## Irrigation

Adequate soil moisture during attainment of fruit maturity is essential because of rapid increase in fruit size. Water stress during growing season may result in less terminal growth and reduced winter hardiness. Excess irrigation causes injury to roots and depletion of plant nutrients.

Phosphorus Deficiency

Potassium Deficiency

Iron Deficiency

Zinc Deficiency

Depending on topography, soil type and water availability, commonly used irrigation methods are basin, contour basin, sprinkler and drip.

## Training and Pruning

### *Initial Pruning and Training*

Peaches are usually trained to an open-center system. Cut back newly planted trees to about 90cm high, just above a lateral branch or bud. If the tree is branched when it comes from the nursery, select three or four laterals with wide-angle crotches (greater than $45^0$) spaced evenly up and around the trunk for the permanent scaffold limbs. The lowest limb should be about 45cm and the highest about 90cm from the ground. If no desirable laterals are available, head the tree to the desired height and cut out all side branches to one bud. A number of shoots develop during the season that can be selected for scaffold limbs.

### *Pruning*

Peach requiree heavy and regular pruning because fruiting occurs laterally only on previous season's growth which bears only once in its life time. Pruning should be done so as to produce 30-70 cm growth under subtropical condition, annually, which is sufficient for maintaining optimum productivity. During early bearing the selected shoots should be lightly pruned by removing 20-30 % linear growth after corrective pruning of the tree. The best time of pruning low chilling peaches is in mid winter.

### *Pruning Older Trees*

To maintain open-center, remove any vigorous upright shoots developing on inside of tree, leaving smaller shoots for fruit production. Prune out poor quality fruiting wood, such as shoots less than pencil-size in diameter or branches that hang downward and are shaded. The desirable wood left for production should be about the diameter of a pencil and from 36-54 cm length. If the length exceeds 72cm, cut off about one-third of this fruiting branch.

Never prune a bearing peach tree heavily enough to make eventual thinning of the fruit unnecessary. Such heavy pruning drastically reduces the crop as well as the size of the tree. Lightly head back terminal growth on the scaffold limbs to outward growing laterals to maintain the open center or bowl-shaped tree. The objective is to open up the tree to allow sunlight penetration and air movement and to improve spray coverage. When the tree is well grown, pruning consists mainly of moderate thinning and heading cuts back to outward-growing laterals to keep the tree low and spreading. A height of 2.5-3.0m is preferred.

## Crop Regulation

Heavy flowering and fruiting are the characteristic features of low chilling peaches resulting in poor quality, small sized fruit and reduction of flowering in subsequent season. Therefore, fruit thinning is very much essential in peach for production of quality fruits. The criteria for fruit thinning in peach are based on leaf to fruit ratio. 30-40 leaves / fruit is the appropriate ratio. Various mechanical thinning techniques are developed, but in recent years, chemical sprays have become widely used for thinning. Application of ethephon (50 to 100 ppm) to peach fruitlets caused effective thinning and also improves the fruit quality (Chahil *et al;* 1980). Spraying of DNOC (1000-2500ppm), prior to full bloom causes effective thinning (Martin, 1988). Spraying thiourea (2 to 3 %) or 2, 4, 5-T (40 to 100 ppm) cause effective thinning (Erez, 1975). Carbaryl spraying at 3 (600 ppm) weeks after full bloom and $GA_3$ (100 to 200 ppm) at full bloom stage increased fruit drop and growth of remaining fruits (Rana and Sharma, 2002).

## Growth and Development

Peach starts bearing 2 to 3 years after planting. Fruit buds are born laterally on one-year-old wood and on short spur like twigs. Generally two flower buds and one leaf bud develop at a node. Low chilling peaches grown in cold regions produce early blooms, which normally suffer from frost injuries. In areas where inadequate chilling conditions prevail, partial flowering can be achieved by spraying $GA_3$ (100 ppm) during autumn. Low chilling peach cultivars are self- fruitful, setting good crop without cross-pollination. Fruit growth pattern in peach follows a double sigmoid curve with initial and final phases of rapid growth separated by a middle phase of comparatively slow growth (Westwood, 1978).

## Harvesting and Yield

Peaches are climacteric therefore, harvested treeripe fruits. Various indices for judging fruit maturity used are days after full bloom, fruit size, fruit firmness, pit discoloration, freeness of pit, ground colour, and TSS: acid ratio (Smith, 1985). Ground colour variation in conjunction with flesh firmness is one of the best maturity indices commonly adopted for peach harvesting. For distant market, fruits are harvested when they attain good colour with hard skin. Whereas, for local consumption, ripe peaches are harvested by twisting with hand. The peak harvesting period of peach in sub tropics areas is last week of April to May.

On an average, 15-20 kg fruits/tree can be harvested in 3 years old tree, but mature productive trees generally produced 7-10 tonnes of fruits / ha.

## Production Problems

*Peach leaf curl aphid (Brachycaudus helichrysi):* It is the most serious pest of low chill peaches. Leaves acquire a characteristics curling as a consequence of sucking of sap from the growing buds during March to May. Floral buds also become weak and result in poor fruit setting and fruits fall off prematurely. Measure injury is caused during the prebloom and blooming period.

Spray 7-10 days before flowering with Methyl demeton 0.025% or dimethoate 0.03% or monocrotophos 0.03% or fenitrothion 0.05%. Repeat spraying, after 15 days, if necessary.

*Peach fruit fly (Dacus spp.):* It is the most destructive pest of low chilling peaches. The flies puncture and lay eggs inside the fruit during maturity to ripening. The infested fruits develop brownish patches due to rotting and drop. In April- May, when the adult flies start appearing on leaves, spray the bait consisting of 0.1% Malathion and 1% sugar/gur (2ml malathion 50EC + 10 g sugar/gur in 1 L water) on foliage.

*Peach green aphid (Myzus persicae):* The colonies of aphid may be seen sucking sap from the young leaves, flower buds and young fruit. It also adversely affects fruit setting. Apart from feeding on foliage, the pest causes great loss by transmitting around 120 plant viruses. It can be controlled by spraying dimethoate 0.03% or monocrotophos 0.03%.

*Flat headed borer (Sphenoptera dadkhani):* In recent years, the flat headed borers have become major pests. Both the grubs and adults are harmful but the grubs make shallow, broad, irregular galleries, which make the bark loose and block the flow of sap thus causing the death of limbs above it. In case of severe infestation, the branches and even entire tree die. For effective control, heading back of severely infested trees and destruction of dead wood should

Peach Leaf Curl

Peach Yellow Leaf role diseases

be done in February. Drench spray of 1000 ml Durmet 20EC (Chlorpyriphos) or 1000 ml Kanodane 20EC in 500 litres of water during February-March and September-October.

## Diseases

*Bacterial canker and Gummosis (Pseudomonas sp.):* This disease is very common in young orchard of low chilling cultivars. Mild winter and moderate summer temperature are the most favorable for disease development. Circular to elongated water soaked gummy lesions similar to borer injury appear on stem/branches and outer bark splits. The bark becomes brown, gummy and sour smelling. To control, clean the wounds and apply Bordeaux paste before the rains start in summer. Continue treating the new lesions, which appear.

*Peach yellow leaf roll disease:* The characteristic symptoms of the disease are yellowing of leaves along with rolling, red spotting, occasional tattering and stunting. Fruits on severely infected trees are shriveled, undersized and drop prematurely. Use disease free bud wood for raising healthy nursery plants. Remove infected plant parts and use oxytetracyclin injection @ 1g/ tree after harvest.

*Peach leaf curl (Taphrina deformans):* Curling, puckering and distortion of young leaves are the typical symptoms. Infected portions develop pink or red bronze colour. Give single dormant spray of either copper oxychloride 0.03% or carbendazim (2ml/L) during late dormancy in January.

*Brown rot and blossom blight (Monilinia laxa):* Small, circular brown spots appear on fruits, which are subsequently covered with ash coloured tufts. The disease extends to the shoots causing wilting and cankering of young branches. Spraying of Captan 50 WP (300g/100 litres water) about 3 weeks before harvest. For controlling blossom wilt or blight, spraying of carbendazim (50/100 litres) first at flowering and repeat after 7-10 days is quite effective.

## Physiological Disorders

### *Split Pit and Gumming*

During pit hardening, some times fruits show splitting at the joint of dorsal and ventral side of the fruits. This splitting is accompanied by exudation of gum, which may fill the entire pit cavity surrounding the kernal. Prolonged dry period followed by sudden rise in moisture content or rain resulting in fluctuation of atmospheric temperature and humidity cause fruit splitting.

### *Sunscald*

Constant exposure to intense sunlight may cause sunscalding on tree trunk, young twigs and branches. To protect plants from sunscalding, painting the exposed surface with lime paste and providing shade by wrapping straw around the trunk and branches may be practiced.

## Constraints

The low chilling cultivars of peach have covered large tracts of the sub-tropical plains of Punjab, U.P, H.P and North East. Generally, peaches are grown in marginal land and fertilizers are not applied according to requirements of trees. The water and fertilizer use efficiency is generally poor. Also, spring frost and hailstorms are adverse weather parameters leading to low productivity of low chilling peaches. Research results have shown that through proper orchard management practices (soil and water conservation and fertilizer application), the fruit yield can be doubled in the existing orchards. The adoption of improved production technology developed by the research system can bring visible and perceptible changes in productivity. Technologies like use of clonal rootstocks, introduction of renewal pruning techniques, adoption of area specific varieties and micronutrient applications have not been transferred and adopted at a satisfactory level.

Peach leaf curl disease is the major plant protection problem in peaches in J&K and H.P, while Uttarakhand and North East are comparatively free from the disease. For checking entries of diseased material in the free areas of U.P. and North-Eastern Hills, strict quarantine and selection of elite disease-free mother plants are very essential. Often it is not followed strictly.

There is no well-organized marketing system. Transportation in the hills itself is problematic. Post-harvest management problems originating from poor harvesting (strip picking) and improper packing system (non CFB boxes) and lack of proper pre-cooling and cold storage facilities result in huge (25-30%) loss of fruits. Capacity of the processing sector is also inadequate. Product diversification, value addition and market infrastructure development would require very substantial investment. The existing processing units are quite old and they require modernization for which substantial investment is required. CA storage trials have shown good promise. Its extension in larger growing areas is needed.

## References

Anonymous . 1993. Horticultural Database. National Horticulture Board.

Anonymous . 2002. Statistical Database. http//www.fao.org.

Chahil, B. S. Grewal, S. S. and Dhatta, A. S. 1980. Effect of thinning on fruit retention and some physico-chemical characteristics of peach. *Punjab Hort. J.* 20 : 70-73.

Childers, N. F. 1983. In: *Modern fruit Science.* Horticulture. Publication. Gainesville, Florida, p.203.

Erez, A. 1975. Thiourea, a new thinning agent for early ripening peaches and nectarines. *Hort.Sci.* 10: 251-253.

Hesse, C. O. 1975. In: *Adv. In Fruit Breeding* (eds). J. Janick and J. N. Moore. Prudue uni. Press, West Lafayette, Indiana, pp. 285-335.

Rana, H. S. and Sharma, L. K. In: Enhancement of Temperate Fruit Production-in changing climate (eds). Jindal, K. K. and Gautam, D. R. Deparment of pomology and Directorate of Research, University of Horticulture and Forestry, Solan (H.P.). p. 399.

Sidhu, B. S., Chahil, B.S., Brar, M. S. and Athwal, B. S. 1980. Effect of soil and foliar application of zinc on yield and quality of peach. *Haryana J. Hort. Sci.* 9(1-2):47-49.

Smith, R.B. 1985. Predicting the details of first commercial harvest of selected Ontario peach cultivar. *J. Amer. Soc. Hort. Sci.* 110 (5): 650-654.

Uppal, D. K., Chopra, S. K. and Chanana Y. R. 1993. Imp. of temp. for subtropical climate. In: *Advances in Horticulture* Vol. I (eds). Chadha K. L. and Pareek O. P. Malhotra publication, New Delhi, pp 445-461.

Voldrick, M. and Kyzlink, V. 1990. *J. Food Sci.* 57: 161

Westwood, M. N. 1979. Temperate Zone Pomology, W. H. Freeman and Company San Francisco, pp. 282.

**Underutilized and Underexploited Horticultural Crops, Vol. 2, 2007**
*K.V. Peter (ed.), pp. 105-116*
New India Publishing Agency, New Delhi (India)
E-mail : newindiapublishingagency@gmail.com
Web: www.bookfactoryindia.com

# 6 Germplasm Resources of Aonla *(Emblica officinalis)*

**ARUN KR. SHUKLA, DHURENDRA SINGH AND ANIL KR. SHUKLA***
Central Institute for Arid Horticulture, Beechwal
Bikaner - 334 006
*Department of Horticulture, MPUAT, Udaipur - 313 001

Plant genetic resources are the key components of crop improvement because of development of precision breeding system through biotechnological approaches as well as conventional system of breeding like selection and hybridization. Breeding system of hybridization is replaced by cybrid technology that facilitates transferring more qualitative and quantitative traits rapidly within incompatible genotypes. Aonla is mainly recognized for its high nutritive, medicinal and therapeutic properties. It contains high amount of vitamin-C (500-1500mg/100g), it is also rich in pectin, iron, calcium and phosphorus. Dried fruits of aonla can be used in curing haemorrhage, diarrhea, chronic dysentery, diabetes, jaundice, dyspepsia and cough. Recently, aonla gained more popularity in arid and semi- arid region of the country because of its hardy nature, prolific bearing capacity and ability to grow under various adversities as a remunerative crop (Shukla *et al.*2002). Aonla is the main ingredient of *Chavanprash* and *Triphala* powder, used for curing different human ailments. Fruit contains a chemical substance gallic acid and leucoanthocyanin, having antioxidant property.

It has significant contribution to improvement of economic status of farming communities through commercial orcharding, nursery production and processing of herbal products/processed fruit products. Production of aonla is affected by physiological/genetic problems viz. frost, variable sex ratio, self-incompatibility, fruit dropping, prolonged fruit bud dormancy, shy and alternate bearing, fruit splitting, post harvest losses, poor quality of value added products and lack of marketing facilities. Keeping in view its vast potential and significance to export trade, there is need to make efforts for efficient management of aonla. To enhance genetic base of aonla and its utilization for crop improvement and product diversification, genetic diversity conservation programme was initiated at different ICAR institutes and SAUs.

## Taxonomy

Aonla (*Emblica officinalis* Gaertn), also known as Indian gooseberry, myrobalan and emblic, belongs to Euphorbiaceous which comprises about 350 (Hooker, 1973) to 500 species (Bailey, 1917) mostly shrubs, a few herbs or trees. It bears two types of shoots. On basis of growth characteristics, these are categorized as long or indeterminate and short or determinate (Bajpai, 1965). Indeterminate shoots are longer and continue to put forth new growth in the season. The determinate shoot appears on nodes of indeterminate shoots and their number at each node varies from 3-5 in different cultivars. These determinate shoots bear small sized leaves (9-14 x 2-4 mm), which are of simple type. Flowering in aonla takes place during March-April under hot arid conditions and in February in Uttar Pradesh on newly emerged determinate shoots. There are two prominent cropping seasons in South India i.e. July - August and April - May (Naik, 1963). Fertilized ovary of aonla, unlike those of other fruit plants, remains dormant for 3.5 months and resumes growth following division in endosperm and zygote nuclei in August. Fruits mature in December under hot arid ecosystem of Rajasthan. Somatic chromosome number is $2n = 2x = 18$ (Bajpai and Shukla, 1985). Other related species *Phyllanthus acidus* commonly known as star gooseberry or otaheitegooseberry is grown for ornamental purposes. In arid and semi arid conditions, aonla is deciduous in nature due to abscission of determinate shoots during January- February. Generally it bears two types of shoot i.e. determinate and indeterminate. Indeterminate shoots are longer and continue to put new growth in the season while flowers are borne in axil of leaves on determinate shoot as axillary cymules. In North India, senescence / leaf shedding of aonla plants begins in February and complete shedding of determinate shoots takes place in February to mid of March. New shoots emerge out during first week of April (Bajpai, 1965, Ram, 1971 and Anon., 2001). Flowering period varies in different varieties from 17-26 days. In South India, flowering occurs twice a year, February-March and June-July (Naik, 1963). In aonla, male flower appears first in the form of cluster at basal part of determinate shoot followed by female flower in axil of leaves at distal end of same shoot. Percentage of female flower varies from 1.12 to 8.33 depending upon varieties. According to Dhar (1979), ratio of female to male flowers varies from 1:109 to 1:501 depending upon cultivars. Sex ratio in different cultivars varies from 1:12 to 1: 89 under hot arid ecosystem of Rajasthan (Anon., 2001). Time of anthesis in aonla is between 5.30 -7.30 pm with peak anthesis taking place at 6.10 to 7.30 pm, after 8.00 pm flowers do not open. Anther dehiscence takes place 10-15 minutes after anthesis. Female flowers open in different stages taking about 70hours. Stigma becomes receptive on 3rd day of anthesis and remains so for 48 hrs. Aonla is highly cross pollinated crop due to its monoecious nature and pollination takes place through wind, honey bees and gravity (Reddy and Janki Bai, 1979). Common visiting hours of honeybees is late evening and morning (Bajpai, 1957).

Since there is no self incompatibility in aonla, causes of poor fruit set seem to be high percentage of staminate flowers (Bajpai, 1968). Fruit setting can be increased up to 18-28% by hand pollination. Pathak and Pathak (1993) did not notice any fruit set for a number of years in isolated trees of Banarasi aonla even after profuse flowering. It indicates presence of self-incompatibility in this cultivar. A gene controlled natural incompatibility system is held responsible for unfruitfulness in some aonla cultivars. Information related to narrow and broad based heritability of desirable traits are still lacking. Preliminary observations indicated that yield and yield attributing traits are polygenically inherited as crosses between large and small-fruited types failed to produce any set segregation patterns in progenies with respect to fruit size and related attributes.

## Distribution of Genetic Resources

Aonla is found in dry deciduous forests of India except in extreme North West ascending to 1450m in Himalayas. It is also found in Sri Lanka USA, Cuba, Iran Pakistan, Iraq, China, Malaysia, Thailand, Philippines, Vietnam, Trinidad and Japan. This crop has extensive adaptability to grow under diverse climatic conditions ranging from non tropical plains to humid sub tropical mid elevation hills. Forests are traditional sources of aonla for medicinal uses estimated 41000 tonnes. Out of this, about 35000 tonnes of aonla are collected from forests of Madhya Pradesh. Increasing demand from industry, especially for Ayurvedic formulations and appreciation of its nutraceutical and medicinal properties for house hold consumption resulted in growing of this tree as cultivated crop. Initially, it was grown as seedling tree having large variability coupled with long gestation period. Aonla has great potential for export to meet growing health consciousness. To commercialize aonla crop in different agro-ecological regions, genetic resources are essential for evolving improved cultivars, selections, hybrids, cybrids and transgenics using different techniques of crop improvement.

Genetic resources of crops include landraces and primitive cultivars, obsolete and old released varieties, recently released varieties, genetic stocks with non desirable attributes and wild relatives (Singh and Srivastava, 2004).Aonla is a hardy fruit plant, which can be successfully grown under various agroedaphic situations. It is indigenous to tropical south-eastern Asia particularly central and southern India (Firminger, 1947). It is also a native of India, Sri Lanka, Malaysia and China. Aonla thrives well throughout Tropical India. Wild or cultivated species in the region extend from base of Himalayas to Sri Lanka and from Malacca to South China. In India, aonla is widely grown in UP, Gujarat, Rajasthan, Madhya Pradesh and Tamilnadu. The seedling aonla plants are commonly observed in Vindhyan hills, Northern hilly region (Garo, Khasi and Jayantia hills) and also in North- Western Himalayan region up to elevation of 1350 m ASL. Besides, potential areas for aonla diversity is Andhra Pradesh (Vizag), Gujarat (Anand, Nadiad), Madhya Pradesh (Panna), Maharashtra (Ajuntha, Akola, Bhandara, Buldhana, Chandrapur, Jalgaon, Kolhapur, Ramtek, Satara, Yoetmal), Tamil Nadu (Annamalai, Azhakiapandia-puram, Khrayamttels, Krishnagiri, Nilgiri hills, Panpozhil, Salem, Siruwani hills), Uttarakhand (Kumaon hills), Uttar Pradesh (Azamgarh, Bareilly, Pratapgarh, Raibareli, Sultanpur, Varanasi).

In aonla, variability exists in seedling population for fruit size, maturity period and yield/ plant. In fact, seed germination is not a problem, therefore, great genetic variability is observed for desirable economic traits in aonla growing areas of the country. Recently, an exploration was conducted in parts of hill ecosystem of Himachal Pradesh, where it was found that there was ample scope for selection of frost tolerant/resistant types, although fruit size was smaller but plants are more hardy than existing commercial varieties (Shukla and Dhandar, 2003). Further, Rajasthan, Gujarat, Uttar Pradesh, Madhya Pradesh, Jharkhand, and NEH region are hunting grounds for aonla variability.

Organizations involved in Survey and collection of aonla germplasm

| Centres | No. of collections |
|---|---|
| ANGRAU, RRS Anantpur | 15 |
| TNAU RRS Arrupukottai | 15 |
| CCS, HAU, RRS Bawal | 06 |
| CIAH, Bikaner | 43 |
| NDUAT, Faizabad | 22 |
| CAZRI, Jodhpur | 03 |
| SKN College Jobner | 06 |
| MPKV, Rahuri | 08 |
| S.K.Nagar | 12 |

*Sources:* Status of genetic resources of aonla at AICRP centers (Saroj and Awasthi, 2004)

Major areas/regions of genetic diversity of aonla in India

| State/Region | Diversity areas |
|---|---|
| Andhra Pradesh | Vizag, Anantpur |
| Gujarat | Anand, Nadiad, Surendra nagar |
| Rajasthan | Ajmer, Jaipur, Kota, Hanumangarh, Alwar and Udaipur |
| Maharashtra | Akola, Bhandra, Buldhana, Chandrapur, Jalgaon, Kholapur, Satna, |
| Tamil Nadu | Salem, Dindigul, Tuticorin. |
| Uttar Pradesh | Vindhyan hills, Bundelkhan, Allahabad, Mirjapur, Sonbhadra, Mathura, Agra, Etawah, Mainpuri. |
| Madhya Pradesh | Bhopal, Gwalior, Shivpur, Indore, Sheopur, Betul, Hoshangabad, Balaghat, Satna, Sidhi, Panna. |
| North Western Himalaya | Jammu & Kashmir, Himachal Pradesh, Uttarakhand up to 1500m above sea level |
| North Eastern Himalaya | Assam, Meghalaya, Manipur, Tripura, Khasi hills, Mizorum. |

## Organizations Involved in Collection of Aonla Germplasm

- Central Institute of Arid Horticulture, Bikaner
- Narendra Dev University of Agriculture and Technology, Faizabad
- Gujarat Agriculture University, Anand
- Central Institute on Subtropical Horticulture, Lucknow
- Central Horticultural Experiment Station, Godhra
- SKN College, Jobner

Satus of some elite genotypes collected by different institutes

| S.No | Organizations | Accession No. | References |
|---|---|---|---|
| 1 | Dry land Agril. Research Station Anantpur, A.P. | 12 | Pareek, 1998 |
| 2 | Agril. Research Station, Arrupukottai | 05 | Pareek, 1998 |
| 3 | Dry land Agril. Research Station, S.K.Nagar, Gujarat | 12 | Pareek, 1998 |
| 4 | NDUAT, Faizabad | 21 | Pareek, 1998 |
| 5 | GBPUAT, Pantnagar | 10 | Ram, 1993 |
| 6 | CIAH, Bikaner | 43 | Shukla *et.al,* 2004 |
| 7 | NBPGR, New Delhi | 05 | Anonymous,1999 |
| 8 | CISH, Lucknow | 12 | Pathak *et.al,* 2003 |
| 9 | MPKV, Rahuri | 06 | Karle, 2000 |

Ideotype of aonla germplasm

| | | |
|---|---|---|
| 1. | Plant traits | Dwarf, regularity and precocity in bearing, wider adaptability, Early and late maturity |
| 2. | Fruit yield | High |
| 3. | Fruit quality | Surface colour, high vitamin-C |
| 4. | Seedlessness | Small sized |
| 5. | Resistance to physiological disorders | Disorders like necrosis and misshapen fruits |
| 6 | Biotic and abiotic stresses | Resistant/tolerant |

## Conservation Techniques of Genetic Resources

Conservation of available genetic resources in fruit crops particularly which are clonally propagated through vegetative means are done by collecting scion wood or vegetative bud of desired genotypes and budded onto rootstocks grown at ex-situ site in field repository. Aonla is propagated through budding technique that involves growing root stocks and subsequently performing budding onto root stock. Further, there are usually two techniques of plant multiplication i.e. either growing root stocks in nursery bed or growing root stocks in field repository at proper spacing for budding operations. Season of budding varies from region to region.

## Utilization of Genetic Resources

Natural plantations and landraces are mainly raised from seed, which is the main cause of high genetic variability among existing populations. Some named cultivars like Banarasi, Francis (Hanthijhool), Chakaiya and Bansi Red are natural selections, which got acclimatized in specific agro climatic zones and they possessed merits and demerits. For instance, Banarasi is early maturing but a shy bearer (less number of female flowers/branch let) and prone to heavy dropping of fruits with poor shelf life. Francis suffers from severe incidence of fruit necrosis and Chakaiya fruits are fibrous, small in size with tendency to bear heavy crop in alternate years. New selections are identified at NDUAT, Faizabad from Chakaiya and are released for commercial cultivation. These are Kanchan -NA4 and Amrit –NA6 (selection from Chakaiya), Krishna and Neelam-NA7 (selection from Banarasi) and Balwant or Agra Bold-NA10 (selection from Banarasi). Similarly in Gujarat, Anand1, Anand2 Anand3 from seedling population available in the area are the new promising types (Pareek and Samadia, 1999).

In early years, there were no improved cultivars of aonla. Popular types were named after locality of their occurrence, plant type and fruit colour. Banarasi, Chakaiya, Hathijhool, Green Tinged, Red Tinged, White Streeked, and Bansi red, are a few among them. Genetic erosion of these resources are high due to severe deforestation, natural calamities and adoption of a few popular cultivars in selected pockets. Enormous variability in aonla still remains unexploited and awaits proper attention on exploration, collection and maintenance to conserve genetic diversity available in nature.

Germplasm survey, collection and evaluation of aonla were initiated at Narendra Dev University of Agriculture and Technology, Faizabad during 1982-83. Ninteen germplasm including 10 in fruiting stage are being maintained at this University. Considerable variability in fruit shape (round, flattened, round, oval, round and triangular); length (3.91-2.90cm); width (4.50-3.50cm); fruit weight (50.55-26.91g); number of segments (6); fruit colour (whitish-green to apricot-yellow, yellowish green, light green with pink tinge, pink); seed size( small to large); flesh colour( whitish green to pink); fibre (more to less); TSS (8-14%); acidity (1.9-2.5%) and ascorbic acid (422-600 mg/100g) in pulp are observed within available genotypes. Diversity in vegetative growth (upright, spreading, drooping); flowering (February end to first week of April); sex ratio (1: 28- 1:355); fruit maturity and productivity were observed with these genotypes.

Important merits and demerits of aonla cultivars

| S.No. | Variety | Breeding method | Merits | Demerits |
|---|---|---|---|---|
| 1 | Banarasi | Seedling selection | Bold size, upright growth, high vitamin-C | Less number of female flowers, self-incompatibility, shy bearer |
| 2 | Francis | Seedling selection | Drooping growth habit, large sized fruits | Susceptible to necrosis |
| 3 | Chakaiya | Seedling selection | Profuse fruiting | More fiber content |
| 4 | Krishna (NA-4) | From Banarasi | Large fruit | Shy bearer |
| 5 | Kanchan (NA-5) | From Chakaiya | Regularbearer, preferredby industries | Small size fruits |
| 6 | NA-6 | From Chakaiya | Less fiber content | Fruit cracking/mis hapen fruit |
| 7 | NA-7 | From Francis | Free from necrosis, prolific bearer | Limb breakage |
| 8 | NA-10 | Banarasi | Bold fruit | Moderately susceptible to necrosis |

### *Challenges ahead (Shukla et.al, 2004)*

- Since, aonla is a highly heterozygous plant, large size of population is required for selection.
- It has long generation cycle i.e. 2-8 years, depending upon species and varieties.
- Lack of recombination.
- Long juvenile phase prohibiting early assessment of genotypes.
- Presence of self-incompatibility.
- Frost Susceptibility.
- Lack of knowledge on inheritance pattern.

## Objectives

- To breed varieties having wider geographic adaptability.
- To develop varieties suitable for export.

- To evolve colored varieties based on market demand.
- To breed varieties resistant to frost.
- To develop varieties resistant to biotic (e.g. rust disease) and abiotic stresses (e.g. drought hardiness).

Exploitation of available hybrid vigour (heterosis) for yield and quality attributes.

- To breed varieties having high yield with good quality fruits.
- Varieties with less fibre content.
- Good pollinating variety.
- Varieties with high sex ratio with more female flowers.

## Important Promising Selections

### *Banarasi*

It is developed through seedling selection from Varanasi district of U.P. It is shy bearing in nature due to less number of female flowers and presence of self-incompatibility. It is rich in vitamin-C. It is an early maturing variety and mildly susceptible to necrosis.

### *Francis*

It is a seedling selection popularly known as Hathijhool. Branches have dropping habit, moderate bearer, and greenish yellow fruit, nearly fibreless and highly susceptible to necrosis.

### *Chakaiya*

Seedling selection having spreading growth habit. It contains fibrous fruits with six segments. It is a late maturing variety and free from necrosis.

### *Krishna*

A chance seedling of Banarasi from Pratapgarh, it is early maturing with shy bearing habit, pinkish green flesh and less fibrous.

### *Kanchan*

It is a chance seedling of Chakaiya with spreading growth habit and bears profusely; fruit is yellowish in colour and suitable for pickle preparation.

### *NA-6*

It is a chance seedling of Chakaiya having semi-spreading growth habit, light green colour of fruit, good keeping quality, late maturing, moderate bearer and free from necrosis

### *NA-7*

It is a seedling selection from open pollinated strain of Francis. It is precocious and prolific bearer, mid season, free from necrosis and suitable for preserve making.

### *NA-8*

It is a chance seedling of Chakaiya. It has upright growth habit, late maturing, moderate bearer, susceptible to necrosis and good keeping quality.

### *NA-9*

It is a seedling selection from open pollination strain of Banarasi. It has semi- spreading growth habit, moderate keeper, early shy bearer and susceptible to necrosis.

### *NA-10*

It is a chance seedling developed from Banarasi locally known as Agra Bold. It has semi spreading growth habit, yellowish green with pink tinge, highly astringent, good keeping quality, early, moderate bearer and moderately susceptible to necrosis.

### *Balwant*

It is developed through selection from Banarasi at RBS College, Agra, UP. It is a prolific bearer, semi spreading growth habit, moderate bearer and good keeping quality.

### *Anand–1*

It is developed at GAU, Anand and is a moderate bearer

### *Anand-2*

It is also developed from GAU, Anand. It is a prolific bearer with fair keeping quality.

### *Laxmi-52*

It is developed from CISH, Lucknow

## Prospects of Germplasm Utilization

- Development of self fruitful varieties.
- Development of coloured fruiting type varieties.
- Exploitation of hybrid vigour.
- To evolve variety(s) with very small stone size.
- Development of variety(s) suitable for export purpose.
- Development of transgenic aonla through biotechnological tools.
- Breeding for resistance to biotic and abiotic stresses (particularly for low and high temperature).
- Identification of rootstocks for salinity resistance, dwarfism and strong tap root system.
- Categorization of germplasm for product diversification.

### References

Bajpai, P.N. 1965 Studies on vegetative growth and development of male and female gametophyte in aonla (*Emblica officinalis* Gaertn) *Agra Univ.J.Res.*(Sci),14: 167-186

Bajpai P.N 1965 Studies on vegetative growth and development of male and female gametophyte in aonla (*Emblica officinalis* G.). *Agra Univ. J.Res.*(Sci.), 14:167-186

Dhandar D.G. and Shukla A.K. 2003 Varietal Improvement in aonla. Paper presented in National Seminar on production and utilization of aonla, 8-10 August,2003 at Salem,

Ferminger T. A. 1947 Ferminger manual Tamil Nadu.of gardening in India 8th Edn. Thaker Spink & Co. Ltd, Calcutta

Kester, D.E. 1965 The inheritance of time of bloom in certain progenies of almond. *Proc. Amer. Soc. Hort. Sci.*, 87: 214-21

Meghwal P.R. and Azam, M.M. 2004 Performance of some aonla cultivars in arid region of Rajasthan *Indian J. of Hort.*, 61(1): 87-88

Naik, K.C.1963 South Indian Fruits and their Culture. P.Varadachary and Co. Madras.

Pathak, R.K and Pathak R.A. 1983 Improvement of minor fruits In: Advances in Horticulture K.L.Chadha and O.P.Pareek (eds) Malhotra Publishing House, New Delhi.

Pathak, R.K. and Srivastava A.K. 1994 Flowering and fruit set studies in Indian Gooseberry. 24th International Hort. Congress, Japan.

Ram,S. 1971 Studies on the physiology of fruit growth in aonla (*Emblica officinalis* Gaertn) thesis submitted to Kanpur University, Kanpur.

Ram, S. 1971 Studies on physiology of fruit growth in Aonla (*Emblica officinalis* G) Ph.D thesis submitted to Kanpur University, Kanpur

Shukla Arun Kumar, Samadia D.K. Shukla, Anil Kumar, and .Dhandar D.G 2005 Genetic resources of aonla *Indian J. PGR* 18(2)188-193

Shushak Kshetra Mein Aonla Utpadan, *Krishi Chayanika,* 3: 20-22

Shukla Arun Kumar, Shukla, Anil Kumar, Saroj, P.L. and .Dhandar D.G 2004 Studies on floral biology of aonla under arid conditions. *Annals of Agri. Res.*, (Accepted)

Shukla Arun Kumar, Dhandar D.G. , Shukla, Anil Kumar, Meena, S.R. and Saroj, P.L. (2004) Germplasm resource conservation of aonla (*Emblica officinalis* L.) Paper presented In National Conference on Resource Conserving Technology for Social Upliftment New Delhi 7-9 December, 2004

Shukla A.K. and Dhandar D.G 2003 Genetic diversity of aonla in hill ecosystem of Himachal Pradesh Paper presented in National Symposium on Agroforestry and sustainable production 7-9 November 2003 at IGFRI, Jhansi

Shukla, Arun Kumar, Shukla, Anil Kumar, Awasthi, O.P. and Vashishtha, B.B. 2002 Shushak Kshetrya Mein Aonla Utpadan, *Krishi Chayanika,* 3: 20-22

Srivastava A.K. and Pathak, R.K. 1993 Studies on sex ratio, pollen viability and suitable pollinizer in aonla (*Emblica officinalis* Gaertn) .In: Absract of Paper presented in Golden Gublee Symposium held at IIHR, Bangalore.

Saroj, P.L. and Awasthi, O.P. 2004 Genetic Resources and Conservation Stretegies in Arid Horticulture In Advances in Arid Horticulture (eds. Saroj *et.al*) IBDC, Lucknow.

Teaotia, S.S, Singh D.B., Singh R.D. and Singh R.N. 1968 Studies on some important varities of aonla (*Phyllanthus emblica* L.) of eastern U.P. *Punjab Hort. J.,* 8:241-244.

**Underutilized and Underexploited Horticultural Crops, Vol. 2, 2007**
*K.V. Peter (ed.), pp. 117-127*
New India Publishing Agency, New Delhi (India)
E-mail : newindiapublishingagency@gmail.com
Web: www.bookfactoryindia.com

# 7 Ber

**ANIL KUMAR SHUKLA*, ARUN KUMAR SHUKLA****
**D K SAROLIA* AND D. SINGH****
*Department of Horticulture, MPUAT, Udaipur - 313001
**Central Institute for Arid Horticulture, Beechwal, Bikaner - 334006

Ber (Chinese date, Chinese fig or bore) (*Ziziphus xylopyrus*) is an important minor fruit of India. It is a hardy fruit tree cultivated all over India and also known as poor man's fruit. It is widely cultivated in Punjab, Haryana, Rajasthan, Uttar Pradesh, Madhya Pradesh, Bihar, Gujarat and Maharashtra. Ber does well even under marginal growing conditions and provides quality yields at low cost (Shukla, 1996). Fruits are eaten fresh but the other forms like dried, candied and squash can also be prepared from ber. Stem bark, root and leaves have medicinal values. Leaves of ber are used as fodder in dry regions. Ber tree is a host plant for rearing the *Tachardia laccad,* a lac insect (Sharma and Kore, 1990).

## Centre of Diversity

*Ziziphus mauritiana* L. is a native of India, while *Ziziphus jujuba* Mill. is native to China. Liu and Cheng (1995) reported that Indo-Malaysia region is the centre of both evolution and distribution of the genus *Ziziphus*. De Candolle (1886) stated that Myanmar (Burma) and India are the homes of ber. In India a number of ber cultivars are developed largely by growers through selection in different regions. Maximum variability of ber is observed in Rajasthan, Gujarat, Haryana, Punjab and Madhya Pradesh. In Gujarat, Mehsana, Anand, Panchmahal, Patan and Sabarkatha districts have rich diversity of ber (Shukla *et al.*, 2003). It is also reported that smaller fruits of Umran are known as

Chameli in Gujarat. Further, there are many unexplored areas of ber variability from where promising types could be collected for further characterization and evaluation. Historically central part of the country is the richest area in wild ber germplasm.

## Germplasm Resources

Since ber is a cross pollinated and heterozygous fruit crop, growing of the plants through seeds results in maximum variability. Collection and evaluation of ber germplasm are in progress at various places. More than 320 genotypes of ber are conserved in field gene bank of CIAH, Bikaner (Anon., 2001). Hisar and Bawal in Haryana, Bangalore in Karnataka, Udaipur, Jobner and Jodhpur in Rajasthan, Rahuri in Maharashtra, Bahadurgarh and Ludhiana in Punjab, Varanasi and Faizabad in Uttar Pradesh, S.K. Nagar and Godhra in Gujarat and Arupukottai in Tamil Nadu are also maintaining field genebanks of ber.

## Botany

Genus *Ziziphus* has more than 600 species (Bailey, 1947). Maheshwari and Singh (1965) recognized six economically important species in India namely *Z. mauritiana* L., *Z nummularia* (Burm. f.) wight and Arn, *Z. oenoplia* Mill, *Z. rugosa* Lam and *Z. xylocarpus* Wild. Cultivated Jujube is derived from the *Z. sativa* Gaertn. (Pathak and Pathak, 1993). Ber cultivars, in general are polyploid in nature. Cytological studies revealed that *Z. lotus* and *Z. jujuba* (Chinese jujube) are diploids, *Z.mauritiana* and *Z. oenoplia* are tetraploids and *Z. rotundifolia* is both tetraploid and hexaploid. Nehra *et al.* (1983, 1984) reported that the cultivar Illaichi is an octaploid (2n=96) and Umran is a tetraploid (2n=48). Khoshoo and Singh (1963) reported that 33 cultivars of ber namely Banarsi, Banarsi Pewandi, Chhuhara, Dandan, Desi Alwar, Gola, Gorwa, Kaithli, Kala Gola, Katha Bombay, Kathaphal, Laddu, Mirchia, Nalagarhi, Narikeli, Nazuk, Noki, Pathani, Sandhura Narnaul, Sanaur, Sanaur -2, Sanaur -3, Sanaur- 4, Sanaur -6, Seo, Safeda Selected, 28/1, Umran, Wilayti, ZG 2, ZG-3 and Hoshiarpur were tetraploids (2n=4x=48), cultivars Illaichi and Mithianwali were octaploids (2n=8x=96), and one cultivar Kalianwali was pentaploid (2n=5x=60). Further, Pareek (1998) reported that Umran and Chonchal are tetraploids while Gola rootstock and Boradi are octaploids. According to Qu *et al.* (1986), out of 111 genotypes *Ziziphus sativa, Ziziphus spinosus, Ziziphus jujuba* Var. Tortuosa, *Ziziphus jujuba* Var. logeniformis were diploids (2n=2x=24) while *Ziziphus jujuba* Var. Zanhu-angla was triploid (2n=3x=36).

Chinese jujube is a small upright tree with glabrous bright green leaves, which sheds leaves during winter and starts vegetative growth late in the spring to avoid danger of frost. Flowering period is April to June and fruit matures in autumn (August to October). While Indian jujube is an evergreen spreading tree with drooping branches. It sheds leaves in summer, August to October is the flowering time and fruits start ripening during January to March.

Important species of *Ziziphus* are given below

| Name of the species | Distribution | Remarks |
|---|---|---|
| *Z. rugosa* | In India it is found in U.P. Bihar, M.P. and Western Peninsula, Myanmar and SriLanka | It produces edible fruits |
| *Z. nummularia* | Found in North West India and A.P. | Bark can be used for tanning, produces edible small size fruits, leaves used for fodder. |
| *Z. oenoplia* | North and peninsular India | Bark can be used for tanning and it produces edible fruits. |
| *Z. sativa* | Punjab and Bengal | Diploid in nature, produces edible fruits. |
| *Z. xylopyrus* | M.P. | Host for Lac Insect, can be used for tanning. |
| *Z. heloia* | South Africa | – – |
| *Z. mucronata* | South Africa | – – |
| *Z. joazeiro* | Brazil | – – |
| *Z. mistol* | Argentina | – – |
| *Z. spina-christi* | Egypt, Syria and Palestine | – |

## Floral Biology

In ber, one to ten branchlets develop from nodes of primary woody branches which bear leaves and flowers. Teaotia and Chauhan (1963) and Desai and Patil (1978) reported that flower buds appear both on mature and current season's growth. Flowers are born in axil of leaves on main and lateral shoots and the inflorescence is an axillary cyme (Vashishtha and Pareek, 1978). Ber flower buds in axillary cymes are borne mainly on current season's growth.

Number of flowers/ cluster varies 12 to 14 (.Vashishta and Pareek 1978), and 16 by Teaotia and Chauhan (1963). Vashishtha and Pareek (1978) observed that 17-20 days are required for the smallest bud (0.13 cm diameter) to attain the largest size (0.35 cm) a day before anthesis. Teaotia and Chauhan (1963) and Josan *et al.* (1980) reported that 21 days are required for development of flower passing through eight stages, the growth of first four stages are slower, whereas Desai *et al.* (1986) observed that flower bud development occurred in seven stages and took, on an average 19.2, 19.3 and 19.85 days in Chhuhara, Sanaur 1 and Shamber cultivars, respectively. Godara (1980) reported that number of days required for bud development varied from 12.7 and 30.3 days. Singh and Jindal (1982) based on studies on Gola Gurgaon, Muria, Kaithali, Chhuhara and Umran reported that Umran had the highest number of hermaphrodite flowers (22.2%).

Teaotia and Chauhan, (1963) observed that time of anthesis varied in different cultivars. It occurred between 7.30 and 8.30 hr. in cultivars Seb, Jogia,, Ponda, Aliganj and Illaichi and between 12.00 and 13.00 hr. in Gola and Mundia. In Banarsi Pewandi and Banarsi Karaka, flowers opened at 12.00 hr. Desai and Patil (1978) found that in Umran and Illaichi, anthesis was completed between 5.30 and 6.45 hr. Singh *et al.* (1970) observed that in Banarsi Karaka, anthesis started after 12.00 hr. and stopped at 15.00 hr. being maximum between 13.00 and 14.00 hr. Anthesis was followed by dehiscence with in 1 or 2 hours and continued for 4 to 6 hours.

Stigmatic receptivity was studied in different varieties. Vashishtha and Pareek (1978) observed that stigma receptivity was maximum on the day of anthesis in Gola, Mundia, Seb, Jogia, Ponda, Aliganj and Illaichi cultivars. The flowers, pollinated 24 hours before anthesis completely failed to set fruits whereas, some success was obtained when pollination was done 24 hours after anthesis. Prasad (1964) observed that stigma becomes receptive 4 hrs. prior to anthesis and it remained at peak at the time of anthesis and continued up to 12 hours after wards.

## Pollination

Generally, cross pollination is the rule in ber. As pollens of ber are sticky, they are not transferred by wind. During flowering time, a lot of insect activities are found in ber orchards. Prominent pollinating insect agents are honey bees, house flies and yellow wasp. Life of individual flower is very short and in an inflorescence many flowers remained unpollinated during their receptive period, further unfertilised flowers droped down. It was also observed that branches of eastern side gave maximum set of 8.47 % compared with minimum set of 2.04% on northern side (Yamdagni *et al.*, 1968). Teaotia and Chauhan (1963) reported that pollen size in Banarsi Pewandi and Thornless were 42μ x 24μ and 35μ x 20.3 μ, respectively. Desai *et al.* (1986) reported that yellow and sticky pollen grains remained in anther unless disturbed, and were 29.34, 26.99 and 30.35 μ in size in Chhuhara, Sanaur -1 and Shamber cultivars, respectively with viability of 79.95, 82.40 and 91.46%. Khoshoo and Singh (1963) observed pollen sterility in ber owing to polyploidy and it was up to 90.5 % in the octaploid cultivar Illaichi. Teaotia and Chauhan (1964) observed that Banarsi Pewandi, Thornless and Banarsi Karaka were reciprocally cross incompatible. Godara (1980) found that cultivars Banarsi Karaka, Illaichi, Kakrola Gola, Kaithali, Kathaphal, Mundia Murhara, Reshmi, Sandhura Narnaul, Safeda Selected and Umran were self incompatible. Illaichi x Kakrol Gola gave the highest fruit set, while Umran showed the best combining ability when used as male or female parent.

## Breeding Methods and Achievements

### *Selection*

Many of the common cultivars are results of selection made by local people in different regions. Promising cultivars under commercial cultivation are Gola, Seb, Banarsi Karaka, Banarsi Peondi, ZG.1, Sanour-1, 2, 3, 4, Katha, Umran and Mundia .

Gomakirti is a clonal selection from Umran which is 25 days earlier than Umran. Recently CIAH Sel-1 of ber is released by CIAH through selection from local materials collected from Bhusawar area of Rajasthan. This selection is more juicy, sweet and early bearer. It is resistant to fruit rot (Anon., 2002, Shukla *et al.*, 2003).

### *Hybridization*

At HAU Hisar 72 hybrids are raised which are under evaluation. Work was also initiated at Jodhpur by making crosses of Katha with Seb and Seb x Tikadi (Pareek, 2001). At the CIAH, Bikaner, CIAH hybrid 1 (Seb x Katha) was found promising, precocious prolific bearer, early maturity with good fruit set in at high temperature under arid environment (Shukla *et al.*, 2003). For fruit fly resistance BCF1 (Seb x Tikadi F1 x Seb) is under evaluation (Anon., 2002).

### *Bitechnological Tools*

Use of DNA finger printing to identify genes for resistance to fruit fly, powdery mildew, salinity/ alkalinity, drought and high ascorbic acid content

Characteristics of important cultivars of ber

| Cultivar | Growth habit | Flowering period | | | Fruit set period | Time of Fruit | | Anthesis | Characters |
|---|---|---|---|---|---|---|---|---|---|
| | | Initiation | Peak | End | Initiation | Peak | End | | |
| 1 | 2 | 3 | 4 | 5 | 6 | 7 | 8 | 9 | 10 |
| Ajmeri | Semi erect | August | September | October | Early September | Late September | Late October | 1 to 2.30 PM | Late in maturity, TSS 18° Brix, Fruit weight 32g Fruit size 4.7x3.5 cm, fruit shape oval, Styler end round. |
| Akhorata | Spreading | End of August | Mid September | Late October | Early September | End September | End of October | 1 to 2.30 PM | Mid season cultivar, fruit weight 20g, fruit size 3.4 x 3.4 cm, TSS 20° Brix, fruit shape obovate, styler end flat. |
| Aliganj | Spreading | End of August | First week of September | End of October | First week of September | Mid of September | End of October | 1 to 2.30 PM | Mid season crop, fruit weight 21g, fruit size 3.8x2.9 cm, TSS 17° Brix , Fruit obovate, styler end round with depression. |
| Bhagwadi | Spreading | End of August | Second week of September | Mid of October | First week of September | Mid of September | Mid of October | 1 to 2.30 PM | Mid season cultivar, fruit size 3.68x3.15 cm, fruit weight about 20g, TSS 15° Brix, fruit oval styler end round. |
| Bahadurgarh | Semi erect | End of August | Mid of September | End of October | Second week of September | End of September | End of October | 1 to 2.30 PM | Mid season cultivar, fruit size 3.93 x 2.70 cm, fruit weight 17g, TSS 23° Brix. |
| Banarsi Karaka | Erect | End of August | First week of September | Mid October | First week of September | End of September | Mid of October | 1 to 2.30 PM | Mid season cultivar, fruit size 5.0 x 3.0 cm, TSS 17° Brix, fruit weight 28g, fruit oval, |

*Contd...*

| | | | | | | | | | |
|---|---|---|---|---|---|---|---|---|---|
| | | | | | | | | | styler end obliquely pointed. |
| Banarsi Pebandi | Semi spreading habit | Mid August | Early September | First week of October | End of August | Mid September | First week of October | 1 to 2.30 PM | Mid season cultivar, oblong fruit, styler end obliquely pointed necked, fruit weight 35 to 40 g, fruit size 5.5 x 3.8 cm to 6.0 x 4.0 cm, TSS 16° Brix. |
| Chhuhara | Erect | End of August | Mid September | End of October | First week of September | End of September | End of October | 7 to 8 AM | Mid season cultivar, oval fruit, styler end round, stem end broad, fruit weight 22g, fruit size 4.5 x 2.9 cm, TSS 23° Brix. |
| Chonchal | Semi erect | Early September | Mid September | End of October | End of September | Mid of October | End of October | 7.0 to 8.0 AM | Mid season cultivar, fruit oblong, fruit weight 25g, fruit size 5.4 x 3.3 cm, TSS 20° Brix. |
| Dandan | Semi erect | End of August | Mid September | End October | Mid September | Early October | End of October | 7.0 to 8.0 AM | Mid season cultivar, fruit oblong, styler end bluntly tapering with ridges, fruit size 5.65 x 3.45 cm fruit weight 27g, TSS 17° Brix. |
| Gobindgarh Selection | Erect | End of August | Mid September | Mid October | First week of September | Mid September | Early October | 1 to 2.30 PM | Mid season cultivar, fruit oblong, styler end obliquely beaked, fruit size 4.2 x 2.5 cm, fruit weight 18g, TSS 19° Brix. |
| Gola | Spreading | End of August | Early September | End of October | Early September | Mid September | End of October | 12 to 2.30 PM | Early cultivar, fruit round, styler end flat, |

*Contd...*

| | | | | | | | | | |
|---|---|---|---|---|---|---|---|---|---|
| | | | | | | to first week of October | | | stem end round, fruit size 3.8 x 3.6 cm, fruit weight 28g, TSS 22° Brix. |
| Goma Kirti | Semi erect | End of August | Mid September | Mid October | End of August | First week of September | Mid October | 1 to 2.30 PM | Maturity is early than Umran, Fruit oval TSS, 18° Brix, fruit weight 35g. |
| Illaichi | Erect | End of August | Mid September | First week of November | Mid of September | Early October | First week of November | 7.0 to 8.0 AM | Octaploid ber, mid season cultivar, fruit oblate, styler end flat ridged, stem end round, fruit size small (2.1 x 2.2 cm), fruit weight 8.5g, TSS 20° Brix. |
| Jhajjar selection | Spreading | Early August | Early September | End of October | End of August | Mid September | End of October | 7.0 to 8.0 AM | Mid season cultivar, fruit obovate, styler and stem end round, fruit weight 21g, fruit size 3.9 x 3.1 cm, TSS 17° Brix. |
| Kaithali | Erect | First week of September | End of September | First week of November | Mid September | Mid October | First week of November | 1 to 2.30 PM | Mid season cultivar, fruit is oblong, styler end round, stem end shallow, mid season cultivar, fruit size 4.8 x 3.5 cm, fruit weight 30 to 35g, TSS 18° Brix. |
| Katha | Semi erect | Mid August | Mid September | End of OCtober | First week of September | End of September | End of October | 1 to 2.30 PM | Late cultivar, oval fruit, styler end round, stem end round and riged, fruit weight 39g, size 5.2 x 4.3 cm, TSS19° Brix. |

*Contd...*

| | | | | | | | | | |
|---|---|---|---|---|---|---|---|---|---|
| Kathaphal | Spreading | End of August | Mid September | End of October | First week of September | End of September | End of October | 1 to 2.30 PM | Fruit obovate, styler end round, stem end shallow, fruit size 3.2 x 2.8cm, fruit weight 15g, TSS17° Brix, mid season cultivar. |
| Mundia | Semierect | End of August | End of September | First week of November | First week of September | End of September | First week of November | 1 to 2.30 PM | Fruit oblong, styler end round, stem end narrow with shallow notch, early in maturity, fruit weight 40g, fruit size 5.8 x 3.6 cm, TSS 22° Brix. |
| Seb | Erect | End of July | Mid August | End of October | Mid August | Mid of September | End of October | 7.0 to 8.0 AM | Mid season cultivar, fruit is round, fruit size 3.8 x 3.4 cm, fruit weight 29g TSS 17° Brix. |
| Umran | Semi erect | End of August | Mid September | End of October | First week of September | End of September | End of October | 1 to 2.30 PM | Late season cultivar, fruit shape oval, styler round, stem end round and ridged, fruit size 4.8 x 3.5 cm, fruit weight 45g, TSS 19° Brix. |

and their transfer in the recipient parent through somatic hybridisation is another approach. Standardization of callus culture and organogenesis from mature and juvenile plant parts of *Ziziphus mauritiana* (Mathur *et al.*, 1993) and *Ziziphus jujuba* (Cheong and Kim, 1984) are attempted and are successful.

### *Other Important Cultivars*

Maharwali, Mehrun, Narikeli, Narma, Noki, Nimaj, Ponda, Pathan, Safed Rohatak, Sanaur-2, Sanaur-3, Tikadi, Thornless, Triloki, Vikas-2, ZG-3, Madhuri and Mirchia.

## References

Anonymous, 2001 Annual Report. CIAH, Bikaner.

Anonymous, 2002 Annual Report. CIAH, Bikaner.

Bailey, L.H. (1947) The Standard Cyclopaedia of Horticulture. Macmillan and Co., New York, pp. 3547-3548.

Cheong, S.T. and Kim, J.S. 1984 Characteristics of the fruit as affected by the developmental stages, embryo germination and in vitro propagation of seedling *Ziziphus jujuba* Mill. *J.Kotean Soc. Hort. Sci.*, 25 (3) : 241-249.

De Candolle, A. 1886 Origin of cultivated plants. Bibliotheque Scientifique Internationale, 43, Paris.

Desai, U.T. and Patil, A.V. 1978 Studies on floral biology of ber (*Ziziphus mauritiana* Lamk.) cultivars Umran and Illaichi. *Udyanika* 2:26-32.

Desai, U.T., Randhawa, D.B. and Warhal, K.N. 1986 Floral biology of ber. *J. Maharashtra, Agric. Univ.*, 11:76.

Godara, N.R. 1980 Studies on floral biology and compatibility behaviour in ber (*Ziziphus mauritiana* Lamk.) Ph.D. Thesis. Haryana Agricultural University, Hisar.

Josan, J.S., Jawanda, J. S., Bal, J.S. and Singh, R. 1980 Studies on the floral biology of ber. Flowering habit, flower bud development, time and duration of flowering, floral morphology, time of anthesis and dehiscence. *Punjab Hort J.*, 20:156-161.

Khoshoo, T.N. and Singh, N. 1963 Cytology for North Western Indian trees. *Ziziphus jujuba* and *Z. rotundifolia*. *Silvae Genetica*, 12 Heft :141-180.

Liu, M. J. and Cheng, C.Y. 1995 A taxonomic study of the genus *Ziziphus*. *Acta Hort.*, 390

Maheshwari, P. and Singh, U. 1965 Dictionary of economic plants in India. ICAR, New Delhi.

Mathur, N., Ramawat, K.G. and Sonie, K.C. 1993 Plantlet regeneration from seedling explant of *Ziziphus* and silver nitrate and nutrient requirment for callus and morphogenesis. *Gartenbavwissenschaft*, 58(6):255-260.

Nehra, N.S., Chitkara, S.D. and Singh, K. 1984 Studies on morphological characters of some wild forms and cultivated varieties of ber. *Punjab Hort J.*, 24 (1/4): 49-59.

Nehra, N.S., Sareen, P.K. and Chitkara, S.D. 1983 Cytological studies in genus *Ziziphus*. *Cytologia*, 48: 103-107.

Pareek, O.P. 1998 AICRP (arid fruits) In: 50 years of horticulture research in India (Eds. S.P. Ghosh, P.S. Bhatnagar and N.P. Sukumaran), Division of Horticulture,ICAR, New Delhi, pp 150-158.

Pareek, O.P. 2001 ber. International Centre for Underutilised Crops, Southampton, UK.

Pathak, R.K. and Pathak, R.A. 1993 Improvement of minor fruits. In: Advances in Horticulture Vol. I (Eds. K.L. Chadha and O.P. Pareek) Malhotra Publishing House New Delhi, pp 412-416.

Prasad, A. 1964 Studies on floral biology of jujube (*Ziziphus mauritiana* Lamk) *Agra Univ. J. Res.*, (Science ) 13:41-48.

Qu, Z.Z., Wang, Y.H., Lu, Z.R. and Yan, G.J. 1986 Studies on the chromosome number of chinese jujube. *Acta. Horti. Sini.* 13 (4):232-236.

Sharma, V.P. and Kore, V.N. 1990 Ber. In: Fruits Tropical and Subtropical (Eds.T.K. Bose and S.K. Mitra), Naya Prokash Publishing House Calcutta, pp 592-615.

Shukla, A.K. 1996 Efficacy of drip and mulching on young plantation of aonla + ber cropping model in sodic soil. M. Sc (Ag) Thesis NDUAT, Faizabad.

Shukla, A.K., Dhandar, D.G., Vashishtha, B.B. and Umamaheswari, C. 2003 New promising ber varieties from CIAH. Accepted in *Indian Horticulture*.

Shukla, Anil Kumar, Shukla, Arun Kumar, Dhandar, D.G. and Patel J.I. 2003 Genetic variability of ber and aonla in Gujarat. National symposium on Agroforestry and Sustainable Production Jhansi 7-9 November2003.

Singh D., Bakhshi, J.C. and Singh, K. 1970 Flowering and fruiting behaviour of ber, variety Banarsi Karaka (*Z. mauritiana* Lamk.). *Punjab Hort. J.*, 10:21-28.

Singh, D. and Jindal, P.C. 1982 Studies on flowering and sex ratio in some ber (*Z. mauritiana* Lamk) cultivars. *Hariyana Agric. Univ. J. Res.*, 12:292-294.

Teaotia, S.S. and Chauhan, R.S. 1963 Flowering, pollination, fruit set and fruit drop studies in ber (*Ziziphus mauritiana* Lamk) 1. floral biology. *Punjab Hort. J.* 3 : 60-70.

Teaotia, S.S. and Chauhan, R.S. 1964 Flowering. pollination, fruit set and fruit drop studies in ber (*Z. mauritiana* Lamk) II pollination, fruit set , fruit development and fruit drop. *Indian J. Hort.*, 2:40-45.

Vashishtha, B.B. and Pareek, O.P. 1978 Flower morphology, fruit set and fruit drop in some ber (*Ziziphus mauritiana* Lamk) cultivars. *Annals Arid Zone*, 18:165-169.

Yamadgani, R., Bajpai, P.N. and Mishra, R.S. 1968 Pollination, fruit set and fruit development studies in ber (*Ziziphus mauritiana* Lamk) *Labdev. Sci. Tech.*, 6:101-103.

# Underutilized Vegetables

**Underutilized and Underexploited Horticultural Crops, Vol. 2, 2007**
*K.V. Peter (ed.), pp. 129-151*
New India Publishing Agency, New Delhi (India)
E-mail : newindiapublishingagency@gmail.com
Web: www.bookfactoryindia.com

# 8 Genetic Resources Management in Cucurbits

**UMESH SRIVASTAVA**
Indian Council of Agricultural Research
Krishi Anusandhan Bhawan-II, Pusa, New Delhi - 110 012

The Cucurbitaceae family is characterized by long training viny stems. There is variation for flower size and colour among the species but all of them have similar general morphology. The species are open pollinated and self compatible. Most of them are either monoecious (eg. *Cucurbita* spp.) Or andromonoecious (e.g. common in melon). Other species show gynomonoecy, androgynomonoecy, androecy, gynoecy, hermaphrodite or dioecy. Chow-chow is the only member of the Cucurbitaceae that is viviparous, contains single seed and propagated by planting whole fruit. The Indian gene centre has rich diversity in genetic resources of ridgegourd (*Luffa acutangula*), bottlegourd (*Lagenaria siceraria*), bittergourd (*Momordica charantia*), ashgourd (*Benincasa hispida*), snakegourd (*Trichosanthes anguina*), pointed gourd- parval (*Trichosanthes dioica*), pumpkin (*Cucurbita* spp.) and roundgourd (*Praecitrullus vulgaris*). Cucurbits are of tremendous economic importance and are cultivated throughout the world from tropical to sub-temperate zones. China, Turkey, India and Iran are important cucurbit growing countries. In India, this diversity is concentrated in Indo-Gangatic plains, north eastern region, north western Himalayas, the Western and Eastern Ghats and sporadically in the tribal dominant belts of central India. More diversity occurs in *Cucurbita* spp. in the north east as also for ashgourd, bottlegourd while for *Luffa*, it is more concentrated in the eastern peninsular tract. In Cucumis *melo* and roundgourd, it is more confined to north-western and Indo-Gangatic plains. In pointed gourd (parval), diversity is concentrated more in eastern part of Indo-Gangatic plains particularly in West Bengal and adjoining Bihar along eastern Uttar Pradesh. *Coccinia cordifolia* (growing wild throughout India, raw fruits used as vegetables), *Cucumis sativus* (distributed throughout India, in Himalayas as well), *Lagenaria siceraria* (African origin but domesticated throughout India, tender fruits used as vegetables), *Luffa cylindrica* (indigenous to India whereas *acutangula* found in western, central and southern India and is regarded as wild form of cultivated species - tender fruits used as vegetables), *L. acutangula* var. amara (occurs in peninsular India and is wild relative of cultivated spongegourd), *L. echinata* (western Himalayas, central India Gangatic plains) and *L. graveolans* (considered as wild progenitor of L. *hermaphrodita* in Bihar and Sikkim) are potential species. In addition, in *Momordica, M. balsamina* occurs in semi dry north western plains, northern parts of Eastern and Western Ghats, *M. dioica* and *M. cochinchinensis* occur wild/semi wild in Gangatic plains. *Trichosanthes* has 21 species occurring in India and the major zone of species concentration are (a) along the Malabar coast in Western Ghat, (b) low and medium elevation zones in Eastern Ghat and NEH region. *Citrullus colocynthes* exhibits much variation in north western plains. In *Cucumis sativus,* variation is concentrated in central and south-western Rajasthan and Madhya Pradesh, southern India, along foothills of Himalayas and north-eastern region, for *C. hardwickii, C. trigonus,* much variability occurs in Himalayas. *C. hystrix* extends its range from eastern plains to north eastern hills in Assam, Tura range of Meghalaya and Mishmi Hills (Srivastava, 1993).

Some cucurbits also have other uses, such as, mature and dry fruit of bottle gourd is used for making vessels and musical instruments, fibrous material of *Luffa* is used for scouring and packing, and some species have ornamental and medicinal uses. Cucurbits are modest sources of calorie, minerals and vitamins also. There are a few exceptions, for example, bitter gourd is rich in vitamin C and iron, pumpkin in carotene, *kakrol* in protein, and chow-chow has fairly high calcium.

## *Origin, Domestication and Extent of Genetic Diversity*

The Indian sub-continent is the centre of origin for a number of wild and cultivated cucurbits. Chakravarty (1982) reported that out of 110 genera and 640 species in the world, 36 genera and 100 species are found in India, which includes 38 endemic species. The region-wise distribution of cucurbits in India is given elsewhere. Many cultivated and wild.

Distribution of cucurbitaceous species in India

| Species | Distribution |
|---|---|
| *Cucumis hardwickii* Royle, *C. trigonus* Roxb., *Luffa graveolens* Roxb., *Trichosanthes multiloba* Miq., *T. himalensis* Clarke | Eastern Himalayas |
| *Cucumis trigonus* Roxb., *Luffa graveolens, Neoluffa sikkimensis* Chakr. | North-Eastern region |
| *Cucumis hystrix* Chakr., *C. trigonus, Luffa graveolens, Momordica cochinchinensis, M. macrophylla* Gage, *M. subangulata* Bl., *Trichosanthes anguina* L, *T. dioica* Roxb., *T. dicaeleosperma* Clarke, *T. khasiana, T. ovata* Cogn., *T. truncata* Clarke | Gangetic plains |
| *Luffa echinata* Roxb., *Momordica cochinchinensis,* M. *cymbalaria* Fenzl, M. *dioica* | Indus plains |
| *Citrullus colocynthis, Cucumis prophetarum* Linn., *Momordica balsamina* L | Western peninsular tract |
| *Cucumis setosus* Cogn., *C. trigonus* Roxb., *Luffa raveolens, Momordica cochinchinenis, M. subangulata* Bl., *Trichosanthes anamalaeiensis, T. bracteata* (Lam.) Voigt., *T. cuspidata* Lamk., *T. nervifolia* L , *T. perotteliana* Cogn. | Western Himalayas |
| *Cucumis hystrix, C. setosus, Luffa acutangula* (L) Roxb. var. *amara, L. graveolens, L. umbellata* (Kleir.) Roem., *Momordica cochinchinenis, M. cymbalaria, M. denticulata* Bl., *M. dioica* Roxb., *Trichosanthes bracteata, T. cordata* Roxb., *T. himalensis, T. multiloba* C.B. Clarke | Eastern peninsular tract |

*Source:* Ram and Srivastava (1999)

Species that date back to pre-historic times are associated with man's culture. Almost all the cucurbits are of tropical origin (Africa, America, and Asia) and these are distributed all over the world in different ecological zones. Apart from cultivated species, enormous diversity occurs in the form of semi-domesticated and wild types. The semi-domesticated species and cucurbits of minor economic value have been selected routinely by local tribes to meet their vegetable requirements and these play an important role in their life. Many of these species are fairly tolerant to biotic and abiotic stresses and can grow on soils, which are not suitable for other vegetables. Some of these species are nutritious and some have medicinal or other uses (Ram and Srivastava, 1999).

Origin and geographical distribution of cultivated cucurbits of global importance*

| Botanical name | Common name/ Vernacular name | Origin | Chromo-some no. (2n) | Geographical distribution | Uses |
|---|---|---|---|---|---|
| *Citrullus lanatus* | Water melon, *Tarbooj* | Africa | 22 | India (Bihar, Rajasthan, Gujarat, Madhya Pradesh), tropical and sub-tropical Africa | Fruit as dessert |
| *Cucumis melo* | Melon, musk-melon, *Kharbooja* | Africa | 24 | India (Uttar Pradesh, Bihar, Rajasthan, Madhya Pradesh), South-West and central Asia, China | Fruit as dessert |
| *Cucumis sativus* | Cucumber, *Kheera* | India | 14 | All over India | Fruit as salad, vegetable and Pickle |
| *Cucurbita maxima* | Winter squash | South America | 40 | Northern Argentina, Bolivia, southern Peru, northern Chile | Fruit and flower as vegetable, seed edible |
| *Cucurbita moschata* | Pumpkin, *Sitaphal* | Central Mexico | 40 | Southwards from Mexico city through central America into northern Colombia and Venezuela | Carotene rich, fruit and flower as vegetable, seed edible |
| *Cucurbita mixta* | Winter squash | Central Mexico | 40 | Central Mexico, and the Yucatan peninsula to Costa Rica | Fruit as vegetable, seed edible |
| *Cucurbita pepo* | Summer squash | Central Mexico | 40 | Northern and Central Mexico | Fruit as vegetable, seed edible |

Origin and geographical distribution of cultivated species of cucurbits of regional or local importance*

| Botanical name | Common name (s) | Origin | Chromosome no. (2n) | Geographical distribution | Uses |
|---|---|---|---|---|---|
| *Benincasa hispida* | Wax gourd, ash gourd, white gourd, *petha* | South-eastern Asia | 24 | Tropical Asia | Fruit, young leaf and bud as vegetable |
| *Citrullus colocynthis* | Colocynth | Africa | 22 | Tropical Afri North-western India | Seed is edible and used to make flour and cooking oil |
| *Coccinia indica* syn. *C. grandis* | Little gourd, ivy gourd, *Kundru* | Tropical Africa | 22-48 | Tropical Asia and Africa | Fruit and young shoot as vegetable |
| *Cucumeropsis mannii* | Egusi melon | Tropical Africa | 22 | Tropical Africa | Edible oilseed |

*Contd...*

| | | | | | |
|---|---|---|---|---|---|
| *Cucumis anguria* | West Indian gherkin | Africa | 24 | Brazil, West Indies | Young fruit as vegetable and pickle |
| *Cucurbita ficifolia* | Fig leaf, malabar gourd | Central Mexico | 40 | Central Mexico to Chili | Mature fruit candied, seed is edible |
| *Cucurbita foetidissima* | Buffalo gourd | Mexico, S West USA | 20 | SW USA & Central Mexico | Oil & protein rich seed |
| *Cyclanthera pedata* | Stuffing cucumber | Tropical America | 32 | Andean countries | Fruits as vegetable |
| *Hogsonia macrocarpia* | Chinese Lard plant | South Asia | 18 | South & SE Asia | Fruits as vegetable' seed as cooking oil |
| *Lagenaria siceraria* | Bottle gourd *Lauki* | Africa | 22 | Trop. & Sub-trop. areas, Africa | Immature fruit & shoot as vegetable |
| *Luffa acutangula* | Ridge gourd, *Dhari tori* | India | 26 | India, East asia | Young fruits as vegetable |
| *L. cylindrica* | Sponge gourd *Chikni tori* | India | 26 | Throughout tropics, India | Young fruits as vegetable, mature for fibre |
| *Momordica charantia* | Bitter gourd *Karela* | Tropical Africa | 22 | Throughout tropics, India, S & SE Asia | Young fruit, leaf as vegetable, seed as condiment |
| *M cochinchinensis* | Cochinchin gourd, sweet gourd | India | 28 | South Asia | Immature fruit as vegetable, young leaf, flower and seeds are edible |
| *M. dioica* | *Kaksa, kakrol* | Southern tropical Asia | 28 | Tropical Asia and Africa | Fruit, shoot and leaf as vegetable |
| *Praecitrullus fistulosus* | Round melon, Indian squash, *tinda* | India | 24 | North-West India | Fruit as vegetable |
| *Sechium edule* | Chayote, chow-chow, mirliton | Mexico, Guatemala | 24 | Throughout tropics especially Latin America | Young fruit and root (tuberous) as vegetable |
| *Sicana odorifera* | Casabanana | Brazil, Peru | 40 | Central America, northern South America | As fruit |
| *Telfairia occidentalis* | Fluted pumpkin | West Africa | 22 | Tropical Africa | Leaf and shoot as vegetable, seed is edible |

*Contd...*

| | | | | | |
|---|---|---|---|---|---|
| *T. pedata* | Oyster nut | Africa | 22 | Tropical Africa | Potential substitute for almonds or Brazil nuts |
| *Trichosanthes anguina* syn. *T. cucumerina* | Snake gourd, *chichinda* | India | 22 | South and South-East Asia | Immature fruit, shoot and leaf as vegetable |
| *T. dioica* Roxb. | Pointed gourd, *parwal* | India | 22 | India | Fruit as vegetable |

*Sirohi *et al.*, 2005

Summer squash (*C. pepo)* and winter squash (*C. maxima);* and two lesser-known species, namely, fig leaf (*C. ficifolia)* and *C. mixta* are described (Andres, 1990). Robinson and Decker-Walters (1999) renamed *C. mixta* as *C. argyrosperma*. Cultivated *Cucurbita* originated in two centres in the new world. Central Mexico is the centre of origin for *C. ficifolia, C. mixta, C. moschata* and *C. pepo,* while southern Peru, Bolivia and Northern Argentina are the centres of origin for *C. maxima.* The areas of diversity are northern Mexico for *C. pepo,* from central Mexico and the Yucatan peninsula to Costa Rica for *C. mixta,* central Mexico through the high plateau of central America and along the Andes to central Chile (1000-2000 m elevation) for cold tolerant *C. ficifolia,* southwards from Mexico City through central America into northern Colombia and Venezuela for *C. moschata,* in northern Argentina, Bolivia, southern Peru and northern Chile for *C. maxima.*Whitaker and Bemis (1975) assigned a key role to *C. moschata* as the nearest putative ancestor of the genus. They suggested *C. moschata* as a link between the wild and cultivated species and that all the five cultivated species have been independently domesticated. In India, *C. moschata* is more widely distributed than other species. According to Decker (1988), archaeological evidences indicate the occurrence of wild populations of *C. pepo* in Mexico and eastern USA around 10,000 and 30,000 years ago, respectively. Domestication of this species took place independently in these two areas. *Cucurbita argyrosperma* was apparently domesticated in southern Mexico where archaeological evidences of its domestication date back to 5200 BC.. Wild population of subspecies *sororia* that occurs in Mexico and central America, gave rise to the domesticated subspecies *argyrosperma*. Landraces of subspecies *argyrosperma* var. *argyrosperma* may have been selected first, the other two group of cultivars, namely, var. *callicarpa* and var. *stenosperma* probably evolved from northern and southern landraces of var. *argyrosperma,* respectively (Merrick, 1990). *Cucurbita ficifolia* was cultivated in the Andes Mountains of Peru in pre-historic times. As the species spread northwards, a large number of perennial xerophytic types evolved adapted to the dessert and plateau regions of northern Mexico and the South-western USA. Attempts have been made to domesticate another xerophytic species, namely, *C. foetidissima.* This species is perennial and grows wild in the arid deserts of South-western USA and Mexico. Its ability to withstand drought has attracted interest in its commercialization. It produces high yield of seed that is rich in protein and oil, and that of fleshy tuberous roots. (Sirohi *et al.*, 2005).

Important wild *Cucurbita* species, their major areas of distribution and Growth habit*

| Species | Natural major area of distribution | Growth habit |
|---|---|---|
| *Cucurbita andreana* | Argentina | Annual |
| *C. californica* | USA (southern California, Arizona) | Perennial |
| *C. cordata* | Mexico (Baja California) | Perennial |
| *C. cylindrata* | Mexico (Baja California) | Perennial |
| *C. digitata* | USA (New Mexico, Arizona), Mexico (Baja California) | Perennial |
| *C. foetidissima* | USA (Nebrask USA (Nabaraska, Kansas, Colorado, Utah, Nevada, Texas, Arizona, Missouri, New Mexico, California), Mexico (Guanajuato) | Perennial |
| *C. fraterna* | Mexico (Tamaulipas) | Annual |
| *C. galeotti* | Mexico (Oaxaca) | Perennial |
| *C. gracilor* | Mexico (Mexico State) | Perennial |
| *C. kellyana* | Mexico (Jalisco) | Annual |
| *C. lundelliana* | Guatemala, Belize | Perennial |
| *C. martinezii* | Mexico (Veracruz) | Annual |
| *C. mooreii* | Mexico (Hidalgo) | Perennial |
| *C. okeechobeensis* | USA (Florida) | Annual |
| *C. palmata* | USA (southern California) | Annual |
| *C. palmeri* | Mexico (Culiacan, Sinaloa) | Perennial |
| *C. pedatifolia* | Mexico (Quaretarao) | Perennial |
| *C. radicans* | Mexico (around Mexico City) | Perennial |
| *C. scabridifolia* | Mexico (North-eastern Mexico, southern Tamaulipas) | Perennial |
| *C. sororia* | Mexico (Guerrero, Oaxaca) | Annual |
| *C. texana* | USA (Central Texas) | Annual |

*Sirohi *et al.*, 2005

*Cucumis* has more than 30 species of which cucumber *(Cucumis sativus)* and melon (*C. melo)* have major economic importance. *Cucumis sativus* is reported to have originated in India where its close relatives with same chromosome number *(2n* = 14) occur. Among all these species, *C. sativus* and to a limited extent *C. callosus* are cultivated while C. *hardwickii,* C. *hystrix,* C. *prophetarum* and *C. setosus* are wild (Chakravarty, 1982; Arora and Nayar,1994). Rich variability is found in foothills of North-West Himalayas and southern hills for *C. hardwickii;* North-East Himalayas (Meghalaya, Mizoram, Assam) for *C. hystrix;* Rajasthan (Sirohi district) and parts of Gujarat, Maharashtra and Tamil Nadu for *C. prophetarum* and parts of Maharashtra for *C. setosus.* Cucumber is cultivated for at least 3,000 years in India and 2,000 years in China. China is considered as a secondary centre of diversity where it occupies second largest area among vegetables after Chinese cabbage. The centre of origin for C. *melo* is still not clear although the evidences point to Africa where wild species of *Cucumis* with same chromosome number *(2n* = 24) frequently occur. However, domestication may have occurred independently in East Asia and South East Asia including India. Today, the primary centre of diversity for this extremely polymorphic species is in South-West and central Asia, mainly Turkey, Syria, Iran, Afghanistan, North and central India, Trans-Caucasian, Turkmenistan, Tajikistan and Uzbekistan; and there is secondary centre of diversity in China and Republic of Korea (Sirohi *et al.*, 2005). Munger and Robinson (1991) proposed seven horticulturally important groups of melons based on fruit characteristics and uses, rather than phylogeny.

- *Cucumis melo var. cantaloupensis (musk melon):* It is usually andromonoecious and has medium-sized fruit with netted, warty or scaly surface; usually orange but sometimes green flesh, and aromatic or musky flavour.
- *C. melo var.flexuosus syn. var. utilissimus (snake melon, long melon):* It is monoecious and has long and slender fruit which is consumed immature.
- *C. melo var. momordica (snap melon):* It is monoecious. It has ovoid to cylindrical fruit and white to pale orange pulp with low sugar content. The smooth surface of the fruits cracks at maturity.
- *C. melo var. inodorus (winter melon):* It is mostly andromonoecious and has white or green flesh that lacks musky odour. Fruit shelf-life is longer than musk melon and is indehiscent at maturity.
- *C. melo var. dudaim (mango melon):* It is pickling type and has small, globular fruit, having white flesh with acid flavour.

- *C. melo var. conomon (pickling melon):* It has small fruit with smooth tender skin, white flesh, usually with a little sweetness and odour. It is andromonoecious, often pickled and also eaten fresh and cooked.
- *C. melo* var. *agrestis (kachri):* It is mostly wild but is cultivated to a little extent. It has slender vine and small fruit.

Two other species of some importance in *Cucumis* are *C. anguria* (West Indian gherkin) and C. *metuliferus* (African horned cucumber). *Cucumis anguria* produces extraordinarily large number of fruits and has potential for wider use in gherkin production (for example, as small pickled fruit). For many years, it was thought to be of American origin, but sexual compatibility and electrophoretic studies have shown that it is a cultigen, which descended from a non-bitter mutant of a bitter African species C. *longipes*. *Cucumis metuliferus* is also of African origin and has commercial potential (Sirohi *et al.*, 2005).

In *Citrullus,* three species have been recognized. These are *C. lanatus, C. colocynthis* and *C. ecirrhosus*. Wild form of several *Citrullus* species is found in dry sandy region of Kalahari Desert. *Praecitrullus fistulosus* (round melon), once included in genus *Citrullus,* is its closely related species. It is believed to have originated in India. The genetic variability in fruit morphological traits is concentrated in North-western India (western Uttar Pradesh, Haryana, Punjab), but the variability is rather limited suggesting its partial or recent domestication. *Luffa* has nine species out of which seven (*L. acutangula, L. cylindrica, L. echinata, L. graveolens, L. tuberosa, L. umbellata)* are found in India. This genus has a long history of cultivation in the tropical Asia and Africa. The probable centre of origin and the primary gene centre of *Luffa* is India. *Luffa acutangula* (ridge gourd) and *L. cylindrica* (sponge gourd) are grown throughout India in tropical and subtropical climates. *Luffa acutangula* has three varieties: var. *acutangula* is grown in South-eastern Asia and other tropical areas; var. amara, a wild form, is confined to peninsular India, while var. *forskohllii*, another wild form, is confined to Yemen. *Luffa echinata* grows in natural habitat in western Himalayas, central India and Gangetic plains. *Luffa graveolens* is a wild species distributed in parts of Northcentral India, and *L. hermaphrodita (satputia),* that bears fruit in clusters, is cultivated in eastern Uttar Pradesh and Bihar (Sirohi *etal.,* 2005).

*Lagenaria,* previously considered to be monotypic genus, has six species which are of African origin. *Lagenaria siceraria* (bottle gourd) is the only cultivated species. It is cultivated throughout the tropical and subtropical regions in Asia, Africa and South America. Based on morphological differences, the cultivars and landraces of African and New World origin are recognized as *L. siceraria* ssp. *siceraria* and those of Asia as *L. siceraria* ssp. asiatica. Other three species, namely, *L. abyssinica, L. breviflora* and *L. sphaerica* are wild perennials. In *Trichosanthes,* 41 species have been reported, and 21 of these have originated in India, parts of tropical Asia or Indo-Malayan region. Among these, snake gourd *(T. cucumerina* var. *anguina* syn. *T. anguina*) and pointed gourd *(T. dioica)* are cultivated throughout India. The major diversity is distributed in North-central and North-eastern India for *T. dioica;* whereas *T. anguina* is distributed in warm humid areas throughout the country but rich diversity is observed in North-eastern region, West Bengal, Karnataka and Kerala. *Momordica* has 45 species, native to Africa. Of these, 7 are found in India. Bitter gourd ( *M. charantia*), a popular vegetable, is grown all over tropical and sub-tropical regions. It has striking variability in India. Sweet gourd *(Momordica cochinchinensis)* is popular in Tripura, Assam and West Bengal whereas, *kakrol (M. dioica)* is cultivated in West Bengal, Assam, parts of Bihar and adjoining areas. There are several other *Momordica* species of minor importance, which include *M. foetida* and *M. grosvenori* which has, potential for the extraction of sweet glycoside component.

Ash gourd *(Benincasa hispida)* is reported to be native of Indonesia and Japan. It was domesticated in India during prehistoric times. It is widely grown all over the country in tropical and subtropical regions and possesses large variability for morphological and quality characters. Its fruits have long storability. It is a potential crop of tropical Africa and America. *Coccinia* genus has nearly 30 species, which are confined to tropical Africa except for ivy gourd (c. *grandis),* which occurs in cultivated and wild forms in South-East Asia in addition to Africa. It is also grown in Latin America, Australia and other tropical areas. In India, it is cultivated in warm humid areas. *Sechium,* formerly considered being monotypic genus, now eight species are recognized. Cho-cho *(Sechium edule)* is believed to be native of Mexico and Guatemala and has been under cultivation for centuries. The wild forms of this species and its close relative, S. *compositum,* are found in that region. *Cyclanthera* has about 30 species and all of these are native to the *Cyclanthera* has about 30 species and all of these are native to the tropics of South America. Stuffing cucumber *(Cyclanthera pedata* var. *edulis)* is cultivated in tropical America and Asia. It is a potential crop for wide cultivation in high altitude areas. Its closely related species *C. brachystachia* is also cultivated for its edible fruits in Nepal and South Taiwan. There are several endangered cucurbitaceous species. These are endangered due to several reasons, such as fragmentation of habitats, deforestation, over exploitation of land and water resources, soil degradation and urbanization.

## Augmentation of Germplasm

The National Bureau of Plant Genetic Resources (NBPGR), New Delhi, has made concerted efforts to augment the plant genetic resources from indigenous and exotic sources. Several explorations have been made in India and a large number of accessions of various cucurbitaceous species have been collected. The important introductions of cucurbits have been made from the USA, Mexico, Japan, France, Denmark, China and UK. Besides cultivated species, some wild species of cucurbits have also been introduced. Some of the prominent wild species introduced include *Citrullus rehmi, Cucumis aculeatus, Cucumis africanus, Cucumis dipsaclus, Cucumis meeurei, Cucumis metuliferus, Cucumis myriocarpus, Cucumis pustulatus, Cucumis ragittatu, Cucumis zeyheri, Cucurbita ecuadorensis* and *Cucurbita martinezii*. The landraces, old and modern cultivars and wild forms found in different parts of the world need to be continuously introduced particularly from the centre of origin and diversity of the cucurbits. The important germplasm introduced since 1990 are given elsewhere.

## Germplasm Evaluation

Characterization and evaluation of cucurbit germplasm is being done by NBPGR and crop based institutes and a number of promising accessions have been identified. Twelve genetic stocks having unique traits, have been registered.

## Germplasm Conservation

Conservation of crop plant diversity has assumed considerable significance today as compared to few decades ago. This has been necessitated due to over increasing human population as well due to depleting plant resources. Table 7 summarizes the available information on general variability status and genetic erosion status of different cucurbit species in Indian region.

It is evident that genetic erosion status for cucumber (both cultivated and related wild species e.g., *C. hardwickii, C. trigonus, C. hystrix* etc. and watermelon have been high followed by melons and in sponge gourd. Thus the situation for collection and conservation of cucumber and watermelon from different parts of India is warranting. Recently an attempt has been made to augment and collect available germplasm diversity in cucumber and melon under Indo-US-PGR programme from Rajasthan and Madhya Pradesh. After proper evaluation and seed increase, the material will be conserved in Indian Gene Bank at NBPGR and at National Seed Storage Laboratory (NSSL), Fort Collins, USA. Properly evaluated and documented germplasm lines are conserved in seed gene bank so that they can be utilized in future in crop-improvement work.

Crops and the areas explored for collection of cucurbit germplasm in India

| Crop | Areas explored |
|---|---|
| *Benincasa hispida* | North-eastern hill region, Uttarakhand, Uttar Pradesh, Himachal Pradesh, Rajasthan, Haryana, Orissa, Kerala, Karnataka, Tamil Nadu |
| *Citrullus lanatus* | North-eastern hill region, Uttar Pradesh, Punjab, Rajasthan, Kerala, Tamil Nadu |
| *Coccinia indica* | Arunachal Pradesh, West Bengal, Bihar, Uttar Pradesh, Chhattisgarh, Karnataka, Kerala, Tamil Nadu |
| *Cucumis hardwickii* | Kulhal range in Himalaya along Poanta Sahib, Dehra Dun, Western Ghat areas, Tehri Garhwal and North-West hill region |
| *Cucumis melo* | Punjab, Haryana, Uttar Pradesh, Rajasthan, Chhattisgarh (Bastar region) |
| *Cucumis melo* var. *momordica* | Uttarakhand, Punjab, Haryana, Uttar Pradesh, Rajasthan |
| *Cucumis sativus* | Kashmir valley, Himachal Pradesh, Punjab, Uttarakhand, Uttar Pradesh, Rajasthan, Bihar, Madhya Pradesh, Chhattisgarh, Orissa, West Bengal, North Cachar, North-eastern hill region, Andhra Pradesh, Karnataka, South Kerala, Tamil Nadu |
| *Cucurbita moschata* | North-eastern hill region, central and eastern Uttar Pradesh, Himachal Pradesh, Uttarakhand, Punjab, Haryana, Rajasthan, |
| *Lagenaria siceraria* | Orissa, Kerala, Tamil Nadu |
| *Luffa acutangula* | Mizoram, Assam plains, Meghalaya (Khasi hills), Tripura, Nagaland, Bihar, Jammu and Kashmir, Himachal Pradesh, Uttar Pradesh, Haryana, Rajasthan, Madhya Pradesh, Chhattisgarh |
| *Luffa cylindrica* | Rajasthan, Haryana, eastern Uttar Pradesh, Bihar, West Bengal, North-eastern hill region, Orissa, Chhattisgarh, Andhra Pradesh |

*Contd...*

| | |
|---|---|
| *Momordica charantia* | North-eastern hill region, West Bengal, Bihar, Uttar Pradesh, Uttarakhand, Haryana, Himachal Pradesh, Chha,ttisgarh, Andhra Pradesh, Kerala, Tamil Nadu |
| *Praecitrullus fistulosus* | Assam, Meghalaya, Tripura, West Bengal, eastern Uttar Pradesh, Bihar, eastern Haryana, Maharashtra (Vidarbha region), Southern Kerala, Southern Tamil Nadu, Andaman and Nicobar islands |
| *Trichosanthes dioica* | Punjab, Rajasthan, Haryana |
| | Eastern Uttar Pradesh, West Bengal, Bihar, Jharkhand and Assam Valley |

Germplasm of cucurbits introduced for specific traits

| **Species** | **EC number** | **Trait** |
|---|---|---|
| *Citrullus lanatus* | EC333659-69 | Wilt resistance |
| | EC382753 | *Fusarium* and anthracnose resistance |
| | EC383809 | *Fusarium* resistance |
| | EC393240-43 | Red fleshed, small round fruit |
| | EC402549 | Sweet Scarlet |
| | EC402549 | Anthracnose and *Fusarium* wilt resistance, early maturity |
| *Citrullus* sp. | EC383809 | *Fusarium* resistance |
| *Cucumls anguria* | EC382500-18 | Pickling type |
| | EC382535-60 | Pickling type |
| | EC399866-909 | Pickling type |
| | EC399910-12 | Pickling type |
| | EC382519-22 | Pickling type |
| *Cucumis melo* | EC342751-54 | Heat tolerance |
| | EC350602 | Aphid resistance |
| | EC350603 | Water melon mosaic virus (WMV) resistance |
| | EC350604 | WMV and powdery mildew resistance, sulphur tolerance |
| | EC382727-31 | Male sterile lines |

*Contd...*

| | | |
|---|---|---|
| | EC382732 | White fruit colour |
| | EC382733 | Yellow green fruit colour |
| | EC382734 | *Fusarium* resistance |
| | EC399914 | Cucumber mosaic virus (CMV) resistance |
| | EC399917, EC399924-31 | Downy mildew resistance |
| | EC399923 | *Bemisia tabacci* resistance |
| | EC399932-33 | CMV tolerance |
| | EC399934 | Powdery mildew resistance |
| *Cucumis sativus* | EC320556 | CMV, scab and powdery mildew resistance |
| | EC329300 | Multiple disease resistance, gynoecious line |
| | EC388737-39 | Gynoecious lines |
| | EC398030 | Early determinate type |
| | EC398030 | Early, cluster bearing, determinate type |
| | EC398966-67 | Angular leaf spot resistance |
| | EC398968-70 | Anthracnose resistance |
| | EC398971-73 | Fruit rot resistance |
| | EC398974-90 | Downy mildew resistance |
| | EC398991-399007 | Leaf spot resistance |
| *Cucumis* sp. | EC399914-37 | CMV and powdery mildew resistance |
| *Cucurbita foetidissima* | EC442706-442713 | Drought tolerance |
| *Cucurbita moschata* | EC329323-24 | High sugar |

Sources/Donors identified for various traits in different species of cucurbits

| **Crops/trait (s)** | **Donor (s)** |
|---|---|
| *Cucumis sativus and related species* | |
| Extended shelf-life | IC203838, IC203839 |
| Early and determinate | EC398030 |
| Gynoecious line | EC382739 |

*Contd...*

| | |
|---|---|
| High yield | IC203838, EC237658, VJ/98-176, VJ/98-151 |
| Anthracnose resistance | PI197087, Poinsett |
| Downy mildew resistance<br>Powdery mildew resistance<br>Cucumber scab resistance<br>Angular leaf spot resistance<br>Bacterial wilt resistance | PI197087<br>Poinsett, Yomaki, PI79376<br>Wisconsin SMR9<br>Poinsett, MSU9402,<br>PI169400 PI200815, PI200818 |
| Cucumber green mottle mosaic resistance White fly resistance | *Cucumis anguria,* C. *african us,* C. *ficifolius*<br>*Cucumis asper,* C. *dinteri,* C. *sagittatus* |
| *Cucumis melo and related species* | |
| High yield | EC303044 |
| Early, high yield, powdery mildew, water melon mosaic virus and aphid resistance, sulphur tolerance and free from blight incidence | EC178496 |
| Downy mildew resistance | Seminole, Edisto, Harvest Queen, *Cucumis callosus* |
| Powdery mildew resistance | PMR45, Campo, Edisto, Georgia 47 |
| *Fusarium* wilt resistance | *Cucumis melo* var. *chi to,* C. *inodorus,* C. *flexuosus,* C. *reticulatus* |
| Water melon mosaic virus resistance Gummy stem blight resistance | PI414723, *Cucumis metuliferus*<br>Chilton, Gulf coast |
| *Citrullus lanatus* | |
| High yield | EC28435, K3228, OC92910 |
| *Fusarium* wilt resistance | Citron, Shipper, Smokylee, Summit |
| Anthracnose resistance | African Citron 8, Charleston Grey, |
| Powdery mildew resistance | Congo, Crimson Sweet, Jubilee |
| *Lagenaria siceraria* | |
| Summer type | Arka Manik |
| Rainy season type | BDJ628, BDJ864, IC1612, U8-217, UlO-315, Ul6-102, |
| Early type | BDJ536, BDJ1733, EC176860, IC92904 IC144389, |

*Contd...*

| | |
|---|---|
| Heavy fruit | IC92426, IC94404, NIC13188, NIC13182<br>IC29404, IC92429, IC11O389 |
| High yield<br>*Cucurbita moschata* | EC305378, IC92429, IC92490,<br>IC201150, IC201159 |
| *Luffa cylindrica*<br>Early type | IC92604, IC92779, IC92797, IC93445 |
| High yield<br>*Cucurbita* spp. | NIC597, NICIO23, NIC13236, NICllO235<br>IC90512, IC92499, IC92533, IC92571<br>NICIO170, NICIO174 |
| Cucumber mosaic virus resistance | NIC23288, NIC23291, NIC23292 |
| Water melon virus resistance | EC305688, IC92475, IC92721, IC92727,<br>IC92751, IC201217, IC201229 |
| Powdery mildew, squash and water melon virus resistance | *C. pepo* cv. Cinderella, Gold Rush<br>*C. ecuadorensis, C. joetidissima*<br>*C. ficifolia, C. lundelliana*<br>*C. foetidissima*<br>*C. lundelliana, C. okeechobeensis*<br>*C. martinezii* |

Crop variability and genetic erosion in Indian region

| **Crop** | **Cultivated (c)/ wild (w)** | **General variability status** | **Genetic erosion status** |
|---|---|---|---|
| Spongegourd | C<br>W | Low<br>High | Medium<br>- |
| Rdigegourd | C | Medium | Low |
| Bittergourd | C | Medium | Low |
| Cucumber | C<br>W | Very High<br>High | High<br>High |
| Melon<br>(*C. melo*) | C<br>W | Very High<br>High | Medium<br>Low |
| Watermelon<br>*Trichosanthes cucumerina* | C<br>C&W | High<br>Medium | High<br>Low |

Utilizing the information that *Cucurbita lundelliana* is cross-compatible with otherwise cross-incompatible *C. pepo, C. mixta, C. moschata* and *C. maxima,* Rhodes (1959) developed an inbreeding population among these species. By using this genepool in a series of bridging crosses, he succeeded in transferring bushy habit from C. *pepo* and powdery mildew resistance from C. *lundelliana* to C. *moschata.* Similarly, to develop a root knot nematode tolerant muskmelon variety, *Cucumis metuliferus* was used as resistant source. However, not much work has been carried out on the transfer of various traits from the wild relatives and desirable traits in wild relatives are yet to be commercially exploited (Sirohi *et al.,* 2005).

Floral Biology of Cucurbit species*

| Crops | Anthesis | Dehiscence | Duration of Pollen fertility | Stigma receptivity |
|---|---|---|---|---|
| Bottlegourd | 17.00 - 20.00 hr | 13.00 - 14.00 hr | On the day of anthesis till next morning | 36 hrs before and 90 hrs after anthesis |
| Ridgegourd | 17.00 - 20.00 hr | 17.00 - 20.00 hr | On the day of anthesis to till 2-3 days after anthesis in winter and 1½ days in rainy season | 6 hrs before and 84 hrs after anthesis |
| Spongegourd | 4.00 -8.00 hrs | 4.00 - 8.00 hrs | On the day of anthesis | 10 hrs before and 120 hrs after anthesis |
| Bittergourd | 9.00 - 10.30 hr | 7.00 - 8.00 hr | 5.00 - 12.00 hr | One day before and one day after anthesis |
| Snakegourd | 18.00 - 21.00 hr | Shortly before anthesis | 10 hrs before and 49 hrs after anthesis | 7 hrs before and 51 hrs after anthesis |
| Ashgourd | 6.00 - 7.00 hr | 3.00 - 4.00 hr | 7.00 -16.00 hr | 12 hrs before and 36 hrs after anthesis |
| Cucumber | 4.30 -7.00 hr | 4.30 - 0.00 hr | Up to 14 hrs | 12 hrs before and 6-7 hrs after anthesis |
| Pumpkin | 3.30 - 6.00 hr | 21.00 - 3.00 hr | 16 hrs after anthesis | 2 hrs before and 10 hrs after anthesis |
| Muskmelon | 5.30 -6.30 hr | 5.00 - 6.00 hr | 5.00 - 14.00 hr | 2 hrs before and 2-3 hrs after anthesis |
| Watermelon | 6.00 - 7.30 hr | 5.00 - 6.30 hr | 5.00 - 11.00 hr | 2 hrs before and 3 hrs after anthesis |

*Kalloo, 1988

Over 400 base collections of different cucurbitaceous crops are under long-term conservation in the National Genebank at NBPGR. In addition, the plant breeders at different institutions are maintaining a number of active collections. *In situ* conservation, as a complementary strategy to *ex situ* conservation, needs to be enhanced at least in the areas with subsistence agriculture by promoting cultivation of farmers' varieties /landraces.

## Germplasm Maintenance

Cucurbits, being cross-pollinated species, require controlled pollination for maintenance of genetic integrity of the accessions. Mode of pollination and isolation distance play an important role in maintaining the integrity of the germplasm. The population should also be large enough to retain various alleles in an accession. Potentially valuable alleles may occur at very low frequencies particularly in wild type and landraces. The desirability of preserving these alleles must be weighed against the cost and space required for generations, yet, the alleles will not necessarily be maintained at the same frequency. To maintain the allelic frequencies, biparent crosses should be preferred to multiparent crosses and an equal number of seeds should be taken from each plant. If pollen mixture is used, the pollen should be thoroughly mixed. The common practice of using small number of plants as female and a large number as males saves space but alters the gene frequency by giving more weight to alleles present in few female plants. Hand pollination requires a great deal of labour. To reduce labour costs, some institutes especially in developed countries, use insect pollination within insect proof net house. Several sexually incompatible species are planted within a net house and an uncontaminated beehive is placed inside for pollinating flowers.

Mode of pollination, isolation distance, seed extraction time of some Cucurbitaceous species

| Species | Pollination | Isolation Distance (m) | Seed extraction Time |
|---|---|---|---|
| Cucumber | Partially selfed and cross pollinated | 400-1000 | Fruits pale yellow/ golden |
| Watermelon | Partially selfed and cross pollinated | 400-1000 | Edible maturity |
| Pumkin | Cross pollinated | 400 | Fruits pale yellow |
| Muskmelon | Partially selfed and cross pollinated | 400-1000 | Edible maturity |
| Bottlegourd | Partially selfed and cross pollinated | 600 | Dried fruit stage |

## Germplasm Utilization

The cultivars of cucurbits have been developed either through direct selection in germplasm or recombination breeding and a few through induced mutation. In recombination breeding, desirable traits available in different sources are combined together through hybridization followed by selection in segregating generations. Both indigenous as well as exotic germplasm have been utilized for developing the cultivars. In some crops, hybrid cultivars have also been developed. There are several wild relatives of cucurbits having resistance to different biotic and abiotic stresses. Some of these are cross-compatible with cultivated species; and interspecific and intergeneric hybridization has been carried out by different researchers to transfer desirable traits from wild to cultivated form and some success has been achieved. There are certain constraints in germplasm utilization. Some materials, particularly wild relatives are very difficult to maintain because of low germinability, self-incompatibility etc. Their maintenance requires specific climatic conditions, careful handling and is labour intensive. Further, there are many crossability barriers. Many interspecific hybrids are sterile, and if fertile, are weak, have more morphological features of the wild types, and suffer a loss of reproductive capacity in later generations. Also, the segregation pattern of inter-specific hybrids is erratic and generally has low chance of obtaining desired recombination for commercial utilization.

### *Gene Pyramiding and Multiple Characters Breeding in Cucurbits*

Since the early 20$^{th}$ century, classical breeding for disease resistance in plants has been a major method for controlling biotic stress resistance in crop plants. Varieties that are resistant or tolerant to one or few specific pathogens are already available in many crops. Although conventional breeding had a significant impact on improving crop plants for resistance to important diseases, there are several constraints in terms of selection of the desired resistant progeny that can react adequately to the evolution of the new virulent pathogens. Moreover, several interesting resistances are difficult to use because the diagnostic tests are often hard to develop due to the challenge posed by inoculum production and maintenance. The DNA-based markers have become important tools for genetic analyses and crop improvement since 1980s. DNA marker technology has been used in commercial plant breeding programmes since the early 1990s, and has proved helpful for the rapid and efficient transfer of useful traits into agronomically desirable genotypes. A great deal of work has been carried out in the last two decades to find molecular markers linked to disease resistance genes in several crop plants, including rice, wheat, maize and tomato. Markers linked to disease resistance loci can now be used for marker-assisted selection (MAS) programmes, as it permits selection of desirable genotypes without being dependent on the phenotypic expression. MAS even allows pyramiding of genes conferring resistance to multiple traits at the same genotype. Molecular mapping has also led to map-based cloning and sequencing of several disease resistance genes, thereby providing a better understanding of the molecular basis of resistance. Here, salient components of the molecular marker-assisted gene pyramiding in cucumber and watermelon will be discussed.

### *Introgression of genes conferring resistance to downy mildew, anthracnose, green Mottled virus from Cucumis hardwikii and other sources into cucumber*

Reduced vulnerability to pest and diseases through incorporation of genetic resistance, wide or narrow genetic base of current cultivars is a major pre-occupation of present day cucumber breeders. A large number of varieties have been released, however resistance to downy mildew is still the major constraint affecting the yield in cucumber. The resistance to downy mildew have been reported in ***Cucumis hardwickii***. India has a large amount of variability in this species as the area of its distribution has been increasing with every expedition. Therefore there is a need to exploit the variability for downy mildew resistance in introgression of gene conferring resistance to cultivated *Cucumis sativus*.

*Recurrent Parent:* Swarn ageti

**Donor Species**

| **Species** | **Traits** |
|---|---|
| *C. hardwickii* | Downy Mildew resistance |

Hybridisation scheme for development of hybrids and recombinant inbred lines

*C. sativus* (*Swarn ageti*) x *C. hardwickii* (Sequential fruiting and Multiple disease resistance)

↓

↓

F1

F2 ↓ Advancement up to F8 generation through selfing

Development of RILs for molecular mapping of important traits

*Expected Output:* Development of interspecific lines with multiple desired features lines

Hybridization scheme for development of hybrids and Recombinant Inbred Lines

| | | |
|---|---|---|
| *C. sativus* (*Swarn ageti*) | x | B-184 |
| | | Resistant to DM, Anthracnose, green mottled virus |
| | ↓<br>↓ | |
| F1 | | |
| F2 | ↓ | Advancement up to F8 generation through selfing |

Development of RILs for molecular mapping of important traits, which will be evaluated at more than 4 locations

*Expected output:* Development of RILs for molecular mapping

### *Transfer of Genes Conferring Resistance to Watermelon Bud Necrosis virus from Citrullus colocynthis into Watermelon*

In last four years the production of watermelon is drastically declined to a about 25% of its total acreages in the country due to viral diseases, particularly watermelon bud necrosis virus. The incidences of the disease are between 39-100 per cent with yield loss of 60-100%. Therefore, an effort on exploitation of wild species in genetic improvement in watermelon, particularly for transfer of Watermelon Bud Necrosis Virus (WBNV) is very much needed. The wild *Citrullus colocynthis* have been reported tolerant to both WBNV and drought.

*Recurrent parents:* Arka Manik

*Donor species:* Citrullus colocynthis

*Genomic constitution:* The two are related with same genomic constitution

*Cross-compatibility:* The two are cross compatible

*Phylogenetic relationship with cultivated species: Citrullus colocynthis* is the progenitor of cultivated watermelon

*Any other information (genetics of traits):* The two species crosses in the nature also

### *Hybridisation and Backcrossing Scheme*

Direct hybridisation

Backcrossing with recurrent parent as female

### *Selection/Screening Strategy*

I Year

Screening of wild species for resistance to WBNV

Identification of sources resistance to WBNV

Hybridisation involving identified wild species with cultivated species

II Year

Advancing the interspecific hybrids produced, screening against WBNV and selecting desired recombinants.

Initiation of appropriate biotechnological manipulation for raising hybrids involving wild and cultivated species.

III Year

Advancement of interspecific population to genetic uniformity by back crossing/ intermittent selfing/ screening against WBNV and selecting desired recombinants.

## Expected Out Put

Development of pre breeding lines with resistance or tolerance to WBNV

## Future Perspectives

- Cucurbit is a big group and has large genetic diversity across the country in different crops which need to be collected, characterized, evaluated, conserved and utilized.
- Large number of cucurbitaceous species are still un-exploited and under-utilized. If exploited, these can play an important role in diversification of agriculture. Work on these species should get greater attention.
- There is a need to introduce diverse germplasm of cucurbits with an emphasis on biotic and abiotic stresses, desirable yield and quality characters. During acquisition, the traits preferred in foreign markets should also get due attention.
- NBPGR has developed exhaustive Descriptor list of large number of vegetable crops including several cucurbitaceous species with the help of vegetable experts. Cucurbits germplasm including wild relatives assembled at NBPGR and other centres across the country should be

characterized and evaluated in different agro-climatic zones using these descriptors.

- Molecular techniques may be utilized for identification of germplasm and study of interspecific differentiation in cultivated species and wild relatives. These tools also need to be deployed to overcome crossability barriers and to enhance utilization of wild species.
- Large genetic variability present in cucurbitaceous crops needs to be utilized with greater vigour for the development of high yielding varieties and FI hybrids with superior fruit quality and multiple resistances to stresses.
- Registration and conservation of promising and unique germplasm should get greater focus.

## References

Arora R.K. and E.R. Nayar 1994 Wild Relatives of Crop Plants in India. Science Monograph No.7, National Bureau of Plant Genetic Resources, New Delhi, India, 90p.

Chakravarty H.L. 1982 Fascicles of Flora of India, Fascicle II, *Cucurbitaceae.* Botanical Survey of India, Howrah, India.

Decker DS 1988 Origin evaluation and systamatics of *Cucurbita pepo* (Cucurbitceae). *Econ. Bot.* 42: 4-15.

Kalloo G. 1988 Vegetable Breeding Vol. -I, CRC Press, Inc. Boca Raton, Florida, p 24-27.

Munger HM and RW Robinson 1991 Nomenclature of *Cuucmis melo* L. *Cucurbit Genetic Cooperative Report* 14: 43.

Ram, D. and Umesh Srivastava 1999 Some lesser known miner cucurbitaceous vegetables: their distribution, diversity and uses. *Indian J. Pl. Genet. Resour.* 12: 307-316.

Rhodes AM 1959 Species hybridization and inter-specific gene transfer indigenous *Cucurbita. Crop. Am. Soc. Hort. Sci.* 74: 546-551.

Sirohi P.S., A.D. Munshi, Gunjeet Kumar and T.K. Behera 2005 Cucurbits. *In*: Plant Genetic Resources: Horticultural Crops. B.S. Dhillon, R.K. Tyagi, S. Saxena and G.J. Randhawa (Eds) p 34-58, Narosa Publishing House Pvt. Ltd., New Delhi.

Srivastava Umesh C. 1993 Exploration and evaluation of cucurbitaceous crops, p 57-67. *In:* Exploration, Evaluation and Conservation of Plant Genetic Resources- Lecture notes. R.S. Rana, Bhag Singh, U. Srivastava, P.N. Mathur, Mathura Rai, B.N. Singh, S. Kochhar and I.S. Bisht (Eds), NBPGR, New Delhi.

Whitaker TW and WP Bemis 1975 Origin and evaluation of the cultivated *Cucurbita. Bulletin Torry Botanical Club* 102: 362-368.

**Underutilized and Underexploited Horticultural Crops, Vol. 2, 2007**
*K.V. Peter (ed.), pp. 153-174*
New India Publishing Agency, New Delhi (India)
E-mail : newindiapublishingagency@gmail.com
Web: www.bookfactoryindia.com

# 9 Arid Vegetables

**D. K. SAMADIA**
Senior Scientist (Horticulture)
Central Institute for Arid Horticulture
Bikaner - 334 006

Indian arid zone is one of the largest sub-tropical deserts of the world, of which 20 % is arid and the remaining semi-arid with varied habitats. The hot arid region is spread over 31.7 million ha mainly in Rajasthan (19.6 m ha), Gujarat (6.22 m ha), Andhra Pradesh, Punjab and Haryana. Rajasthan is the largest state in the country covering approximately 34.2 m ha geographical area out of which 27.1 m ha (79.08 % of total area) is laying in arid and semi-arid climate, spread over 21 out of 32 districts. Western part of Rajasthan is under hot arid climate (Thar Desert) and it extends from $24^0$ 37′ to $30^0$ 10′ N latitude and $69^0$ 29′ to $76^0$ 05′ E longitude. Barmer, Bikaner, Churu, Ganganagar, Hanumangarh, Jaisalmer, Jalore, Jhunjhunnu, Jodhpur, Nagour, Pali and Sikar districts of Western Rajasthan account for 63.4 % area of hot arid realm of India. Apart from these, about 10 % area each of Sirohi, Ajmer and Jaipur districts are also in arid region. Punjab and Haryana bound it in North and Gujarat in South. The Aravali hill ranges provide a natural boundary with semi-arid Eastern Rajasthan. In the West, it has a long international boundary with Pakistan.

Arid zone is characterized by extremes of high temperatures, and low, uncertain and variable precipitation. These factors predominantly limit scope for diversification and choice of crops and their varieties and also productivity. Growing of most hardy vegetables should be encouraged. Drier conditions of arid region greatly favour development of quality in vegetables particularly cucurbits, chillies and spices. This region also offers excellent opportunity for quality seed production in vegetable crops. Besides, the arid situations also help in low incidence of diseases and insect-pest infestations (Samadia *et al.*, 2004).

### *Agro Climate – Necessary Tool for Planning*

Arid zone eco-system of Rajasthan is very fragile and is prone to serious imbalances even with the slightest disturbance due to mismanagement of resources. Among the eastern margin of the desert, the mean annual rainfall is 500 mm, while in the western most part, it is 100 mm. The rainfall is largely monsoon driven. It comes between June and September (in 9-21 short spells). About 90 %of rainfall is received in July and August. Delayed onset and early withdrawal of monsoon, the main source of rain, is quite common. Large coefficient of variation (40-60%) and erratic distribution during the monsoon are characteristic features of rainfall, leading often to prolonged drought and failure of rainfed crops.

Winter is the period of light wind (3.4-8.3 km/h) from North and Northeast. From March onwards the wind peaks up and heralds in change of season. Strong to very strong winds with sand storm are experienced during May-July when the southwest monsoon sets in. The land is subjected to strong wind during hot summer months, which leads to blowing away of sand and other finer particles from parts of landscape and deposition of sediments at other localities. Major threat to agriculture from wind erosion is injury that is caused to crops through sand blasting of tender seedlings, stem, buds, flowers and tender fruits during sand storms. In areas of depositions, smaller seedling or plants get buried under sand.

Mean maximum air temperature in May-June varies from $42^0$ C in the west to $40^0$ C in the east, but it is common to experience 48-$50^0$ C. The mean minimum air temperature is recorded during December-January ($6^0$-$10^0$ C), where the temperature sometime dips down below freezing point. Occurrence of frost is frequent in the arid tracts. Here, March to October months are hot and shiny with a continuous high temperature conditions ($35^0$-$38^{0C}$) for the longer period of time. Erratic and variable rainfall, extremes of temperature, high evapo-transpiration, acute water stress and high and hot wind conditions are the restrictions imposed by nature in the Indian arid zone which necessitates micro level planning for successful vegetable cultivation. By and large field soils are coarse textured and sandy, and poor in fertility and water retention capacity

Since in arid region, rainfall is low and erratic to a few short spells leading to both soil and atmospheric water deficit exceeding 24-30 mm during hot months. In western India, 54 % of area receives less than 250 mm, 26.5 % of the area 250-300 mm and only 17 % areas receives 350-600 mm rainfall. Under rainfed conditions very few crops like *mateera*, snapmelon, *kachri, tinda* and cluster bean are grown for vegetable production during *kharif* season. Among adaptable crops, cucurbits and brinjal, cluster bean and Indian bean are drought hardy. In these crops, there is need to select the most hardy and short duration genotypes for rainfed production. Besides, the primary need is to develop viable systems of vegetable production by incorporating native species having horticultural potential. This will not only assist in creating favourable microclimate in production sites but also insuring better utilization of available resources (Samadia, 2004).

### *Vegetable Farming – A Way Towards Prosperity*

In arid areas, mixed cropping of *bajra* with legumes like cluster bean, moth bean or cowpea dominated conventional farming systems under rainfed conditions. Here, indigenous cucurbits particularly *mateera, kachri,* snapmelon and *tinda* are rescue crops and provide subsistence income to farmers. Besides, a number of indigenous species such as *khejri, kair, lasora, kumat, khimp, phog, guarpatha* and *khumbi* have horticultural potential ensuring nutritious vegetables to desert dwellers under extremes of environmental stresses. Presently, systematic production of vegetable crops is rare in arid regions mainly because of lack of suitable crop genotypes and appropriate production technologies suited to prevailing agro-climatic conditions.

Based on surveys, studies , research and development work conducted over more than one decade (1995-2004) in arid and semi arid regions, the extent of opportunities and constricts were assessed in respect of horticulture based farming and recommendations were carved out for vegetable cultivation in Rajasthan. Besides, a road map was suggested for cultivation of vegetable crops under rainfed, partially and/or assured irrigated conditions in different sub agro climatic zones of Western Rajasthan. Perennial species having commercial significance and found potential as vegetables in fresh, dehydrated or processed forms have also been highlighted to develop resource use efficient conventional or intensive systems of crop production. By incorporating these underutilized floras in production sites as a component of plantations for windbreak, boundary side, bio fencing, hedgerows, field dividers, strip blocks or in production models help in development of favourable microclimate for vegetable production and ensures continuation of income from these species. (Samadia *et al.*, 2004).

Existing low productivity of arid vegetables can be increased by adopting and improving production technology and innovative approaches. In this region, special attention is needed to exploit native genetic resources of underutilized horticultural species for development of commercial farming systems. Systematic breeding approaches for targeted crops is required on high priority basis to develop better genotypes with multiple uses and attributes including processing qualities and wider adaptability under abiotic stresses. In this part of the country, vegetable production and productivity largely depend upon availability and management of limited irrigation and rainwater. Quality of produce and productivity of arid vegetables can be improved manifold, if due consideration is given to *in situ* run-off harvest, moisture conservation, and adoption of improved and innovative techniques in a strategic manner- "Horticulture based crop production site management approach" (HBCPSMA, Samadia 2004) as an integrated concept.

Indian arid zone grows a large number of vegetables compared to any hot arid desert in the world. Production of vegetables in arid areas poses bio-physical constraints limiting extent and type of crop productivity. Major restrictions are related to scarcity of water, high temperature, excessive evaporation, and soil conditions. Potential of growing vegetable crops is recognized particularly in the Indian desert where conditions are not as extreme as in desert of Sahara and Arabia. Some of the well established and underutilized vegetable crops prospectively suitable for arid regions are given elsewhere. As a result of concerted vegetable improvement programmes in the country, a large number of high yielding varieties and hybrids suited to different agro-climatic conditions are developed. Owing to variable and to some extent manageable agro-climatic conditions, availability of limited irrigation water and potentials for different types of crop cultivation systems (rainfed, partially or assured irrigation) in the vast arid areas, a careful selection of crops and their varieties for specific micro region could be a gainful boon for success in vegetable production.

Peculiar arid agro climatic conditions impart unique quality in watermelon (*mateera*), muskmelon, snapmelon, *kachri*, chilli and spices. Indigenous products from *khejri* (*Prosopis cineraria*), *kair* (*Capparis decidua*) and *kachri* (*Cucumis melo* var. *callosus*) can be exploited for preparation of value added products for domestic and export market. Dry vegetable products have a lot of market potential. High temperature and low humidity help in solar drying of vegetables. The practice is already common for drying of *khejri* (*sangri*), cluster bean, *methi* and fruits of *kachri*, snapmelon, roundmelon and chillies.

Potential vegetables for hot arid region of India

| Group | Vegetables |
|---|---|
| Cucurbits | Watermelon [*mateera*](*Citrullus lanatus*), muskmelon (*Cucumis melo*), snapmelon (*Cucumis melo* var. *momordica*), *kachri* (*Cucumis* spp.), *kakri* (*Cucumis* spp.), roundmelon (*Praecitrullus fistullosus*), bottlegourd (*Lagenaria siceraria*), ridgegourd (*Luffa acutangula*), spongegourd (*Luffa cylindrica*), bittergourd (*Momordica charantia*), *kundru* (*Coccinia grandis*) |
| Legumes | Clusterbean (*Cyamopsis tetragonoloba*), cowpea (*Vigna unguiculata*), Indian bean (*Lablab purpureas*), pea (*Pisum sativum*), sword bean (*Canvalia gladiata*) |
| Solanaceous crops | Chilli (*Capsicum annuum*), brinjal (*Solanum melongena*), tomato (*Lycopersicon esculentum*) |
| Cole crops | Cabbage (*Brassica oleracea* var *capitata*), cauliflower (*Brassica oleracea* var.*botrytis*), knol-khol (*Brassica oleracea* var. *caulorapa*) |
| Bulbs | Onion (*Allium cepa*), garlic (*Allium sativum*) |
| Root crops | Radish (*Raphanus sativus*), carrot (*Daucus carota*) |
| Okra | Okra (*Abelmoschus esculentus*) |
| Leaf/herbal | *Methi* (*Trigonella foenum graecum*), *palak* (*Spinacia oleracea and Beta vulgaris*), coriander (*Coriandum sativum*), chaulai (*Amaranthus* spp), *bathua* (*Chenopodium spp.*), *sowa* (*Anethum sowa*) |
| Perennial species | *Khejri* [*Sangri*](*Prosopis cineraria*), drum-stick [*Sehjana*](*Moringa oleifera*), Indian Aloe [*Guarpatha*](*Aloe barbadensis*), *phog* (*Callgonum polygonoides*), *khimp* (*Laptadenia pyrotechnica*), *Kair* (*Capparis deciduas*), *Lasora* (*Cordia myxa*), .*Kumat* (*Acacia senagal*) |

Under exploited crop products used commercially

| Crops | Form of utilization |
|---|---|
| Kachri | Dried fruits for vegetables, pickles, chattnies, curry powder |
| Snapmelon | Dried fruits (*Khelera*) for vegetables, seed kernels for magaz |
| Tinda | Dried fruits (*Phophalia*) for vegetables |
| Mateera | Seeds for extraction of kernels for magaz |
| Khejri | Dried tender pods (*Sangri*) for vegetables |

*Contd...*

| | |
|---|---|
| Cluster bean | Dried tender pods (*Guar phali*) for vegetables |
| Indian bean | Dried tender pods and seeds (*Sem phali*) for vegetables |
| Kumat | Dried seeds for vegetables |
| Lasora | Dried tender fruits for vegetables, pickles |
| Kair | Dried immature fruits for vegetables, pickles |
| Methi | Dried leaves (*Methipani*) and seeds (*Methidana*) for vegetables |
| Coriander | Dried leaves |
| Mint | Dried leaves |
| Chillies | Dried fruits, whole or powder |

## *Research and Development in Underutilized Arid Vegetables*

Farmers of arid zone are resource poor and carry out subsistence farming. Primary requirements of a farmer like food and fodder are largely met from monsoon based harvests and are mostly uncertain. Proper production systems should be required which support livelihood depending upon land and water resources. Vegetable crops are highly remunerative for replacing subsistence farming and also to provide higher employment opportunity and high potential of value addition. Native genetic resources of potential species having horticultural significance are quite drought hardy, adapted to adverse situations and ensuring production under low input management systems of arid areas.

Realizing importance of vegetable crops particularly underexploited and less popular, and having commercial production potentials in arid and semi arid regions, systematic research work on germplasm collection, conservation and utilization was started in 1994 at CIAH, Bikaner. Intensive crop specific surveys in targeted variability pockets and explorations were undertaken in arid and semi arid regions and a large number of land races, semi-cultivated and popular types of *mateera*, *kachri*, snapmelon, beans and a few perennial horticultural species of vegetables were made over years for systematic evaluation, characterization and conservation of indigenous germplasm. As a result of systematic purification followed by selection and hybridization at CIAH, Bikaner, high yielding and high quality genotypes are developed and recommended for commercial cultivation. Promising varieties/selections are in *mateera* (AHW 19, AHW 65 and $F_6$/a), *kachri* (AHK 119 and AHK 200), snapmelon (AHS 10 and AHS 82), *kakdi* (AHC 2 and AHC 13), bottle gourd (AHLS Round 1 and AHLS Long 1), round melon (AHRM 1), cluster bean

(AHG 13), Indian bean (AHDB 3 and AHDB 16) and sword bean (AHSB 1) (Pareek and Samadia, 2002 and Samadia, 2004). Besides, intensive research and development work for commercialization of underutilized horticultural crops like *khejri*, *kair*, *lasora*, *shejana*, *guarpatta* and *phog* was undertaken by identifying superior genotypes, developing agro-techniques for systematic establishment of orchards and also for product diversification (Samadia,2004).

Selection of production sites and the most potential horticultural crops and their genotypes are the key factors in resource based vegetable crop planning. For successful vegetable cultivation under adverse arid agro-climatic situations, a large number of production technologies are developed, improved or modified at CIAH, Bikaner adopting innovative crop production site management approaches.

## Cucurbit Culture in Arid Region

Cucurbits form a big group of vegetable crops cultivated extensively in India. The group consists of a wide range of crops, used either as salad or cooking, pickling, as dessert fruit or in processed form. Cultivation of underutilized cucurbits like *mateera, kachri*, snapmelon and *tinda* are very popular under rainfed conventional farming systems in arid regions. Owing to their drought hardy nature, these crops are not only adapted but also exhibit high productivity even under extreme stress conditions of hot arid environment. The existing low productivity of these underutilized vegetables could be increased manifold simply by adopting high yielding varieties and appropriate agro-techniques under integrated crop production and management concept (Samadia, 2006).

## Mateera

*Mateera* is an indigenous type of watermelon (*Citrullus lanatus*) grown extensively in North - Western India particularly in arid and semi arid areas of Rajasthan. Mouth appeal of mateera fruit is attributed to sweetness, taste, aroma and texture of the edible flesh (pulp). Its ripe fruits have sweet, juicy, refreshing and cooling pulp. Immature, small, green and tender fruits of about 100 g weight (*loiya*) contain protein, carbohydrate, fat, crude fibre, calcium and phosphorus in appreciable quantities and are used extensively as vegetable for making *rayta* and *curries*. Seeds are protein rich (25 – 32 %) and are roasted and eaten as snacks. Seeds also yield very nutritive oil (30-40 %). Seed kernels (*magaz*) are extracted on large scale and are used in sweets, bakery and ice creams. Rind of ripen fruit is pickled or turned in to a sugar candy (*tuty fruity*).

Quality of edible fruit flesh (sweetness, taste, firmness and colour, and seed size and content) is the important aspect of consumer's preference and its wider acceptability. Sweetness (TSS) varies from 5 -15%. Sweetness is mainly due to glucose, fructose and sucrose which increase with fruit maturity, ripening and temperature (dryer climatic conditions). Several workers reported nutritional composition of fruit flesh, but there are variations in the quantity of composition and this might be because of variation in cultivars, climatic conditions and growing season and also stages of ripened fruits taken for analysis. As per information available in literature, approximate nutritional value per 100 g fruit pulp is moisture (91 - 96 %), protein (0.2 - 0.3 %), carbohydrate (3.3 - 5.5 %), fibre (0.20 - 0.25 %), fat (0.2 - 0.3 %) and total minerals (0.2 - 0.3 %), calcium (7 - 9 mg), phosphorus (10 - 12 mg), iron (6.5 - 7.9 mg), potassium (140 - 160 mg), sodium (25 - 27 mg), copper (0.05 - 0.06 mg), sulphur (40 - 42 mg), magnesium (11 - 13 mg), oxalic acid (9 - 11 mg), and ascorbic acid (1 - 5 mg). The rind of mature fruit constitutes 30 - 35 % of whole fruit by weight and it is rich in carbohydrate (3.5 - 3. 8 %) and minerals (0.2 - 0.3 %).

Why *Mateera* cultivation in arid region? For production of *mateera* during *kharif*, farmers broadcast its seeds between crops of *bajra* or arid legumes as mixed crop with advent of monsoon. During good rainfall years, arid zone farmers harvest *mateera* fruits from October - November *i.e.* at the time of harvesting of *bajra* or guar as a rainfed crop and provides income as a bonus. In recent past, its systematic cultivation is becoming popular both as a sole or mixed crop with *bajra*, guar, groundnut or in fruit orchards. It can also be cultivated in summer as irrigated crop under extremes of arid conditions. In the absence of better and standard varieties of drought hardy *mateera*, farmers of this region are forced to use heterogeneous and mixed seeds collected from open pollinated crops only on the basis of fruit sweetness. Obviously, this results in poor fruit quality, yields and no asssurance on better fruit quality to consumers. Intensive studies conducted on farmer's field on fruit quality parameters of *mateera* reveal that hardly five % fruit are of better quality and no assurance of fruit quality even before cut. Absence of standard varieties in *mateera* type watermelon for stressed conditions in arid region leads to poor fruit productivity and quality. Consequently it fetches low market price leading to poor returns to growers. Improved watermelon varieties and hybrids (Durgapura Meetha, Arka Manik, Sugar Baby, Charleston, Arka Jyothi) require high irrigation water (15 times), have low yield potential and suffer from severe fruit cracking under extremes of aridity and high temperature conditions

Mateera is a drought hardy crop in Indian desert with potentials to extent its period of availability (April to November) and diversified utilities (tender fruits "Loiya" as vegetable, ripe fruit pulp as dessert for thrust quenching and refreshing, edible oil rich seeds for kernals and roasting, besides non-edible rind for processing and animal feed). The crop is attracted for systematic and concerted research for commercial exploitation. A research plan on mateera was developed during 1994-95 at CIAH, Bikaner to develop standard varieties for uniform and quality production. Intensive surveys in arid and semi arid areas of Rajasthan were under taken during 1994 to 1995 and resulted build up of large genetic materials in *Citrullus lanatus* at CIAH. The collected *mateera* (watermelon) diversity was evaluated over years (1995-2001) and is being maintained as *ex situ* gene reserve. Considerable variability in *mateera* is available with regard to growth, maturity, flowering and fruiting, fruit yield and quality characters and seediness, besides possessing a number of genes for biotic and abiotic stresses.

*Mateera* genotypes were categorized on the basis of economically important traits and promising genotypes - AHW 18, AHW 19, AHW 65, AHW 108 and AHW 140- were identified. As a result of purification, two high yielding and better quality varieties - AHW 19 and AHW 65- were developed and released in 1998 for cultivation in arid region. These recommended varieties are now widely adopted by growers for commercial cultivation in arid region. These varieties are suitable for cultivation as rainfed crop in *kharif* and as irrigated crop in summer in arid region (Parrek et al., 1998, Samadia and Pareek, 2001 and Samadia, 2004). Characteristics of these varieties are given below.

### *AHW 19*

A vine produces 3.2-3.4 fruits and first harvest starts 75-80 days after sowing. Fruit is oblong with green to dark green stripes and does not crack under arid agro-climate. Fruit is 30-33 cm long, 60-65 cm in girth and has average weight of 4.0 kg. Flesh is dark pink, solid (firm) and has good eating quality, taste and sugar content (8.0 - 8.4 % TSS). Seeds are big and bold. It is a heavy yielder (425-486 q/ha).

### *AHW 65*

It is an early maturing (72 days after sowing) variety having medium sized (2.5-3.0 kg) round fruits with green stripes. It is andromonoecious in sex form and thus gives more fruit set. Harvesting at early tender stage (100g fruit size) can be done for vegetable purpose (*Loiya*). A vine yields 3.7-4.2 mature fruits giving a yield of 382-401 q/ha. Fruits are 25-29 cm in length and 58-61 cm in girth. The flesh is delicious, pink and solid (firm) having TSS (8.0-8.5%).

## Kachri

*Kachri* (*Cucumis melo* var. *callosus/agrestis*) is a drought hardy underexploited vegetable crop growing abundantly during rainy season in arid and semi arid regions particularly in the *Thar* desert of Rajasthan. Rural people collect *kachri* fruits grown with rainfed traditional mixed crops and receive income as bonus. It is also in cultivation as sole or as mixed crop with other *kharif* crops like *bajra, guar* and groundnut or in fruit orchards. It can also be cultivated in summer as an irrigated crop. In the absence of standard varieties, fruit productivity and quality are highly variable and poor and consequently fetch low returns to growers (Samadia, 2004).

Mature fruits are usually cooked with various vegetable preparations, *chutneys*, pickles and are also used for garnishing vegetables or *salad*. Mature fruits are also dehydrated for off-season use. *Kachri* has great significance in social, cultural and religious occasions of desert dwellers. Fresh fruits of *kachri* are used in *Diwali Pooja* and a special gift pack, *Dibadi* is send to kith and kin which contains dehydrated conventional vegetables/fruits like *kachri*, *ker, sangria* and *tinda*. *Kachri* is one of the components of delicious vegetable popularly known as *Panchkuta* in desert districts of Northern-Western India.

Mature fruits are peeled, dried whole or sliced and can be stored as such or in powder form. The dehydrated fruits are sold @ Rs.30-50/ kg, and thus have great potential for exploitation as an industrial crop. *Kachri* powder is used as souring agent in combination with chillies, coriander, turmeric, cumin, *methi*, and other spices to manufacture various kind of curry powders. *Kachri* powder is also used in combination with salts, black pepper, cumin, ginger and sugar to prepare *Churan*. This is commonly used to cure stomach pain, nausea, vomiting and constipation. The dehydrated *kachri* is coughicide, vermicide, cooling refrigerent, diuretic and gastric stimulant. It is useful to treat facial paralysis, cough, debility, coryza, bile, obesity, piles and constipation. The fresh fruit of *kachri* (per 100 g weight basis) contains 88.2 % moisture, 7.45 % carbohydrate, 0.28 % crude protein, 1.28 % fat, 1.21 % crude fibre, 1.46 % total ash, 0.09 mg calcium, 0.0029 mg phosphorus, 0.182 mg iron, 0.0046 mg copper, 0.052 mg zinc, 0.058 mg manganese and 29.81 mg vitamin C. The fresh *kachri* fruit provides 47.24 K cal energy / 100 g weight basis.

Realizing importance of this drought hardy minor cucurbit in Indian desert for subsistence income, potentialities to extent its period of availability (April to November), market demand and diversified uses, systematic research work was initiated during 1994 at CIAH, Bikaner. The primary objective of the research programme was to develop varieties for different uses such as vegetable, *salad* and processing (dehydration) to make it a profit-making crop for conventional and intensive production systems and possibilities to have crop based processing industry in this region. Systematic research programme was implemented and efficient production technology was standardized at CIAH for commercialization of this crop. Two promising varieties have now been developed by CIAH, Bikaner and have already become popular in the region (Samadia and Pareek, 2000 and Pareek and Samadia, 2002).

### *AHK 119*

It is an early, high yielding selection from land races of *kachri*. Fruits are suitable for processing (dehydration). Fruits are small, egg shaped having 50-60 g weight, 5-6 cm length and 4.0-4.5 cm diameter. It flowers very early, male flower appearing about 28 days and female flower at 40 days after sowing. Fruits are ready for harvest in 70 days from sowing. Vines are medium long (1.7-2.5 m) with 5-7 primary branches. On an average, 22 fruits are borne / vine giving a yield of 94-99 q/ha. It is drought hardy and can be grown under high temperature conditions. It can be sown in February-March as an irrigated summer crop and in June- July as a rainfed crop.

### *AHK 200*

It is a very early maturing variety possessing drought tolerance characteristics. It was selected from land races growing in arid region. Fruits are suitable for garnishing vegetables or as *salad*. Big sized fruits are oblong in shape and developed tough, light yellow skin at maturity. Fruits are 100-120 g in weight, 7.5 cm in length and 5.1 cm in diameter. The vines are medium in length (1.75-2.5 m) with 5-6 primary branches. Anthesis of first male flower starts 26 days and female flower 32 days from sowing. Fruits are ready for harvest in 65 days from sowing. On an average, 20 fruits can be harvested from a single vine giving a yield of 100-113 q/ha. The variety is suitable for sowing in summer and rainy seasons.

## Snapmelon

Snapmelon or *phoot kakdi* (*Cucumis melo* var. *momordica*) commonly known as k*akdia*, is a popular vegetable widely grown in western India, particularly in arid and semi arid regions of Rajasthan (Samadia, 2003). Its mature but unripe fruits are used to prepare *rayta*. Ripe fruits are used as dessert or as *salad*. Ripe fruits are also dehydrated for off-season use. These are locally called *khelera*. Seed kernels are extracted commercially and are extensively used in sweets and bakery products. Fruits of snapmelon are rich in micro nutrients. It contains moisture (79.93 g), crude protein (0.37 g), fat

(1.12 g), carbohydrate (15.60 g), crude fiber (1.34 g), and total minerals (1.64 g), calcium (0.76 mg), phosphorus (0.088 mg), iron (0.843 mg), copper (1.04 mg), zinc (0.202 mg), manganese (0.097 mg) and vitamin C (18.6 mg) and 79.29 K cal energy / 100 g fresh weight basis.

Snapmelon is cultivated as mixed or self-sown crop with other rainfed *kharif* crops. It can also be cultivated in summer as an irrigated crop. In the absence of standard varieties, fruit productivity and quality are variable and poor. Consequently, it fetches low market price resulting low returns to growers. With the aim to develop better quality and high yielding drought hardy varieties of snapmelon for cultivation under arid region, intensive surveys in rich genetic diversity pockets were made during 1994 and 1995 in Rajasthan. Ninety collections were made and evaluated over the seasons (1995 to 1997) at CIAH, Bikaner. These collections showed wide variation for agro-morphological traits, viz., plant growth, flowering and fruiting behavior, maturity, and fruit yield and fruit size, shape and quality characters. Genotypes were characterized for important attributes like earliness, fruits / plant, fruit size, flesh quality and taste. Genotypes AHS 6, AHS 10, AHS 50, AHS 54, AHS 64 and AHS 82 were promising (Pareek *et al.*, 1998).

After purification, two promising selections namely AHS 10 and AHS 82 were developed and recommended for cultivation in 1998. Existing low productivity could be increased by adopting improved production technologies and use of high yielding varieties both under rainfed and irrigated farming systems. These technologies are now widely accepted for commercial cultivation in arid regions (Samadia *et al.*, 2005). Characteristics of these varieties are given below.

### *AHS 10*

It is an early, high yielding selection from local land races of arid region. Fruit is oblong and medium in size, 800-950 g in weight, 17-20 cm in length and 9.7-10.5 cm in diameter. Edible flesh thickness is 2.1-2.6 cm and fruit cavity is 5.5-6.1 cm wide. The whitish pink flesh is sweet having 4.5-5.0 $^{0}$Brix TSS. Male flowers appear in 26 days and female flowers in 38 days after sowing. Ripe fruits can be harvested 67 to 120 days after sowing. Vines of this variety are medium in vigour with 2.5 m length and 8.5 branches. It is highly productive giving 4.0-4.5 fruits/plant and upto 202-222 q/ha.

### *AHS 82*

This selection from local genetic material yields uniform, medium sized fruits. Male and female flowers appear in 28 and 35 days from sowing, respectively. First harvesting can be done 67-70 days after sowing. Plants are vigorous with average vine length of 2.25 m and 6.5 branches. Each vine bears 4.5-5.0 fruits giving a yield of 225-248 q/ha. Fruits are 800-950 g in weight, 20-25 cm in length and 9-10 cm in diameter. Edible flesh thickness is 2.1-2.4 cm with 5.4-6.0 cm wide fruit cavity. Flesh is sweet (4.3-4.9 $^{0}$Brix TSS), tasty and light pink in colour.

## Traditional Methods of Culture of Cucurbits

### *Mixed Farming*

Mixed cropping of cucurbits (*mateera, kachri,* snapmelon and *tinda*) with a major proportion of arid legumes (cluster bean, moth bean and cowpea) and *bajra* is very popular and helps in subsistence income under rainfed condition of arid region. For production of *mateera, kachri, tinda* and snapmelon during *kharif,* farmers broadcast seeds with advent of monsoon with bajra, *til* and arid legumes (guar, moth or cowpea) the crop becomes ready for harvest with other rainfed crops during October-November. During good rainfall years, these drought hardy cucurbits ensure good returns. Systematic production of cucurbits is negligible in vast rainfed arid region farming system.

### *Bari Cultivation*

In the arid tract around Bikaner, *mateera* is cultivated on barren sand dunes and sandy undulating wastelands. Local people have developed an ingenious technique for its cultivation on conserved moisture under the extremely harsh agro climate. For this a holding (locally called *bari*) is divided in two parts. While one part is put under cultivation during March to November, the other part is kept fallow and weed-free to conserve rainwater for the next year's sowing. Before sowing in March the *bari* is fenced with thorny twigs of *jharber* (*Ziziphus nummularia*) and a deep trench is dug all along its outer side to protect the cultivated areas from animals and to catch the creeping sand from the adjoining sandy wasteland. The area is then divided into beds of 1.5 - 2 m x 1.5 - 2 m size (catchments) by fixing the dry brushwood of the local weeds such as *Crotolaria burhia* (*sania*) and *Aerva pseudotomentosa* (*bui*). In each bed, two pits of 15 - 20 cm diameter and depth are dug in each of which 5 - 6 seeds of *mateera* are sown. To take care of the risk in monoculture, some farmers also mix the seeds of snapmelon (*Cucumis melo* var. *momordica*) while sowing in these pits. At the time of sowing about one litre water is poured in each pit. The seeds germinate and plants grow very slowly until middle of June on the conserved moisture in the dune. With the advent of rains, growth rate picks up and fruiting begins. Immature fruits (*loiya*), harvested during the summer, are sold as vegetable. The ripe fruits are harvested during September-October.

## Improved Cultivation Technology

For successful cultivation of underutilized crops in the arid region, improved production technologies are to be adopted after right selection of crops and their genotypes. The existing poor quality and low productivity of underutilized vegetable crops could be increased to a great extent by adopting recommended agro techniques under strategic and conceptual approaches *i.e.* Horticulture based crop production site management for the best use of available resources and sustainable vegetable based conventional or intensive farming under the fragile eco-systems of arid areas. Owing to variable patterns of micro agro-climate, availability of water and rainfall distribution, systems of crop cultivation (rainfed, partially or assured irrigated) and available resources and market, a careful recommendations of crop combinations for each production sites or micro level locations (specific sub zone/regions) is essential for viable and resource use crop production in environmentally stressed arid areas.

As a result of germplasm utilization and improvement programme in horticultural crops in India, a large number of high yielding selections, varieties and hybrids suited to different agro climatic conditions were developed and recommended under national and regional net work. In recent past at CIAH, Bikaner, a large number genotypes in commercially potential and under-exploited crops were evaluated and suitable selections, varieties or lines developed and recommended for the resource based crop cultivation under arid conditions. These improved cultivars also improved productivity of high quality produce if used in the traditional systems of '*Bari* cultivation' and mixed cropping. The agro-techniques developed or improved for maximizing higher returns / unit area in vegetables under environmentally stressed areas are related to site selection, field micro-climate management, seed treatment, method and time of sowing, maintenance of plant population, soil-water conservation, irrigation systems and scheduling, foliar feeding and crop protection measures. Besides, post harvest management, on farm value additions, organic and hi-tech farming and marketing were also taken up for remunerative cultivation of arid vegetables.

### *Development of Production Sites*

The region encounters strong winds during hot summer, which blow away sand and other finer particles that are deposited at other localities. In such areas the fertile topsoil is blown away leaving behind a coarser substrate poor in nutrients, thus, affecting production potentials of the fields land. At the sites where deposition takes place the coarse sand grains veneer the surface and bury productive soil. The problem becomes more serious in vegetable crop fields which have to be ploughed before or during the summer, and thus remain without any protective plant cover. The vegetable crops also get damaged from winds through sand blasting of tender seedlings, stem, buds, flowers and tender fruits during sand storms. The smaller seedlings or plants get buried under sand. The plants thus get mutilated, and cannot make normal growth leading to crop failure or reduce productivity. Besides, this region experiences extremes of high temperature from March to October (38 - 40 $^0$ C). Here days are bright, hot and scorchy. Due to high aridity conditions, the evapo-transpiration rate from plants and soil is very high. The hot and dusty winds (*Loo* and *Andhi*) drastically affect vegetable crops. Therefore, it is essential to select proper landscape and fields for the development of production sites for the vegetable farming.

For successful vegetable cultivation under adverse situations, a complete land use plan of production sites should be the prerequisite considering large sized land holdings of arid zone farmers (4-10 ha). Therefore, based on research and development of horticultural production systems and experiences in vegetable farming in the arid areas some innovative approaches are recommended for development and management of horticulture based crop production sites (HBCPSMA of Samadia, 2004). In this concept, major emphasis is on development of favourable microclimate and protection of crop fields in production site. Besides, adoption of a series of developed and recommended agro-techniques for specific crop combinations are key input factors for cultivation. Field should be prepared as per topography and physical soil conditions. Ploughing or harrowing during mid June-July and November-December helps to obtain weed free field, pre and post monsoon rainwater harvesting and *in situ* moisture conservation. For better plant growth, fruit set and higher yields in vegetable crops, it is strongly recommended that surrounding microclimate of the fields "*Bari*" should be developed by planting suitable indigenous trees or shrubs as a system of crop production in the form wind breaks, hedge-rows or field divider plantations and bio-fence plants supported by thorny twigs (or barbed wire fence) all-around in production sites. This will not only improve and create favourable microclimate in production sites but also prevents entry of wild animals in it. At CIAH, based on available resources and potentialities, vegetable and indigenous horticultural crop based production site management approaches have been suggested for better resource utilization and income to farmers. Khejri based crop diversification combinations under rainfed situation of arid region like Organic *Panchkuta* production (*Sangri, kair, lasora, kumat and kachri*), Khejri and arid zone cucurbits, cluster bean and lesser-known crops.

### Field Preparation

In field preparation, due consideration should be given to moisture conservation and watershed management approaches. Ploughing or harrowing followed by planking should be done during end of June-mid July and November-December. This is to keep fields weed free and conserve rainwater. In light sandy soils, field should be fertilized @ 200-250 q/ha farmyard manures at about 15 days before sowing. For rainfed crop cultivation, adopting production site management approaches, the selected field plots for vegetable crops should be kept ready well in advance to get advantages of the limited monsoon rains.

### Growing Season

The underutilized cucurbits are grown extensively as rainfed crops in the arid areas. Due to ease in cultivation and tolerance of extremes of aridity conditions, these crops are also grown right from village house steads (*Bari*), banks of river or water reservoir or pond, and in farm land near town and big cities with supplemental irrigation and slightly modification in sowing time. In arid region of western India, the best time of sowing is February-March for summer and July for rainy season crops. Sowing can be done with onset of monsoon in *kharif* season as rainfed crops.

### Sowing Techniques

In trials conducted at CIAH, Bikaner, cultivation in channel system was the most practicable as it saves losses of both water and nutrients, and helps in better care of young seedlings and ease in cultural operations like plant protection sprays and weed management. This system also reduces pre-harvest losses of fruits by restricting water flow only in channels. The vine spread area remains to be kept dry. During rainy days, it is observed that in the channel or furrow of 60-75 cm width water collection within the channel or furrow and in nearby area is more than in remaining field area making maximum water harvesting and availability to root zone of the sown crop.

Prior to 15 days of crop sowing, the selected field plot should be cross-ploughed two times followed by planking. Then 60-75 cm wide and 20-25 cm deep channels are made at appropriate distance (1.5-2.5 m), depending upon season, crop and cultivars. The seeds are sown in the inner northern slope of each channel at 50 to 75 cm spacing depending on crop and varieties. Before sowing, seeds should be soaked in fungicide solution having 2 g bavistin / litre of water. This is done for about 6-8 hours for snapmelon, *kakri* and *kachri* and for 10-12 hours for *mateera*, *tinda*, and gourds. For early germination in bold seeded cucurbits viz., *mateera*, *tinda* and gourds, the soaked seeds should be wrapped in wet gunny cloth and buried for 10-12 hours about 30 cm deep in soil or in FYM pit, prior to sowing. Two-three seeds are sown at each point. When the germinated seedlings are 8-10 cm in height, thinning should be done keeping one or two plants at each point. Proper care should be taken to protect young seedlings from birds, wild lizards, squirrels and insects pest particularly beetles.

Recommendations for under utilized cucurbit vegetables.under arid conditions

| Crop | Sowing period | Channel to channel distance (m) | Seed rate (kg/ha) | Basal dose (kg/ha) | | | Top dressing (kg/ha) |
|---|---|---|---|---|---|---|---|
| | | | | N | P | K | |
| Mateera | February-March<br>June-July | 2.0-2.5 | 2.5-2.5 | 40 | 40 | 40 | 50 kg urea<br>25 DAS<br>50 kg urea<br>40 DAS |
| Snapmelon | February-March<br>June-July | 2.0-2.5 | 2.0-2.5 | 40 | 40 | 40 | 50 kg urea<br>25 DAS<br>50 kg urea<br>40 DAS |
| Kachri | February-March<br>June-July | 1.5-2.0 | 1.5-2.0 | 40 | 40 | 40 | 50 kg urea<br>25 DAS<br>50 kg urea<br>40 DAS |
| *Tinda* | February<br>June-July | 1.5-2.0 | 2.0-2.5 | 40 | 40 | 40 | 50 kg urea<br>25 DAS<br>50 kg urea<br>45 DAS |

### *Water Management*

Common method of irrigation is by flow (check basin, ring basin, furrow or channel) since it is simple and low cost but is very inefficient causing great conveyance and percolation water losses. Studies conducted at CIAH, Bikaner clearly demonstrated that even in conventional method about 33 % water saving can be made if channel system of cultivation is practiced. Under channel system of irrigation water losses are lesser than in check basin. Labour and time requirement are also lower and water distribution in root zone is more uniform. Less number of irrigations is required for *mateera, kachri, tinda and* snapmelon as compared to watermelon, muskmelon and gourds. During summer, irrigations are required at an interval of 6-10 days depending upon stage of crop growth and extremeness in weather situations. During very early stage of crop growth irrigation should be done more frequently. During rainy season, irrigation is minimum except in long dry spell particularly under rainfed situations.

Modern methods including sprinkler, bubblers (hose pipe and channel) and drip can be used for high water use efficiency in water scarce areas. Among controlled irrigation, drip system of irrigation is the most suitable for cucurbits under arid conditions. At CIAH, a series of experiments were conducted to assess feasibility of different combinations under drip and micro sprinkler systems in cucurbits. In conclusion, single lateral lines (12-16 mm size) at 1.5 - 2.0 m distances with on-line drippers (4 lph) at 50 cm distance was the most suitable for kachri, snapmelon and mateera. Increased fruit yield of about 25-30 % is obtained in these crops in comparison of channel system of irrigation under arid conditions. Based on seasonal agro climatic and weather situations, crop potentiality and available resources, a complete production technology adopting drip irrigation was developed and recommended for commercial cultivation of arid zone cucurbits. Drip irrigation is the most efficient method of irrigation when it comes to water saving. Since drippers deliver water directly to soil adjoining to the root system, which absorbs water immediately. Efficiency of drip irrigation systems in Indian arid zone would only be at the highest level if it is planned, laid out and maintained after understanding prevailing climatic situations, crops under cultivation and competency in resource management.

### *Nutrition Management*

Although it is highly impracticable to make general recommendation of specific dose for different soil conditions and crop production sites, those based on experimental results at CIAH, Bikaner can be adopted after due modifications depending upon soil fertility status, system and intensity of cropping. Recommended basal doses of nitrogen, phosphorus and potassium should be given at the time of preparation of channels. The remaining half dose of nitrogen should be applied in two split doses, first at the time of vine spreads 25-30 days after sowing and the other at 40-50 days after sowing. Foliar application of urea at vine spread stage @ 20 g/litre of water maintains good plant growth and vigour and results in higher fruit yields in snapmelon, *kachri*, *tinda*, and *kakdi*. In snapmelon, foliar feeding with urea (2%) at 25 and 35 days after sowing improves fruit sets and significantly higher yields by 45 % over control (5.05 kg/plant). Arid region also offers good potentialities for organic farming (vermicompost, farm waste manure etc. in Bari system of cultivation) in under exploited cucurbits owing to their low input requirements and the least incidence of insect pests and diseases.

### *Inter Culture*

To keep the crop free of weeds, inter-culture operations are started 20-25 days after sowing and depending upon the season, 2 to 3 weedings and hoeing operations may be required. Mulching with locally available materials like hay, straw, weeds etc. improves fruit yield in cucurbits. In summer season crop of bottlegourd, mulching with dry hay (or *khimp*) resulted increase in fruit set and there by 30 % higher yields under hot arid conditions of western Rajasthan. Mulching reduces water evaporation, delays drying of soil and reduces soil thermal regime during daytime under extremely high temperature conditions. It also checks weed population and improves microbial activity of soil by improving environment around active root zone. Continuous use of organic mulches (*bui, keemp* or dry hays) is helpful in improving organic matter content of soil, which in turn improves water holding capacity of sandy soil. In arid zone cucurbits, mulching should be done at 30-35 days after sowing in vine spared areas. Mulching with locally available dry hays and weeds are given significant improvement in quality fruit yield by 58 % over control (5.18 kg /plant) in snapmelon cultivar AHS 82. Among bio-regulators imposed for the improvement in fruit set and high quality early yields under irrigated or rainfed mixed crop cultivation practices, spraying of $GA_3$ @ 20 ppm at 2 - 4 leaf stage (18 - 20 days after sowing) significantly increased yield potential by 63 % over the control (5.05 kg/plant) in snapmelon.

### *Crop Protection*

In arid zone cucurbits, noteworthy results are obtained on integrated crop protection under Horticultural crops based production site management approaches. The important techniques for crop protection measures include field selection and sanitation, summer ploughing, clean cultivation, seed treatment and sowing, foliar feeding, prophylactic sprays, protection measures against wild lizards, squirrels, rats, birds, wild animals and regular monitoring in the production sites.

In arid region, common insect pests attacking cucurbits are red pumpkin beetle, epilachna beetle, aphid, mites and fruit fly. Red pumpkin beetle and epilachna beetle cause devastating losses at early stage (cotyledon to one or two true leaf stage) resulting in very poor plant stand. Spraying of dimethoate @ 1 ml/litre of water or sevin @ 2 g/litre for controlling beetles and metasystox or monocrotophos @ 1 ml/litre for the control aphids and fruit fly are useful.

Dusting of methyl parathion (2 % dust) mixed with wood ash (in 1:50 ratio) at 2-4 leaf stage of crop growth during morning hours was the most effective and economical to manage red pumpkin and epilachna beetles. For effective insect pest management, four sprays one each of sevin, monocrotophos, metasystox or dimethoate at an interval of 12-15 days are recommended. The first spray should be given at 2-4 true leaf stage in all the cucurbit vegetables.

Disease occurrence is very limited in arid region. The most common diseases are powdery mildew, wilt, downy mildew and viral complexes. Powdery mildew can be controlled by one or two sprays of karathane (1 ml/ litre of water). Spraying of Dithane M 45 (2 g/litre) gives good control against downy mildew. For effective disease management, preventive spray of Dithane M 45 (0.2 %) and bavistin (0.2 %) at 15 days interval is recommended. First spray should be given about 20 days after sowing.

### *Harvesting*

In *mateera*, immature tender fruits are harvested 45-50 days after sowing. Fruits for vegetable purpose (*Loiya*) are plucked daily to collect more number of uniform grade fruits (100 g weight) as it fetches more price. Retaining only 2-3 fruits/vine improved fruit size and marketable yield under arid conditions. The crop will be ready for harvesting in about 75-80 days after sowing depending upon cultivars, season and method of cultivation. Maturity in *mateera* can be judged from change in fruit colour of ground spot to yellow and by thumping test. On thumping, mature fruit gives a dull sound. Yield potential of improved mateera varieties ranged from 280 to 350 q/ha under varying level of adoption of production technology. The over-mature, deformed and damaged fruits can be collected for extraction of oil rich seeds

*Kachri* is a short duration crop and becomes ready for harvesting in 65-70 days after sowing. Harvesting continues upto 110-120 days after sowing. Fully mature fruits are harvested at regular intervals. Grading of *kachri* fruits according to size at field level could result in higher profits. Therefore, fruits should be graded in three size categories *viz.*, small, medium and big. In general, big size *kachri* fruits are relished for making special vegetables, garnishing the vegetable salad whereas smaller and medium sized fruits are used for processing particularly for dehydration. Average yield of *kachri* varieties AHK 119 and AHK 200 is 85 and 102 q/ha, respectively. Yield also depends on whether crop is rainfed, partially irrigated or completely irrigated, or whether it is sole or mixed crop. Considering a market price of *kachri* fruits @ Rs 3.00/ kg in the wholesale market, the returns / ha would range from Rs. 25500 - 30600 /crop.

Snapmelon is ready for harvesting in 65-70 days after sowing. Harvesting continues upto 110-120 days after sowing. Fully mature fruits at turn of ripening are harvested for distant market. Ripe fruits are harvested for domestic markets. At farmer's fields, yield potential of improved snapmelon varieties ranges from 200 to 250 q/ha under different level of adoption of production technologies and cultivation practices. Considering all the factors, if a farmer is able to dispose snapmelon fruits @ Rs. 150 per quintal in the wholesale market, the returns / hectare of the produce would range from Rs. 30,000 to 37,500/ crop.

## References

Pareek, O. P. and D. K. Samadia 2002 Promising Indigenous Cucurbit Varieties. *Indian Horticulture* 47 (2): 15-18.

Pareek, O. P., B. B. Vashishtha and D. K. Samadia 1998 Genetic diversity in drought hardy cucurbits from hot arid zone of India. IPGRI APO Regional Newsletter for Asia, the Pacific and Oceania. 28: 22-23

Rai, Mathura 2006. Research and development in vegetable crops : Present status and future strategies. *Indian Horticulture* 51 (3):4-7.

Samadia, D. K. 2006 Technology for cucurbitaceous crop cultivation in arid regions. *Intensive Agriculture* 44 (1-2):3-7.

Samadia, D. K. 2005 Improvement for high quality yield in arid vegetables. National Seminar on Commercialization of Horticulture in Non-traditional areas, CIAH, Bikaner. February 5-6, 2005. pp 30-31.

Samadia, D. K. 2004 Genetic resource management for sustainable horticultural development in arid regions of India. Sustainable Resource Management. (Eds.. B.B.S. Kapoor and Ahemad Ali) Madhu Publications, Bikaner (India). pp 163-173

Samadia, D. K. 2004 Protected Cultivation Technology for Arid Regions of India. *Advances in Arid Horticulture* Vol. I (Eds., P.L. Saroj, B.B. Vashishtha and D. G. Dhandar). International Book Distributing Co Lucknow (India). pp 267-289

Samadia, D. K. 2004 Technology for the production of *mateera* in arid and semi arid areas. *Intensive Agriculture* 42 (3-4): 22-26.

Samadia, D. K. 2004 Improvement of Arid Vegetables. *Advances in Arid Horticulture* Vol. I (Eds., P.L. Saroj, B.B. Vashishtha and D. G. Dhandar). International Book Distributing Co., Lucknow (India). 185-201 pp.

Samadia, D. K. 2003 Snapmelon production technology for arid and semi arid areas. *Intensive Agriculture* 40(1-2): 27-29.

Samadia, D. K. and O. P. Pareek 2000 Two *Cucumis* (*kachri*) varieties of arid region *ICAR NEWS* 6 (2):15-16.

Samadia, D. K. and O. P. Pareek 2001. AHW 19 and AHW 65: New *mateera* varieties. *Indian Horticulture* 46 (2): 8-9 & 21

Samadia, D. K,. O. P. Pareek and B. B. Vashishtha 2005. AHS 10 and AHS 82: New snapmelons. *Indian Horticulture* 50 ((2): 10-11.

Samadia, D. K., B. B. Vashishtha and G, B, Raturi 2004. Prospects of vegetable production in the arid zone of India. Impact of vegetable research in India (Eds. Sant Kumar, P.K.Joshi and Suresh Pal), National Centre for Agricultural Economics and Policy Research (ICAR), New Delhi. 101 –114 pp.

**Underutilized and Underexploited Horticultural Crops, Vol. 2, 2007**
*K.V. Peter (ed.), pp. 175-193*
New India Publishing Agency, New Delhi (India)
E-mail : newindiapublishingagency@gmail.com
Web: www.bookfactoryindia.com

# 10 Cucurbits

**D.K. SINGH**
Associate Professor, Department of Vegetable Science
G. B. Pant University of Agriculture and Technology
Pantnagar-263 145, Uttarakhand

The Cucurbitaceae family consists of nearly 100 genera and over 750 species. Although most of them have Old World origin (Whitaker and Davis, 1962), many species originated in the New World and at least seven genera have origins in both hemispheres (Esquinas-Alcazar and Gulick, 1983). There is considerable genetic diversity within the family and the range of adaptation for cucurbit species includes tropical and subtropical regions, arid deserts and temperate locations. A few species are adaptable to production at elevations as high as 2000 m. The genetic diversity in cucurbits extends to both vegetative and reproductive characteristics. There is considerable range in the monoploid ($x$) chromosome number (Jeffrey, 1990) including 7 (*Cucumis sativus*), 11 (*Citrullus* spp., *Momordica* spp., *Lagenaria* spp., *Sechium* spp., and *Trichosanthes* spp.), 12 (*Benincasa hispida, Coccinia spp., Cucumis* spp. other than *C. sativus*, and *Praecitrullus fistulosus*), 13 (*Luffa* spp.), and 20 (*Cucurbita* spp.).

Archaeological evidence indicated that cucurbits were present in ancient and prehistoric cultures. *Lagenaria* was associated with man as early as 12,000 BC in Peru (Esquinas-Alcazar and Gulick, 1983). Archaeological expeditions in the Oaxaca region of Mexico have reported that *Cucurbita pepo* was associated with man as early as 8500 BC and cultivated by 4050 BC (Esquinas-Alcazar and Gulick, 1983). Written Chinese records describing the use of cultivated cucurbits have been found as early as 685 BC (Herklots, 1972).

American Indians cultivated squash in pre-Columbian times (Whitaker and Davis, 1962) and chayote was a common vegetable among the Aztecs prior to the Spanish conquest (Herklots, 1972). Depending upon the species, virtually all parts of the plant can be used for food including leaves, shoots, roots, flowers, seeds and immature and mature fruits. Starch can be extracted from roots and the seeds are a rich source of oils and proteins (Jacks *et al.*, 1972). In addition, some cucurbits are used for ornamental purposes (e.g., gourds), for utensils (e.g., bowls, ladles, sponges, boxes, birdhouses, musical instruments) and for fuel and pharmacological uses in certain areas of the world.

Out of large variability of cucurbits, some species are cultivated locally and are not very popular despite of their high yield potential, resistance to biotic and abiotic stresses and nutritive value. These cucurbits are categorized as Minor Cucurbits or Under-utilized cucurbits. Some of these under utilized cucurbits have been described as follows

List of underutilized cucurbits

| Name | Botanical name | Origin | Chromosome No. (2n) |
|---|---|---|---|
| Stuffing cucumber (Pepino cucumber) | *Cyclanthera pedata* | Latin America | 32 |
| Chayote (Chow-Chow) | *Sechium edule* (Jacq.) Swartz | Mexico and Central America | 24 |
| Bitter melon (Kachari) | *Cucumis callosus* | India | 24 |
| Snake gourd | *Trichosanthes anguina* | India | 22 |
| Ivy gourd | *Coccinia grandis* | India | 24 |
| Buffalo gourd | *Curcurbita foetidissima* | South west America and Mexico | - |
| Long melon | *Cucumis melo* L. var *utilissimus* | India | 24 |
| Round melon (Indian squash) | *Praecitrullus fistulosus* | Indo Burma | 24 |
| Snap melon | *Cucumis melo* var. *momordica* | India | 24 |
| Sweet gourd | *Momordica cochinchinensis* | South East Asia | 28 |

## *Stuffing Cucumber (Pepino cucumber)*

### Botanical Name

*Cyclanthera pedata*

It is locally known as *Meetha karela* or *Korila* and is relatively cold tolerant and adapted to elevations up to 2,000 m. It is also easy to cultivate in the tropics and subtropics. It is currently cultivated in the Caribbean and in Central and South America. It is distributed in warmer parts of South Asia. Sometimes it is found as an escape in Northern Himalayas, Bhutan and Nepal. In India, this vegetable grows in backyards during warm season.

### Botanical Description

The plant is an annual climber and vine length goes up to several meters. The stem is branched, glabrous; leaves are on glabrous petioles 1-8 cm long, blades broadly ovate to orbicular, pedately lobate, margins serrulate to deeply dentate and essentially glabrous. Staminate flowers are in panicles with a few-several (rarely many) flowers confined to upper half or less of peduncle, receptacle to 3.5 mm wide, sepals obscure or to 2 mm long, corolla obscurely papillate, whitish, 3-6 mm across, filament column to 0.75 mm long, androecial head to 1.5 mm wide. Pistillate flowers are sub sessile, stigmas sessile, to 2.5 mm broad, with a depressed center; fruiting peduncles 0.5-2 cm long. Fruits greenish white, striate, lacrimiform and somewhat gibbous, not explosively dehiscent, 15 cm long and 6 cm in diameter, setae none or few, confined to abaxial part of the fruit. Seeds are black, more or less turtle-shaped, 12-16 mm long (Dieterle, 1976). The foliage is glabrous and odoriferous. Fruits are pale green, flattened, and mostly hollow. The seed cavity is spongy, and the seeds are attached to a single placenta.

### Edible Use

Young fruits are eaten raw or cooked and have a similar taste to cucumbers though they are not crisp. Older fruits are cooked, they can be stuffed in much the same way as marrows. Seeds are usually removed and the fruits are eaten raw or cooked. They are often used stuffed with meat, fish or cheese, then baked and eaten similar to stuffed peppers. The shoots are also edible. Matured fruits are preserved by sun drying and used as vegetable round the year. Matured plant is fed to the animals.

### Medicinal Uses

The plant has hypoglycemic, hypotensive and lipotropic effect and it is suggested for people suffering with cholesterol and obesity. Analysis of flavonoids from *Cyclanthera pedata* fruits by liquid chromatography/electro-spray mass spectrometry showed its antioxidant activity.

Nutritional composition (per 100 g)

| | |
|---|---|
| Moisture (g) | 82.7 |
| Total ash (g) | 1.1 |
| Crude fat (g) | 0.2 |
| Crude fibre (g) | 4.2 |
| Crude protein (g) | 1.5 |
| Carbohydrate (g) | 10.3 |
| Energy (Kcal) | 49 |

### Cultivation Practices

It requires a very warm, sunny and sheltered position in a rich well-drained soil. It is occasionally cultivated for its edible fruit in S. America. Plants are not very cold-hardy in Britain and are normally grown in a greenhouse. If the plants are started in a greenhouse in the spring and grown on well, it is possible to get reasonable yields in most years outdoors in Britain. Fruiting plants have been seen outdoors on a number of occasions at a few botanical gardens.

In India, it is basically grown in hills during rainy seasons. It is a seed sown crop. Two to three seeds are planted in pits, along with 10-15 Kg FYM/pit. Plants are trained on trellis. Fruits can be harvested within 70 days and harvesting prolongs for 2 months. Individual plant yields 800-1200 g fruits.

## *Chayote (Chow-Chow)*

### Botanical Name

*Sechium edule* (Jacq.) Swartz

Chayote is the most accepted name for *Sechium edule* but it is also *known* as cho cho, chow chow, mirliton and vegetable pear. In China, it is know as Buddha's hand and in India it is known as vilaiti vanga in local language. It is believed to be native of Mexico and Central America where it has been grown for centuries and remains popular today. No archaeological evidences are available to indicate how long *S. edule was* cultivated. Its fleshy fruit, which has a single seed with a smooth testa, does not allow it to be preserved and, as far as is known, no pollen grains or other structure of this species have been identified on archaeological sites. Chronicles from the time of the conquest record that in Mexico, at least, the chayote has been cultivated since pre-Columbian times. As regards linguistic references, the common names of native origin are concentrated mainly in Mexico and Central America. Exploration records concur in the finding that the widest variation of *S. edule* under cultivation is found between southern Mexico and Guatemala (Newstrom, 1990). The geographical distribution of the wild relatives of *S. edule* also testifies to the Mesoamerican origin of this crop. It is believed that *S. edule* is domesticated within the cultural area of Mesoamerica, and specifically in the region lying between southern Mexico and Guatemala. Chayote cultivation is widely distributed in Mesoamerica. It was introduced into the Antilles and South America between the eighteenth and nineteenth centuries. During this same period, the chayote was introduced into Europe from where it was taken to Africa, Asia and Australia, while its introduction into the United States dates from the late nineteenth century. Latin American countries are main producer of chayote from where it is exported to USA, Europe and Asia. Presently this crop is grown in limited scale in USA, Europe, India, China and other East Asian country. In India it is much popular in North Eastern States including Sikkim and West Bengal.

### Botanical Description

The chayote is a herbaceous, perennial, monoecious climber, with thickened roots and slender, branching stems up to 10 m long. The stems are angular, grooved and glabrous. Several stems grow simultaneously from a single root in the cultivated plants. The roots thicken towards the base and appear woody, while towards the apex there are many thin, firm, herbaceous branches. Its leaves are on sulcate petioles of 8 to 15 cm length. They are ovate-cordate to sub-orbicular, measure 8-18 x 9-22 cm, are slightly lobate (with three to five angular lobes) and have minutely denticulate margins and three to five divided tendrils. Both blade surfaces are pubescent when young, later become glabrous. Flowers are unisexual, normally pentamerous, coaxillary and with ten nectaries in the form of a pore at the base of the calyx. The staminate flowers grow in axillary racemose inflorescences that are 10 to 30 cm long, and the groups of flowers are distributed at intervals along the rachis. The calyx is patteliform and 5 mm wide, the sepals triangular and 3 to 6 mm long, the petals triangular, greenish to greenish-white and measure 4 to 8 x 2 to 3 mm. There are five stamens, and the filaments are fused almost along their total length, forming a thickened column which separates at the apex into three or five short branches. The pistillate flowers are normally on the same axilla as the staminate flowers; they are usually solitary but are occasionally in pairs; the ovary is globose, ovoid or piriform, glabrous, inerm and unilocular; the perianth is as in the staminate flowers but has slightly different dimensions; the styles are fused in a slender column and the nectaries are generally less evident than in the staminate flowers. The floral biology of *S. edule* has been studied in detail (Newstrom, 1989). There are various patterns in the structure and sexual expression of the staminate and pistillate flowers, which seem to be determined by genetic, environmental and seasonal factors and by the age of the plants.

Pollination is entomogamous. Among the most efficient pollinators are species of native bees of the genus *Trigona,* chiefly in areas of medium and low altitudes and which are free from pesticides, and the honeybee (*Apis mellifera*) on commercial plantations where the use of pesticides is very frequent. Secondary pollinators include wasps of the genera *Polybia, Synoeca* and *Parachartegus.* The fruit is solitary or rarely occurs in pairs. It is fleshy and sometimes longitudinally sulcate or crestate; it is of very different shapes and sizes, has a number and type of spines; it is white and yellowish, or pale green to dark green with a pale green to whitish flesh that is bitter in the wild plants and non bitter in the cultivated ones. The seed is ovoid and compressed with a soft and smooth testa. The fruit is viviparous, *viz.* the seeds germinate inside the fruit even when it is still on the plant. This characteristic does not occur in any of the wild species, in which the seeds germinate asynchronically after falling to the ground. A possible explanation for the viviparism of the cultivate species is that the process of domestication may have resulted in suppression of the dormancy mechanisms.

## Edible Uses

The chayote is used mainly for human consumption. The fruit, stems and young leaves as well as the tuberized portions of the roots are eaten as a vegetable, both alone and plain boiled, and as an ingredient of numerous stews. Because of its softness, the fruit has been used for children's food juices, sauces and pasta dishes. The fruit and roots are not only used as human food but also as fodder.

The edible parts of *S. edule* have a lower fibre, protein and vitamin content than other plants. However, the calorie and carbohydrate content is high, chiefly in the case of the young stems, root and seed, while the micronutrients and macronutrients supplied by the fruit are adequate. The fruit and particularly the seeds are rich in amino acids such as aspartic acid, glutamic acid, alanine, arginine, cysteine, phenylalanine, glycine, histidine, isoleucine, leucine, methionine (only in the fruit), proline, serine, tyrosine, threonine and valine.

## Medicinal Uses

The chayote also has medicinal uses; infusions of the leaves are used to dissolve kidney stones and to assist in the treatment of arteriosclerosis and hypertension; infusions of the fruit are used to alleviate urine retention. The cardiovascular properties of the infusions of leaves have been tested in modern studies, while their great effectiveness in curing kidney diseases has been known since colonial times on the Yucatán peninsula, where these ailments are very common.

Nutritional Composition of Fruit

| | |
|---|---|
| Moisture (%) | 89.0-93.4 |
| Protein (%) | 0.9-1.1 |
| Fats (%) | 0.1-0.3 |
| Carbohydrates (%) | 3.5-7.7 |
| Soluble sugar (%) | 3.3 |
| Starch (%) | 0. 2 |
| Fibre (%) | 0.4-1.0 |
| Ashes (%) | 0.4-0.6 |
| Ca (mg) | 12.0-19.0 |
| P (mg) | 4.0-30.0 |
| Fe (mg) | 0.2-0.6 |
| Vitamin A (mg) | 5.0 |
| Thiamin (mg) | 0.03 |
| Riboflavin (mg) | 0.04 |
| Niacin (mg) | 0.4-0.5 |
| Ascorbic acid (mg) | 11.0-20.0 |
| Calories (Kcal 100g) | 26.0-31.0 |

*Source:* Aung *et al.,* (1990)

## Cultivation Practices

It is grown in the traditional way on family plots and in backyards and vegetable gardens. The viviparous characteristic of its fruit is familiar to peasant farmers, so that fruit selected for consumption is kept—without being allowed to germinate—by a small cut or puncture made in the embryos, while those selected for seed are simply allowed to ripen until it is decided to plant them.

The normal and most effective form of propagation is from seed. The most widespread sowing practice consists of planting one or more whole fruits. However, on some small holdings the seed is carefully removed and sown in pots or other media that enable it to be handled for subsequent transplantation in the final sowing plot.

In areas of traditional production, the sowing plot is prepared beforehand by making a hollow in the soil that is big enough to allow the roots to attain maximum development. Next to the sowing plots, a frame of wood or other material is commonly erected so that the plant can grow on it quickly. For the same reasons, sowing is also frequently carried out close to a tree.
During the first weeks of development, the amount of care given is relatively high (irrigation, fertilization with animal or chicken manure, etc.), although attention to the root (protection against physical damage and application of organic fertilizers) is considered to be of great importance throughout the plant's life cycle.

Sowing can be done at any time of the year, although it generally takes place at the beginning of the rainy season. The average length of the plants' productive cycle is three years or, in exceptional cases, eight.

On commercial plantations, sowing is carried out using rooted cuttings or selected seed. The plants are sown on permanent beds with trellises and are laid out at distances that allow the easiest possible harvesting, transport to cold-storage rooms and packaging. On the commercial type of plantations, chemical and foliar fertilizers are generally used as well as herbicides and nematicides.

## Indian Research Scenario

Eleven accession of chow-chow collected from East and West Garo Khasi hills of Meghalaya are being maintained in field gene bank of NBPGR, RS, Bhowali, Uttaranchal. These accessions have been characterized and evaluated for 20 agro-botanical characters and considerable diversity has been found for earliness and number of fruits per plant. The resistance to abiotic and biotic factors has also been studied (Verma *et al.*, 2006). Similar studies have been described by Chakravarty, 1990.

## Prospects for Improvement

In spite of the fact that the whole of the *S. edule* plant can be used and with numerous applications (parts of the plant are used for different purposes), in several countries the majority of these uses have not become widespread and ways have not been devised to make them accessible to sectors of the population other than the peasant community.

The most widespread uses at all levels are that of the fruit as a table vegetable and in the preparation of some industrialized foods. Commercial demand requires a morphologically homogeneous production which rules out the possibility of the considerable range of fruit produced under traditional cultivation systems appearing in the market. However, as the standards required for export are very different from those accepted for the product for local consumption, it is not very likely that the usual varieties will be abandoned and that serious genetic erosion will occur in the species. While planning for intensification and diversification of *S. edule* production we have to include the following aspects:

The establishment of permanent gene banks to maintain varietal diversity, wild populations and related congeners. These collections can be used to evaluate resistance to diseases, type of growth and organoleptic characteristics of the fruit. They will allow growers to be supplied with new sowing material and will be used as a basis for genetic improvement.

Programmes for selecting varieties with high root yield or high production of young stem should be initiated. Both are popularly accepted consumer items, with a high nutritional value and potential use as a basic material in agro-industries. The development of vegetative propagation methods will provide growers with sowing material at reasonable prices. Basic studies on the most important diseases *(Ascochyta phaseolorum, Mycovellosiella cucurbiticola, Fusarium oxysporum* and complexes of these and other species), particularly those which attack the fruit and which cause 35 to 40 percent of the rejects in commercial production. Identification of problems in post harvest handling, packaging and storage during the marketing process.

## *Bitter melon (Kachari)*

### Botanical Name

*Cucumis callosus*

*Cucumis callosus* is a feral species of India and found in Rajasthan, Punjab, Haryana, parts of Maharashtra, the North Western Provinces, and the Sind area (now in Pakistan). It is known by several vernacular name *viz., Kachari* in Rajasthan, Haryana, Chipper in Punjab, *Pehtool* in eastern UP, besides several names in other parts of country. It has attracted the attention of muskmelon breeders, as this species is reported to possess genes for resistance to fruit fly and leaf-eating caterpillars. Sambandam and Chelliah (1972) have given evidence of free compatibility between *Cucumis callosus* and *Cucumis melo*. Parthasarthy and Sambandam (2005) conducted an experiment to ascertain the compatibility with other *Cucumis* species. Crosses were made (including reciprocals) with seven *Cucumis* species, namely, *C. metuliferus, C. anguria, C. longipes* (*C. anguria* var. *longipes*), *C. zeyheri, C. myriocarpus, C. dipsaceus* and *C. melo*. The metaphase plates revealed the somatic chromosome number of 24 for *C. callosus*. The crosses with seven *Cucumis* species indicated that C. *callosus* was crossable with only C. *melo*. Parthasarthy and Sambandam (2005) based on the basis of chromosome number and its free compatibility with C. *melo*, and normal behavior of chromosomes during diakinesis, concluded that *Cucumis callosus* does not warrant a separate species status and is nothing but a progenitor of *Cucumis melo*.

## Botanical Description

Vine length remains up to 2-3.0 meter with many primary branches. Its stem is soft-hairy to glabrous, angled, leaves ovate to reniform, shallowly three lobed and rough. Flowers are monoecious in nature and borne solitary on stem. Both staminate and pistillate flowers appear on short stout pedicels. Under open natural system, female flowers to male flowers ratio has been found as 7: 20. Calyx and corolla are 5 in number and yellow in colour. Anther bears abundant yellow colour pollen, however stigma is divided in 3 lobes. Anthesis starts from 6.00 a.m. and continues up to 10:30 a.m. It is partly cross and partly self-pollinated by bees. Fruits are oval in shape, skin is covered by light irregular greenish white stripes; fruits remain light green in colour after ripening.

## Uses

Immature fruits are cooked as vegetable and for *Chutney* preparations. Besides *Chutney*, half cut green fruits are dried in sun for future use with *sarsoan saag* in winter season. It is also used in place of tomato in vegetable for adding sour taste.

## Cultivation Practices

*Cucumis callosus* is essentially a warm season crop. A long period of warm and humid climate is required. It can be grown in mild climate, but is sensitive to frost. The optimum temperature requirement is 24°C-30°C for successful crop growth. Seed sowing time is June-July and fruit harvesting remains up to mid-November. The seed rate for *kachari* is about 2.5 kg/ha. The seeds are sown on raised bed or in furrows or in pits when it is grown as sole crop. It is mostly grown as mixed crop in *kharif season*. The spacing followed is 0.5 m x 0.2-0.3 m. Generally no irrigation is given by farmers when grown as mixed crop with *kharif* crops but it needs at least 1 or 2 irrigations for better yield. Fruit takes 8-10 days from anthesis to harvesting stage. Fruits are harvested in immature stage for domestic uses. With proper care it produces 50-60 q/ ha immature fruits. Crop sold in market for as high as Rs. 10 /kg.

## *Snake Gourd*

### Botanical Name

*Trichasanthes anguina*

Snake gourd is also known as serpent gourd, *chichinda*, and *padwal*. In the South Indian state of Kerala, it is known as *padavalanga*. It is a commonly grown vegetable in India especially in Kerala, Tamil Nadu, Andhra Pradesh and Karnataka. It is widely distributed in humid tropical areas of many countries including South East Asia, China, Japan, Australia, South America and Africa. It is probably a native of India or Indo Malayan region. It is a tropical or subtropical vine, raised for its strikingly long fruit, used as a vegetable and for medicine. *Trichosanthes* is a large genus having up to 44 species out of which 21 are found in India. *T. cucumerina* is supposed to be progenitor of *T. anguina*.

### Botanical Description

It is an annual climber. Flowers are basically monoecious i.e. individual flowers are either male or female, but both sexes can be found on the same plant and are pollinated by insects. The narrow, soft-skinned fruit can reach up to 2 metre long and 10cm thick. Its soft, bland, somewhat mucilaginous flesh is similar to that of the luffa and the calabash.

### Edible Uses

It is most popular in the cuisine of South Asia and Southeast Asia. The shoots, tendrils and leaves are also eaten as greens.Young fruit is cooked as vegetable. It can be used in curries or eaten as a vegetable like green beans. The bright red pulp around the mature seeds is extracted and used in cooking in much the same way that tomatoes are used (Facciola,1990).

### Medicinal Uses

The fruit is considered to be anthelmintic, emetic and purgative and seeds are said to be cooling (Chopra *et al.*, 1986).

### Cultivation Practices

The crop prefers a tropical warm humid climate. A loamy soil having pH range of 6.0-7.0 is the best suited for the crop. The crop cannot tolerate frost. It cannot be grown in the shade. Some of the important varieties in India are Kaumudi, Baby, T A-19, Co-1, Co-2, MDU1, PKM1, APAU Swetha. Seeds are sown during June-July or December-January. Seed rate is about 4.0-6.0 kg/ha. Seeds are sown in a pit. Pits of 60 cm diameter and 30-45 cm depth are made at a distance of 2-2.5 metre. Well rotten FYM and fertilizers are mixed with topsoil in the pit and seeds are sown @ 4-5 per pit. After two weeks unhealthy plants are removed and three plants per pit are retained. FYM @ 20-25 t/ha is applied as basal dose along with half dose of N (35 kg) and full dose of $P_2O_5$ (25 kg) and $K_2O$ (25 kg/ha). The remaining dose of N (35 kg) is applied in several split doses at fortnightly intervals. The seedlings when starts producing tendrils should be staked to thin bamboo poles using banana fibres so as to enable the vines to spread on the pandal. Snake gourd is trained on pandal made up of bamboo stakes and cross wires for spreading the vines. Five days after fruit set, at the tip of the developing fruits, a small stone is tied using a piece of banana fibre otherwise the developing ovary will be coiled. However, short fruited types do not require such training. During the initial stages of growth irrigate at an interval of 3-4 days. Irrigate on alternate days during flowering and fruiting periods. The harvest starts from 78 to 80 days after sowing. The harvest is done at an interval of 5-6 days. A total number of 6-8 harvests can be done. Yield ranges from 18 to 25 tonnes/ha.

## *Ivy Gourd*

### Botanical Name

*Coccinia grandis*

*Coccinia grandis is* also known as *tindora, kundri, kundru, kowai, kovai, tindla,* gentleman's toes (compare lady's fingers), scarlet gourd or *thainli.* Probably this crop was domesticated in India and South East Asia. It is cultivated in India, China, South East Asian countries and Tropical Africa. In India, it is grown mainly in West Bengal, Assam, Bihar, Jharkhand, Chhattisgarh and Southern India. It is a tropical vine grown for its small edible fruits, which may be eaten immature and green, or mature and deep red. The young shoots and leaves may also be eaten as greens. It is perennial and provides vegetables throughout the year except during winter season when crop goes under dormancy due to low temperature.

### Botanical Description

It is a herbaceous perennial climber. The roots are long and tuberous. The leaves are angular having five lobes with serrated margin. The plants are dioecious (individual flowers are either male or female, but only one sex is to be found on any one plant so both male and female plants must be grown if seed is required). Pollination is done by insects. The plant is self-sterile. The fruits are smooth, roundish, long, ovoid or elliptical shape with 2.5-5.0 cm diameter. Immature fruits are normally green. There are vertical stripes of white colour on fruits at maturity in some genotypes. The fruits become scarlet red on ripening.

### Edible Uses

Young leaves and long slender stem tops are cooked and eaten as a pot herb or added to soups. Young and tender green fruits are used as raw in salads or cooked and added to curries etc. Ripe scarlet fruit is fleshy and sweet.

### Medicinal Uses

The juice of the roots and leaves is considered to be a useful treatment for diabetes. The juice of the stem is dripped into the eyes to treat cataracts. The leaves are used as a poultice in treating skin eruptions. The plant has hypoglycemic, laxative, ophthalmic and have poultice properties. It is used internally in the treatment of gonorrhoea. Aqueous and ethanolic extracts of the plant have shown hypoglycemic principles (Chopra *et al.*, 1986).

### Cultivation Practices

It requires warm and humid climate with an ideal temperature of 20-30$^0$C. It continues to produce fruits throughout the year in South India but the plants remain dormant during winter in Northern India. The dormant period provides an opportunity to prune its vines which results in sufficient number of flowers and fruits in next season. The plant can be grown in light (sandy), medium (loamy) and heavy (clay) soils and requires well-drained soil. It cannot be grown in the shade. It is propagated by stem cutting. The planting is done in June-July or Feb-March. The plant population in field should have at least 10% of the male plant. The vines are often trained on bowel or bamboo structure. The flowering starts after 50-60 days of planting and average yield is about 10-15 tonnes per ha.

### Indian Research Scenario

Research work on ivy gourd was initiated during 1995-96 at IGAU, Raipur with the collection of 36 germplasm from remote area of Chhattisgarh. Besides 11 genotypes from Gujarat, 3 from Maharastra and 42 from North Eastern region were also collected and all 93 genotypes were evaluated under Chhattisgarh condition. 2 genotypes *viz.* Ac. No. 5 (roundish fruit) and Ac. No. 35 (long fruit) were found promising to grow in Chhattisgarh state. Both of the varieties have been released by State Varietal Release Committee. After release, genotype Ac. No. 5 is known as Indira Kundru 5, which has 4.30 cm fruit length, 2-63cm fruit diameter and 22.94 kg per plant yield and genotype Ac. No. 35 is known as Indira Kundru-35, which has 6.0 cm fruit length, 2.43 cm fruit diameter and 21.08 kg per plant yield.

## *Buffalo gourd*

### Botanical Name

*Curcurbita foetidissima*

The buffalo gourd is a possible new root starch crop for semiarid regions. This is a xerophytic perennial and its ability to withstand drought has attracted interest for its commercialization It is native to the southwestern USA and Mexico. It is often recognized by its fetid odour and bitter gourds. Despite its odour, various Native American and Mexican tribes have used buffalo gourd for at least nine thousand years. Buffalo gourd grows predominately along roadsides and in disturbed soils. It is an extremely fast growing plant well adapted to arid conditions. The vine crawls along the ground acting as a good soil binder.

### Botanical Description

It is a coarse, trailing perennial, with gray-green, rough leaves, yellow flowers, and a large root. It undergoes dormancy after flowering and fruiting. It has a long white-fleshed perennial taproot with annual stems and leaves, whose lengths can exceed 650 ft. This plant is recognizable by its rough textured triangular leaves that are longer than they are broad. The flowers are monoecious (both sexes are found on the same plant), which are pollinated by insects. The plant is self-fertile. The flowers are bright yellow reaching 4 inches. Gourds that reach 2 to 3 inches long are also found with this plant having green or yellow stripes (Tull, 1999).

### Uses

It has been used traditionally in various ways as a food, cosmetic, detergent, insecticide and ritualistic rattle, to name a few. As research discovers the important resources of buffalo gourd it can be expected that this plant will become a valuable asset to the future. It is used as a vegetable; it can also be dried for later use. The young fruit is used which is bitter and becomes more bitter as it gets older. Buffalo gourd produces a high yield of seeds which are rich in protein and oil. The seeds can be ground into a powder and used as a thickening in soups or can be mixed with cereal flours when making cakes and biscuits (Facciola, 1990). Some caution is advised due to reported toxicity (Coffey, 1993).The foliage can be used for cattle feed and the large tuberous roots are a source of carbohydrates.

### Medicinal Uses

It is laxative, poultice and vermifuge. Buffalo gourd was employed medicinally by many native North American tribes who used it particularly in the treatment of skin complaints (Moerman, 1998). It is still employed in modern herbalism as a safe and effective vermicide (Bown, 1995). The leaves, stems and roots are laxative and poultice (Uphof, 1995).

Nutritional Composition (Berry *et al.*, 1976)

| | Fruit pulp | Seed | Root |
|---|---|---|---|
| Protein (%) | 30.1 | 32.9 | 15.1 |
| Fat (%) | 1.19 | 33.0 | 0.69 |
| Carbohydrate (%) | 26.9 | 0 | 64.6 |
| Acid-detergent fibre (%) | 27.7 | 35.1 | 15.8 |
| Energy (kcal/g) | 3.82 | 6.17 | 4.09 |

### Cultivation Practices

It requires a rich, well-drained moisture retentive soil and a very warm, sunny and sheltered position (Huxley, 1992). Established plants are very drought tolerant. This species is not very hardy. It is usually grown as an annual in temperate climates. The roots can survive temperatures down to about -25°C. It does not hybridize naturally with other members of this genus though crosses have been made under controlled conditions. This crop is sown by seeds early to mid spring under greenhouse in a rich soil. Germination should take place within 2 weeks. Two to three seeds per pot are sown and thinned out to the best plant. If grown outside, plants are first grown inside for sometime and after the last expected frosts, they are planted in open giving them cloche or frame protection for at least their first few weeks.

## *Long melon*

### Botanical Name

*Cucumis melo* var. *utilissimus*

This is locally known as *kakri* or *tar* which is a popular crop of Bihar, Uttar Pradesh and Uttaranchal. It is generally cultivated in river beds or *diara* areas. Its fruits are very much used as refreshing material to get relief from scorching heats during summer season due to high moisture content of fruits. Fruits are commonly sold along the roadside near bus stands and railway station in summer season. Young immature fruits are used for salad purpose.

It is used in *Raita* or cooked form also in Bihar and eastern U.P. Plants are annual climber which is monoecious in nature. Fruit length varies from 20 to 100cm. Short fruits are generally cylindrical and straight but longer fruits may be curved or coiled. The fruit colour ranges from light green to dark green. Longer and thicker fruits with prominent ridges were preferred earlier but nowadays consumers prefer smaller fruits with lesser girth size and smooth surface having undeveloped seed. This is a warm season crop. It prefers sandy loam soil. Seeds are sown in riverbeds during November-December and in Feb-March in up lands. Seed rate is 3-4 kg/ha. Four to five seeds are sown in a hill placed at 1.5-2 m x 60-75 cm. After germination 2 plants per hill are maintained. Farmer usually retains the seed from the previous crop. The fruit from riverbed come early to the market. Immature tender fruits are harvested and sold immediately due to very short post harvest life. The average yield is about 25-30 tonnes/ha.

There has been a limited research on this crop by research institutions and State Agricultural Universities in India. However, few varieties have been developed *viz.* Arka Sheetal, (IIHR, Bangalore) and Pant Kakri-1 (GBPUAT, Pantnagar). Pant Kakri-1 is a long fruited cultivar of long melon which has been developed through inbreeding and selection in the indigenous germplasm. The first green immature fruit picking starts in fifty days after sowing and its yield potential is 30 tonnes/ha.

### *Round Melon (Indian Squash)*

#### Botanical Name

*Praecitrullus fistulosus*

It is locally known as *tinda* which is a popular vegetable in North Indian states of Punjab, Haryana, Rajasthan, Delhi and Western U.P. It is an annual climber or vine crop which are monoecious in nature. Fruits are flattish round with 5-8 cm diameter. Skin colour is light green to dark green and flesh colour is white. Immature fruits are harvested and used as cooked vegetable similar to squash. It requires warm and dry climate with optimum temperature in the range of 27-30$^0$C. It prefers well drained light sandy loam soil with high organic matter. The optimum soil pH is around 6-7. This crop is sown during February-March or June-July in North India and December-January or June-August in western and Southern India. Seeds are usually sown on both sides of furrows. Fifty cm wide furrows are prepared at a distance of 1.5 meter and plant to plant distance is kept at 50-60 cm. The seed rate is about 3-5 kg/ha. Arka Tinda, S-58, Annamalai Tinda and Tinda Tonk are few important varieties of this crop. At the time of field preparation about 20-25 tonnes of FYM is applied per ha. The fertilizer dose is normally 50-75 kg N, 30-35 kg P and 30 kg K per ha. Half dose of nitrogen along with full dose of P and K are incorporated into the soil at the time of land preparation and the rest quantity of nitrogen is given as top dressing at the time of flowering and fruit set. In summer crop frequent irrigation at weekly interval is required. The fruits are ready for harvest after about 80-90 days of sowing. Immature fruits should be harvested after about 5-7 days of fruit set. The average yield is about 8-10 t/ha.

## *Snap Melon*

### Botanical Name

*Cucumis melo* var. *momordica*

This is a group of melon and locally known as phoot which is grown in India and some other Asian countries. This crop may have an Indian origin as it is still found in wild forms. It is cultivated in a limited scale in Bihar, Jharkhand, Uttar Pradesh, Rajasthan, Haryana and Punjab, especially as an intercrop with maize or sorghum. The plants are climber and mostly monoecious in nature. Fruits are ovoid to cylindrical in shape with a dimension of 30-60 X 7-15 cm. Flesh is white or pale orange, low in sugar content or rather sour in taste. The smooth surface of the fruit cracks or bursts as maturity approaches and fruit disintegrates at ripening. The immature fruits are cooked as vegetables and ripened fruits are consumed fresh. The post harvest life of fruit is very short. The seeds are also roasted and eaten in some parts. Snap melon has been very useful in melon breeding programme in USA because of its resistance to aphids and viruses. The crop may be grown in a wide rang of soil but light sandy loam soil with high organic matter is most suited. Seeds are sown during January-March or in June-July. Seeds are sown in hills/pits at a distance of 150 cm X 75 cm. Seed rate is about 2-3 kg/ha. Fruits are generally harvested at half ripe stage or before the bursting of fruit. The average yield is about 15-20 tonnes/ha.

## *Sweet Gourd*

### Botanical Name

*Momordica cochinchinensis*

This is commonly known as sweet gourd of Assam or *kheksa* or *bhat karela*. It is also called as cochinchin gourd or spine bitter cucumber. It may have an Indian origin and domesticated throughout Southeast Asia. In India a large variability is existing in North-Eastern states specially in Tripura and Assam. Sweet gourd is grown in North Eastern States, Assam, West Bengal, Bihar, Jharkhand, Andaman Island and Gujarat. It is a dioecious perennial climber with tuberous roots. Vines are vigorous with large leaves and thick woody stems. The large solitary flowers are cream-coloured to pale yellow, with black area at the bases of the petals. The fruit are avoid measuring 10-20 cm long and are covered with raised conical spines. The green colour of immature fruit turns to orange yellow colour on ripening. Ripened fruits contain numerous flattened black seeds of larger size inside the thick orange red flesh. Immature fruits are cooked as vegetable or prepared in curries. Young leaves, flowers and seeds are also edible. Oil can be extracted from the seeds and used for illumination. The roots contain a triterpenoid saponin and produce froth when rubbed in water which may serve as a substitute for soap. Various parts of the plant have medicinal uses in China, India and elsewhere in southern Asia (Robinson and Walters, 1997). The leaves, seeds and roots of sweet gourd have medicinal property especially for treatments of sores, vicers and bone fracture (Swarup, 2006). This crop is usually grown during the rainy season in the hot humid tropics. The normal method of propagation is through tuberous roots. As it is dioecious in nature, tubers from both male and female plants are collected separately to enable the growers to have a high propagation of female plants. Male population is kept at about 10 percent of the total population. The tubers are planted in nursery for sprouting and thereafter transplanted in the main field at a distance of about 1 m x 1 m. The vines are trained on bowers for obtaining better yield. The new vines start growing in summer and flowering as well fruit set take place during June. It takes about 60-65 days to flower after planting. The immature, tender fruits are harvested during May- September.

## References

Aung l H, Ball A and Kushad M. 1990 Development and nutritional aspects of chayote (*Sechium edule*, Cucurbitaceae). *Econ. Bot.*, 44: 157-164.

Berry J W, Wewber C W, Dreher M L and Bemis W P. 1976 Chemical composition of buffalo gourd, a potential food source. *Journal of Food Science* 41: 2, 465-466.

Bown D. 1995. Encyclopaedia of Herbs and their Uses. Dorling Kindersley, London.

Chakravarty, H. L. 1990 Cucurbits of India and their role in the development of vegetable crops. (In) Biology and utilization of the Cucurbitaceae (eds.) D M Bates, R W Robinson and C Jeffrey, Cornell University Press, Ithaca, N Y. pp 325-334.

Chopra R N, Nayar S L and Chopra I C 1986 Glossary of Indian Medicinal Plants (Including the Supplement). Council of Scientific and Industrial Research, New Delhi.

Coffey T. 1993. The History and Folklore of North American Wild Flowers. Facts on File.

Dieterle Jennie V A. 1976 Curcurbitaceae (In) Nash, Dorothy L., ed. Flora of Guatemala. Fieldiana. *Botany*, 24 (11/4): 336.

Esquinas Alcazar J T and Gulick P J. 1983 Genetic resources of Cucurbitaceae. Int. Board for Plant Genet. Resources, Rome.

Facciola S. 1990 Cornucopia - A Source Book of Edible Plants. Kampong Publications

Huxley A. 1992 The New RHS Dictionary of Gardening. MacMillan Press.

Herklots G A C. 1972 Vegetables in south-east Asia. George Allen & Unwin, London.

Jacks T H, Hensarling T P and Yatsu L Y. 1972 Cucurbit seeds: Characterizations and uses of oils and proteins, a review. *Econ. Bot.* 26:135-141.

Jeffrey C. 1990 Systematics of the Cucurbitaceae: An overview, (In) D.M. Bates, R.W. Robinson, and C. Jeffrey (eds.). Biology and utilization of the Cucurbitaceae. Cornell Univ. Press, Ithaca. pp 3-28.

Moerman D. 1998 Native American Ethnobotany Timber Press. Oregon.

Newstrom L. 1989 Reproductive biology and evolution of the cultivated chayote *Sechium edule:* Cucurbitaceae. (In) G.H. Bock & Y.B. Linhart, eds. Evolutionary ecology of plants, Boulder, Colo., USA, Westview. pp 491-509.

Newstrom L. 1990 Origin and evolution of chayote, *Sechium edule.* (In) D.M. Bates, R.W. Robinson & C. Jeffrey, eds. Biology and utilization of the Cucurbitaceae, Ithaca, N.Y., Cornell University Press. pp 141-149.

Parthasarathy V A and Sambandam C N. 2005 Taxonomy of *Cucumis callosus* (Rottl.) Cogn: The wild melon of India. *Cucurbit Genetics Coop. Rpt.* 3: 66-67.

Robinson R W and Walters Decker D S. 1997 Cucurbits. CAB International, New York, USA. pp 100-101.

Sambandam C M and Chelliah S. 1972 Scheme for the evaluation of cantaloupe and muskmelon varieties for the resistance to the fruit fly. USDA PL - 480 Res. Project. Annamalai Univ. pp 67.

Sarnik D A, Verma Satish K and Sharma G L. 2006 Evaluation of Ivy gourd (*Coccinia grandis* L.) germplasm. (In) Cucurbits Breeding. pp 205-208.

Swarup Vishnu 2006 Vegetable Science and Technology in India. Kalyani Publishers. Ludhiana, India.

Tull Delena. 1999 Edible and Useful Plants of Texas and the Southwest. University of Texas Press, Austin.

Uphof J C. 1995 The Dictionary of Economic Plants. Weinheim.

Verma S K, Negi K S, Muneem K C and Arya R R. 2006 Chow-chow (*Sechium edule* (Jac 1) Swart 2) germplasm maintained at NBPGR Regional Station Bhowali, (In) Cucurbits Breeding and Production Technology, Proceedings of National Seminar on Cucurbits held on 22-23 Sept. 2005, G.B. Pant University Agriculture and Technology, Pantnagar. pp 169-182.

Whitaker T W and Davis G N. 1962 Cucurbits Inter Science Publishers, Inc., New York.

**Underutilized and Underexploited Horticultural Crops, Vol. 2, 2007**
*K.V. Peter (ed.), pp. 195-197*
New India Publishing Agency, New Delhi (India)
E-mail : newindiapublishingagency@gmail.com
Web: www.bookfactoryindia.com

# 11 Alternanthera

**P. INDIRA**
Kerala Agricultural University, P.O.KAU
Thrissur - 680 656, Kerala

*Common names:* Sessile joyweed, Joyweed, Dwarf copperleaf (English), Geluti (Hindi), Ponnanganni green (Tamil)

*Latin names: Alternanthera sessilis* (L) Syn: *Alternanthera nodiflora* R. Br., *Gomphrena sessilis* (L), *llecebrum sessile* (L) Linnaeus belongs to family Amaramlnaceaes.

## Habit and Habitat

Perennial herbs, pantropical, damp shady areas, swamps, pond margins, shallow ditches, road sides, low-lying waste lands, damp pastures, cultivated areas. In New Guinea, «a plant of damp places; ditches, wet headlands, roadsides; sometimes a weed of plantations, particularly at altitudes of 1500 m or higher. From near sea level to over 2000m» (Henty & Pritchard, 1975). In Tonga, «occasional as a waste area weed» (Yuncker, 1959).

## Origin and Distribution

Perhaps Southern Asia (Smith, 1981), widely distributed as a weed. Widespread throughout tropics and subtropics: tropical Africa, Southern Asia to Japan, Southeast Asia, Australia, and the Pacific Islands. In Western hemisphere: the West Indies, Central America, tropical South America including Argentina, and the United States.

## Identification Remarks

*Alternanthera sessilis* is unique in the genus in that its tepals are generally shorter than the utricle. The utricles of other *Alternanthera* species sit well below the tops of tepals.

## Propagation

The plant spreads by seeds, which are wind and water-dispersed, and by rooting at stem nodes.

## Description

*Alternanthera sessilis* is an annual or perennial herb, 10-45 cm tall, stem ascending or creeping, green or somewhat tinged purple, striped, hairy, across nodes with a transverse row of hairs. Petiole 1-4 mm, glabrous or pilose; leaf blade linear-lanceolate, oblong-obovate, or ovate-oblong, 1-8 × 0.2-2 cm, glabrous or pilose, base attenuate, margin entire or slightly serrate, apex acute or obtuse. Heads 1-4, axillary, sessile, at first globose, later cylindric, 3-6 mm in diameter. Flowers dense; rachis densely white hairy, Bracts and bracteoles white, glabrous, apex acuminate; bracts ovate-lanceolate, ca. 1 mm; bracteoles subulate, 1-1.5 mm. Tepals white, ovate, 2-3 mm, glabrous, with a vein, apex acuminate or acute. Stamens 3; filaments ca. 0.7 mm, connate into a cup at base; anthers oblong; pseudostaminodes subulate, shorter than stamens, margin entire, apex acuminate. Style very short; stigma shortly parted. Utricles enclosed in perianth, dark brown, obovoid, 2-2.5 mm. Seeds ovoid. $2n = 34$

It prefers wet conditions, but occurs in both wetlands and uplands and can grow on a variety of soil types. *Alternanthera sessilis* is a weed of rice throughout tropical lands, and of other cereal crops, sugarcane, and bananas. Young shoots and leaves are eaten as a vegetable in Southeast Asia. One hundred grams of leaves contain 5.0g protein, 0.7g fat, 11.6g carbohydrate, 510mg Ca, 60mg P,1.63mg iron and vitamins like Vit-A (1.93mg), B2 (riboflavin .014mg), B3 (niacin 1.2mg ) and vitamin C (17mg). The leaves and tender shoots are clipped, fried and consumed. The first clipping is done 1 ½ to 2 months after planting. (Veeraraghavathatham *et al.*, 1998).

### References

Henty, E. E. & G. H. Pritchard. 1975. *Weed of New Guinea and their control*. 2nd edition. Division of Botany, Department of Forests, Lae, Papua New Guinea. p. 53.

Smith, A. C. 1981. *Flora Vitiensis nova: A new flora of Fiji.* Lawai, Kauai, Hawaii. National Tropical Botanical Garden. 2:291-292.

Veeraraghavathatham, D., Javaharlal , M. and Seemanthini Ramadas. 1998. *A guide on vegetable cultivation* . Suri Associates, Coimatore. pp 244-245

Yuncker, T.G. 1959. Plants of Tonga. *B. P. Bishop Museum Bull.* 220:109.

**Underutilized and Underexploited Horticultural Crops, Vol. 2, 2007**
*K.V. Peter (ed.), pp. 199-217*
New India Publishing Agency, New Delhi (India)
E-mail : newindiapublishingagency@gmail.com
Web: www.bookfactoryindia.com

# 12 *Hibiscus* and *Abelmoschus* Species of Potential Uses

**SURESH BABU, K V**
Department of Olericulture, College of Horticulture
Kerala Agricultural University, Vellanikkara
Thrissur - 680 686, Kerala

Malvaceae, the mallow family, carries herbs, shrubs or trees comprising about 75 genera and perhaps as many as 1,500 species. *Hibiscus* and *Abelmoschus* are two important genuses under this plant family. Several species of plants of potential uses come under these two genera about which limited publications are available.

## *Hibiscus*

*Hibiscus* is a large genus of about 200-220 species of flowering plants in Malvaceae, native to warm temperate, subtropical and tropical regions throughout the world. The genus includes both annual and perennial herbaceous plants and woody shrubs and small trees. The leaves are alternate, simple, ovate to lanceolate, often with a toothed or lobed margin. The flowers are large, conspicuous, trumpet-shaped, with five petals, ranging from white to pink, red, purple or yellow, and from 4-15 cm broad. The fruit is a dry five-lobed capsule, containing several seeds in each lobe, which are released when the capsule splits open at maturity.

The genus *Hibiscus* exhibits great diversity in the production of natural materials, with both edible and industrial applications. Components such as flowers and green tissues may be consumed as specialty foods, while the bast fibers (cellulosic), core fibers (lignitic and cellulosic), and seed oils (primarily poly-unsaturated) are a renewable source of industrial materials and edible oils.

About 200-220 species are known, including:

- *Hibiscus acetosella*
- *Hibiscus arnottianus* - Koki'o
- *Hibiscus brackenridgei* - Ma'o hau hele
- *Hibiscus calyphyllus*
- *Hibiscus cameronii*
- *Hibiscus cannabinus* - Kenaf
- *Hibiscus cisplatinus*
- *Hibiscus clayi* - Hawaiian hibiscus (red)
- *Hibiscus coccineus*
- *Hibiscus diversifolius*
- *Hibiscus elatus*
- *Hibiscus fragilis* - Mandrinette
- *Hibiscus furcellatus* - 'Akiohala
- *Hibiscus fuscus*
- *Hibiscus grandiflorus*
- *Hibiscus hastatus*
- *Hibiscus heterophyllus*
- *Hibiscus indicus*
- *Hibiscus kokio* - Koki'o 'ula
- *Hibiscus laevis* - Halberd-leaved rosemallow
- *Hibiscus lasiocarpos*
- *Hibiscus lavaterioides*
- *Hibiscus ludwigii*
- *Hibiscus macrophyllus*
- *Hibiscus militaris* - Syn. of *Hibiscus laevis*
- *Hibiscus moscheutos* - Swamp Rose-mallow

- *Hibiscus mutabilis* - Cotton rosemallow
- *Hibiscus paramutabilis*
- *Hibiscus pedunculatus*
- *Hibiscus platanifolius*
- *Hibiscus radiatus*
- *Hibiscus rosa-sinensis* - Chinese hibiscus
- *Hibiscus sabdariffa* - Roselle or Omutete or Sorrel
- *Hibiscus schizopetalus*
- *Hibiscus scottii*
- *Hibiscus sinosyriacus*
- *Hibiscus syriacus* - Hibiscus
- *Hibiscus tiliaceus* - Hau
- *Hibiscus trionum* - Flower-of-an-Hour
- *Hibiscus waimeae* - Koki'o ke'oke'o

Many of these species are grown for their showy flowers and many are used as landscape shrubs. *Hibiscus syriacus* (*Mugunghwa* or "Rose of Sharon") is the national flower of South Korea, while *Hibiscus rosa-sinensis* (*Bunga Raya* or "Chinese hibiscus") is the national flower of Malaysia. The *ma'o hau hele* (*Hibiscus brackenridgei*) is the state flower of Hawai.

In temperate zones, probably the most commonly grown ornamental species is *Hibiscus syriacus*, the common garden Hibiscus, also known in some areas as the "Rose of Althea" or "Rose of Sharon" (but not to be confused with the unrelated *Hypericum calycinum*, also called "Rose of Sharon"). In tropical and subtropical areas, the Chinese hibiscus (*H. rosa-sinensis*), with its many showy hybrids, is the most popular hibiscus.

One species of *Hibiscus*, known as Kenaf (*Hibiscus cannabinus*), is extensively used in paper making. Another, roselle (*Hibiscus sabdariffa*) is used as a vegetable and to make herbal teas and jams (especially in the Caribbean). In Mexico, the jamaica drink is quite popular and is made from calyces of the roselle plant. In Egypt and Sudan, roselle petals are used to make a beloved tea named after the plant, karkade, which can be served hot or chilled with ice.

Extracts of some hibiscus species are claimed to have health benefits, including prevention of constipation, bladder infections and nausea, and high blood pressure. The studies that yielded these results are debated. An unspecified hibiscus plant is used to make a herbal tea, typically blended with rosehip.

Hibiscus may reduce cholesterol-An extract from the hibiscus flower could have the same heart health benefits as red wine and tea, researchers suggest. A team from the Chung Shan Medical University in Taiwan says the flower contains antioxidants that help control cholesterol levels. They said animal studies showed the extract could reduce cholesterol in animals, so it may help humans.

The bark of the hibiscus contains strong fibers. They can be obtained by letting the stripped bark sit in the sea for some time in order to let the organic material rot away. In Polynesia these fibers (fau, pûrau) are used for making grass skirts.

Some *Hibiscus* species and cultivars such as 'Texas Star' look superficially similar to marijuana at a glance. This led to a police raid in one instance

Species of *Hibiscus* of economic importance are:

### *Hibiscus cannabinus (Kenaf )*

*Hibiscus cannabinus* (Kenaf) is a species of *Hibiscus*, probably native to southern Asia, though its exact natural origin is unknown. The name also applies to the fibre obtained from this plant. Kenaf is one of the allied fibres of jute and shows similar characteristics. Other names include Bimli, Ambary, Ambari Hemp, Deccan Hemp, and Bimlipatum Jute.It is an annual or biennial herbaceous plant (rarely a short-lived perennial) growing to 1.5-3.5 m tall with a woody base. The stems are 1-2 cm diameter, often but not always branched. The leaves are 10-15 cm long, variable in shape, with leaves near the base of the stems being deeply lobed with 3-7 lobes, while leaves near the top of the stem are shallowly lobed or unlobed lanceolate. The flowers are 8-15 cm diameter, white, yellow, or purple; when white or yellow, the centre is still dark purple. The fruit is a capsule 2 cm diameter, containing several seeds.

Kenaf has long history of cultivation for its fibre in India, Bangladesh, Thailand, parts of Africa, and to a small extent in southeast Europe. The stems produce two types of fibre, a coarser fibre in the outer layer, and a finer fibre in the core. It matures in 100 to 125 days. About 200 cultivars are grown.The main uses of kenaf fiber are the manufacture of rope, twine, coarse cloth (similar to that made from jute), and paper.Because it is very fast-growing compared to most trees, kenaf is a promising alternative to wood pulp for paper making that could free valuable timber for other uses. Kenaf seeds yield a vegetable oil that, while considered inedible, is used for industrial purposes and as oil-lamp fuel.Emerging uses of kenaf fiber include engineered wood, insulation, and clothing-grade cloth. Kenaf is also grown in Mackay, it is still in its trial stages but produce a much higher quality kenaf than in any other climate. Australian kenaf is of the highest standard.

### Description

Woody to herbaceous annual, mostly unbranched, fast-growing, with prickly stems, up to 4.2 m tall; leaves alternate, long-petiolate, shallowly to deeply parted, with 3–7 toothed lobes; flowers solitary, large to 10 cm in diameter, short-stalked, axillary, yellow with purple centers; sepals 5, petals 5; stamens numerous, connate; ovary superior; fruit a many-seeded, hairy capsule about 1 cm long; seeds brown, glabrous, wedge-shaped, 5 mm long, 3 mm wide, weight 25g/1000. Root a deep-penetrating taproot with deep-seated laterals (Reed, 1976).

### Uses

Kenaf is cultivated for its bast fibers which resemble and substitute for jute fibers. Fiber strands, 1.5–3 m long, are used for rope, cordage, canvas, sacking, carpet backing, and fishing nets. It is cultivated secondarily for the seeds which contain about 20% oil, used for: salad, cooking, and lubricant oils. Oil is also used in the manufacture of soap, linoleum, paints and varnishes, and for illumination. In 1968, Florida used 600 ha growing kenaf for bean poles (Whitely, 1981). Recently the pulp has been used in paper-making. Leaves are used as a potherb. A concentrated food for cattle in the form of seed-cake comes from the residue after oil extraction. Africans use soot from the stems as a black pigment. They also use a piece of the stem as a base for drilling fire.

### Harvesting

Highest quality fiber is obtained when plants are harvested during the flowering period, usually in August north of the equator. Planting to harvest is from 90–125 days. Three reaping methods are used: hand-cut, machine-cut, and pulled by roots, as the lower stem contains the best fibers. Hand-cutting is practiced only where labor is cheap and plentiful. A water source is necessary at harvest time since retting from 5–22 days is required to separate the bark from the wood and release the fibers. The fibers are then cleaned by hand and dried in the sun. A skilled worker can strip and wash about 36–45 kg dry clean fiber per day. Equipment essential for growing and harvesting kenaf consists of a tractor, plow, harrow, light grain drill, and a decorticator. A small farmer can generally make money with the smallest hand-fed decorticator and six hectares. However, it is more profitable to plant a larger acreage and use a semi-automatic decorticator. Farmers in Cuba and the United States avoid expensive hand labor by utilizing a self-propelled harvester decorticator, which will produce 45 kg of dried fiber per working hour (Reed, 1976).

### Folk Medicine

Reported to be anodyne, aperitif, aphrodisiac, fattening, purgative, and stomachik, kenaf is a folk remedy for bilious conditions, bruises, fever, and puerperium (Duke and Wain, 1981). Powdered leaves are applied to Guinea worms in Africa. Africans use peelings from the stems for anemia, fatigue, lassitude, etc. In Gambia, the leaf infusion is used for coughs. Ayurvedics use the leaves for dysentery and bilious, blood and throat disorders. Seeds are applied externally to aches and bruises. Medicinally, juice of the flowers with sugar and black pepper is used in biliousness with acidity. Seeds are considered aphrodisiac and fattening.

## *Hibiscus sabdariffa (Roselle)*

*Hibiscus sabdariffa* (Roselle) is a species of hibiscus native to the Old World tropics. It is an annual or perennial herb or woody-based subshrub, growing to 2–2.5 m tall. The leaves are deeply three- to five-lobed, 8–15 cm long, arranged alternately on the stems. The flowers are 8–10 cm in diameter, white to pale yellow with a dark red spot at the base of each petal, and have a stout fleshy calyx at the base, 1.5–2 cm wide, enlarging to 3–3.5 cm, fleshy and bright red as the fruit matures. It is an annual plant, and takes about six months to mature.

It is also known as meœta/meshta on the Indian Subcontinent, bissap in Senegal, the Congo and France, wanjo in the Gambia, zobo in Nigeria, karkade in Egypt, Saudi Arabia, and Sudan, omutete in Namibia, sorrel in the Caribbean and jamaica, in Latin America.

### Uses

Primarily, the plant is cultivated for the production for bast fibre from the stem of the plant. The fibre may be used as a substitute for jute in making burlap.The red calyces of the plant are increasingly exported to America and Europe, where they are used as food colourings. Germany is the main importer. It can also be found in markets (as flowers or syrup) in some places such as France, where there are Senegalese immigrant communities. The green leaves are used like a spicy version of spinach. They give flavour to the Senegalese fish and rice dish *thiéboudieune*. Proper records are not kept, but the Senegalese government estimates national production and consumption at 700 metric tons per year.In Africa, especially the Sahel, roselle is commonly used to make a sugary herbal tea that is commonly sold on the street. The dried flowers can be found in every market. In the Caribbean the drink is made from the fresh fruit, and it is considered an integral part of Christmas celebrations. The Caribbean Development Company, a Trinidad and Tobago brewery, produces a *Shandy Sorrel* in which the tea is combined with beer.

### *Fountain Drink*

With the advent in the U.S. of interest in south-of-the-border cuisine, the calyces are sold in bags usually labelled "Flor de Jamaica" and have long been available in health food stores in the U.S. for making a tea that is high in vitamin C. This drink is particularly good for people who have a tendency, temporary or otherwise, toward water retention: it is a mild diuretic.

### Production

China and Thailand are the largest producers and control much of the world supply. Thailand invested heavily in roselle production and their product is of superior quality, whereas China's product, with less stringent quality control practices, is less reliable and reputable. The world's best roselle comes from the Sudan, but the quantity is low and poor processing hampers quality. Mexico, Egypt, Senegal, Tanzania, Mali and Jamaica are also important suppliers but production is mostly used domestically.

In the Indian Subcontinent (especially in the Ganges Delta region), roselle is cultivated for vegetable fibres. Roselle is called meœta (or meshta, the *œ* indicating an *sh* sound) in the region. Most of its fibres are locally consumed. However, the fibre (as well as cuttings or butts) from the roselle plant has great demand in various natural fibre utilizing industries.

### Theraputic Value

*Hibiscus sabdariffa* L, which is grown in China, India and Taiwan is used as a traditional remedy to treat high blood pressure and liver disorders. The researchers found it contains antioxidants that are known to reduce the dangerous build up of fats inside the arteries. They carried out a study looking at the effects of hibiscus extract on low density lipoprotein, LDL or "bad" cholesterol in rats. They found the hibiscus extract significantly reduced the build up of fatty deposits in arteries and blood cholesterol levels.

According to a study performed by Shan Medical University's Institute of Biochemistry in Taiwan, extract from the hibiscus flower (Hibiscus sabdariffa) lowers both low-density lipoprotein cholesterol levels and the risk of heart disease. These results were published in the September 15th issue of the Journal of the Science of Food and Agriculture. Dr. Chau-Jong Wang and his colleagues suspect that the high antioxidant content contained in the hibiscus extract contributes to the lowered LDL levels noted in the experiment.

For the experiment, Wang and colleagues divided rats up into three groups, according to the atherosclerosis-inducing diet they were to be placed on. One group was placed on a high cholesterol diet, while the second group was placed on a high fructose diet. High glucose levels are thought to increase triglyceride levels, contributing to the fact of why diabetics are more susceptible to higher cholesterol levels and heart disease. Some groups containing each diet were supplemented with hibiscus extract to see whether or not it had any affect on their cholesterol levels. According to the results, both groups saw a surprising decrease in low density lipoproteins upon the administration of hibiscus flower extract. Additionally, the rats consuming a high fructose diet also saw a decrease in triglyceride levels. It is thought that the antioxidant chemicals, such as flavonoids, polyphenolics and anthocyanins, contained in the flower play a large role in preventing the oxidation of low density lipoproteins (the "bad cholesterol"). This oxidation is what contributes to atherosclerosis, the build up of a waxy plaque on the walls of arteries.

Does the news that hibiscus flower extract actually has some therapeutic value surprise scientists? Not really. Hibiscus flower extract has been used in many folk remedies for liver disorders and high blood pressure. However, this is the first piece of research that establishes the healthy benefits of consuming hibiscus flower extract.

## Food Uses

Roselle fruits are best prepared for use by washing, then making an incision around the tough base of the calyx below the bracts to free and remove it with the seed capsule attached. The calyces are then ready for immediate use. They may be merely chopped and added to fruit salads. In Africa, they are frequently cooked as a side-dish eaten with pulverized peanuts. For stewing as sauce or filling for tarts or pies, they may be left intact, if tender, and cooked with sugar. The product will be almost indistinguishable from cranberry sauce in taste and appearance. For making a finer-textured sauce or juice, sirup, jam, marmalade, relish, chutney or jelly, the calyces may be first chopped in a wooden bowl or passed through a meat grinder. Or the calyces, after cooking, may be pressed through a sieve. Some cooks steam the roselle with a little water until soft before adding the sugar, then boil for 15 minutes.

Roselle sauce or sirup may be added to puddings, cake frosting, gelatins and salad dressings, also poured over gingerbread, pancakes, waffles or ice cream. It is not necessary to add pectin to make a firm jelly. In fact, the calyces possess 3.19% pectin and, in Pakistan, roselle has been recommended as a source of pectin for the fruit-preserving industry.

Juice made by cooking a quantity of calyces with 1/4 water in ratio to amount of calyces, is used for cold drinks and may be frozen or bottled if not for immediate needs. In sterilized, sealed bottles or jars, it keeps well providing no sugar has been added. In the West Indies and tropical America, roselle is prized primarily for the cooling, lemonade-like beverage made from the calyces. This is still "one of the most popular summer drinks of Mexico", as Rose observed in 1899. In Egypt, roselle "ade" is consumed cold in the summer, hot in winter. In Jamaica, a traditional Christmas drink is prepared by putting roselle into an earthenware jug with a little grated ginger and sugar as desired, pouring boiling water over it and letting it stand overnight. The liquid is drained off and served with ice and often with a dash of rum. A similar spiced drink has long been made by natives of West Tropical Africa. The juice makes a very colorful wine.

John Ripperton of the Hawaiian Experiment Station maintained that, for jelly and wine-making, it is unnecessary to take out the seed capsule, but neglecting to do so may result in a "stringy" product which would be contaminated with the minute hairs from the surface of the capsule and these hairs are quite likely to be injurious unless carefully filtered out.

The calyces are either frozen or dried in the sun or artificially for out-of-season supply, marketing or export. In Mexico today, the dried calyces are packed for sale in imprinted, plastic bags. It is calculated that 11 lbs (5 kg) of fresh calyces dehydrate to 1 lb (0.45 kg) of dried roselle, which is equal to the fresh for most culinary purposes. However, dried calyces as sold for "tea" do not yield high color and flavor if merely steeped; they must be boiled.

For retailing in Africa, dried roselle is pressed into solid cakes or balls. In Senegal, the dried calyces are squeezed into great balls weighing 175 lbs (80 kg) for shipment to Europe, where they are utilized to make extracts for flavoring liqueurs. In the United States, Food and Drug Administration regulations permit the use of the extracts in alcoholic beverages.

The young leaves and tender stems of roselle are eaten raw in salads or cooked as greens alone or in combination with other vegetables or with meat or fish. They are also added to curries as seasoning. The leaves of green roselle are marketed in large quantities in Dakar, West Africa. The juice of the boiled and strained leaves and stems is utilized for the same purposes as the juice extracted from the calyces. The herbage is apparently mostly utilized in the fresh state though Wester proposed that it be evaporated and compressed for export from the Philippines.

The seeds are somewhat bitter but have been ground to a meal for human food in Africa and have also been roasted as a substitute for coffee. The residue remaining after extraction of oil by parching, soaking in water containing ashes for 3 or 4 days, and then pounding the seeds, or by crushing and boiling them, is eaten in soup or blended with bean meal in patties. It is high in protein.

### *Hibiscus acetosella (Cranberry Hibiscus or False Roselle)*

This species has striking red leaves similar to a Japanese maple. It can be planted in the spring and kept pruned for an attractive annual shrub and may be grown as a temporary hedge. Cranberry hibiscus is nematode and insect resistant and does well in sandy soil. The young tender red leaves have a tart flavor and are an attractive addition to salads, slaws, or stir-fries. In the fall it has pink blossoms. About thirty blossoms can be picked at dusk after they have folded, and blended with lime juice and sugar to make a beautiful and tasty drink. The petals add a bright red color rather than any special flavor.

*Hibiscus syriacus* (The Rose of Sharon) is a deciduous shrub that grows up to 3 metres tall and 2 metres wide, though it can be kept smaller by trimming it, and can be grown as a hedge. It prefers a well-drained humus rich fertile soil in a sheltered position in full sun and succeeds in any soil of good or moderate quality. Plants are hardy to about -20°C but plants only really succeed in the warmer counties of Britain because of their late flowering habit. They are also slightly tender when young and when planted in colder areas of the country, they will need protection for the first few winters. The young leaves can be eaten raw or cooked. They have a very mild, slightly nutty flavour, and though slightly on the tough side they make an acceptable addition to the salad bowl.

*Hibiscus* species with leaves that have reportedly been eaten by people

- *Hibiscus acetosella*
- *Hibiscus bifurcatus*
- *Hibiscus cannabinus*
- *Hibiscus diversifolius*

- *Hibiscus furcatus*
- *Hibiscus rosa-sinensis*
- *Hibiscus sabdariffa*
- *Hibiscus surattensis*
- *Hibiscus syriacus*
- *Hibiscus tiliaceus*

### *Hibiscus species as Potential Nutraceutical*

In the past, perennial and temperate-zone tree-form *Hibiscus* species and hybrids have been relegated to the status of little-known garden plants, at least in the United States. However, their potential as a new source of edible flowers and natural food colorants (all within the anthocyanin class of edible flower pigments) elevates them to the status as—possibly the newest—of the New Crops.

The results of previous investigations of nutritional analyses of dehydrated petals of many *Hibiscus* species indicate substantial yields of dietary proteins, sugars, fiber, and unsaturated fats, as well as most of the major dietary vitamins. These could be recovered from floral harvests on a commercial basis. Applications may include use of whole fresh, dried (dehydrated or freeze-dried), or frozen petals for use in nutraceutical products and specialty gourmet foods, or as a base for extraction of natural food colorants for applications in foods and beverages. Other applications for fresh and processed blossoms and their extractions may include formulation of cellulosic and mucilagenous dietary products, transdermal carriers for non-alimentary drug delivery, and skin-care formulations.

## II Abelmoschus

*Abelmoschus* is a genus of about fifteen species of flowering plants in Malvaceae, native to tropical Africa, Asia and northern Australia. It was formerly included within *Hibiscus*, but is now classified as a distinct genus.

The genus *Abelmoschus* has eight important species including the cultivated or semi cultivated species such as *A. esculentus* (L.) Moench, *A.caillei* (L.)Stewes, *Abelmoschus moschatus* Medic., *A. manihot* (L.) Medic., and wild forms such as *A. ficulneus, A. crinitus A.tuberculatus* and *A. angulosus,* distributed in the South and South East Asia, Africa and in North Australia

The genus comprises annual and perennial herbaceous plants, growing to 2 m tall. The leaves are 10-40 cm long and broad, palmately lobed with 3-

Hibiscus acetosella

Hibiscus sabdariffa

Hibiscus elatus

7 lobes, the lobes very variable in depth, from barely lobed, to cut almost to the base of the leaf. The flowers are 4-8 cm diameter, with five white to yellow petals, often with a red or purple spot at the base of each petal. The fruit is a capsule, 5-20 cm long, containing numerous seeds.

The taxonomy of *Abelmoschus* is complex. Originally cultivated and wild species were grouped within the genus *Hibiscus*, section *Abelmoschus* by Linnaeus (1773), later raised to a distinct genus *Abelmoschus* by Medikus (1787), further re-established as a genus based on the caducous nature of the calyx by Schumann (1908), and later rehabilitated by Hochreutiner (1924) based on the relationship of the calyx to the petal and staminal column aestivation.

Describing the genus, Bates (1968) considered the group to belong to the family Malvaceae, with a number of circumscriptive characters. Van Borssum-Waalkes (1966) and Bates (1968) further reviewed the genus. Van Borssum-Waalkes distinguished only six species: three cultivated (*A. moschantus, A. manihot* and *A. esculentus*) and three wild (*A. ficuleus, A. crinitus* and *A. angulosus*). However, van Borssum-Waalkes's classification did not mention *Abelmoschus manihot* var. *caillei*.

*Abelmoschus manihot* var *caillei* was described by Chavalier (1940) as a West African taxon resembling *Abelmoschus esculentus* and later elevated to a distinct species by Stevels (1988).Thus it is presently known as *Abelmoschus caillei*.

Several species of *Abelmoschus* are edible, with both the young seed pods and the young leaves being eaten as a vegetable. The most important species commercially is Okra (*Abelmoschus esculentus*).

### *Okra (Abelmoschus esculentus)*

*Immature fruit:* cooked on their own or added to soups, they can be used fresh or dried. Mucilaginous, they are commonly used as a thickening for soups, stews and sauces. The fruits are rich in pectin and are also a fair source of iron and calcium. The fresh fruits contain 740 IU vitamin A. The fruit should be harvested while young, older fruits soon become fibrous. The fruit can be up to 20cm long. Seed - cooked or ground into a meal and used in making bread or made into 'tofu' or 'tempeh. The roasted seed is a coffee substitute. Probably the best of the coffee substitutes. The seed contains up to 22% of edible oil. The leaves, flower buds, flowers and calyces can be eaten cooked as greens. The leaves can be dried, crushed into a powder and stored for later use. They are also used as flavouring. Root - it is edible but very fibrous. Mucilaginous, without very much flavour.

## Medicinal Uses

Antispasmodic; Demulcent; Diaphoretic; Diuretic; Emollient; Stimulant; Vulnerary. The roots are very rich in mucilage, having a strongly demulcent action. They are said by some to be better than marsh mallow (*Althaea officinalis*). This mucilage can be used as a plasma replacement. An infusion of the roots is used in the treatment of syphilis. The juice of the roots is used externally in Nepal to treat cuts, wounds and boils. The leaves furnish an emollient poultice. A decoction of the immature capsules is demulcent, diuretic and emollient. It is used in the treatment of catarrhal infections, ardor urinae, dysuria and gonorrhea. The seeds are antispasmodic, cordial and stimulant. An infusion of the roasted seeds has soporific properties.

## A source of Phytate

Okra pod is a source of Phytate( Phytic acid) whiich is chemicaly known as Inositol Hexaphosphate, it has mineral binding properties. Researchers now believe that phytate, found in the crude fiber of okra pod , is a major ingredient responsible for preventing colon cancer and other cancers by reducing oxidative stress in the lumen of the intestinal tract. Phytate rarely appears in soluble fiber. It is found in association with insoluble dietry fibres and hence must be well cooked in order to free phytate from the fiber and enable it to be absorbed in the system.

## Other Uses

A fibre obtained from the stems is used as a substitute for jute. It is also used in making paper and textiles. The fibres are about 2.4mm long. When used for paper the stems are harvested in late summer or autumn after the edible seedpods have been harvested, the leaves are removed and the stems are steamed until the fibres can be stripped off. The fibres are cooked for 2 hours with lye and then put in a ball mill for 3 hours. The paper is cream coloured. A decoction of the root or of the seeds is used as a size for paper.

## *Abelmoschus manihot*

Known as an *edible hibiscus*, this obscure Chinese species is a tropical perennial, *half-shrub* that can reach 6 feet in height. *A. manihot's* flowers are three-five inches, yellow to white in color, and have a dark red eye. The edible leaves have a large ovate shape with five to nine lobes. Several varieties exist of the species in which the lobes of the leaf can vary in depth and leaf color can range from sulfur yellow to very dark green. This plant prefers full sun and rich, moist soil. Propagation occurs through seed sown in the spring or root division. The leaves of *Abelmoschus manihot* are highly nutritious, and are commonly used in salads, stir-fry, stews, and soups. A single leaf can cover a slice of bread in a sandwich and is used like lettuce. The leaves are tender when raw but turn slimy if overcooked. They are high in nutrition. Add compost to the soil before planting to reduce nematode damage. The plant is frost sensitive.

As *Abelmoschus manihot* occurs in tropical Asia and in northern Queensland, it grows best in tropical and subtropical regions since it is frost tender. However, although it is a perennial, it can be grown as an annual in temperate areas, flowering well in the first year and setting seed.

*Abelmoschus manihot* has a single central stem to 2m and short sparse branches. It grows very quickly. Cuttings taken in spring can reach over 2 metres by autumn in sub-tropical areas. It is a hardy plant, which prefers a sunny aspect with rich, moist, well- drained soil.

The plants may naturalize. Plants grow well in an ornamental vegetable garden, and this may not only add to the appearance of the garden but may also make it easier to ensure that the plant does not naturalize.

The plant is suitable for pot culture.The large, up to 15 cm (6 inches) hibiscus-like flowers are a brilliant lemon with deep purple centres borne on long pedicels at the apex of the plant. Flowers are produced in the warmer months. The flowers are pollinated by insects. Petals can be added to salads or cooked.

While the large yellow flowers are very ornamental, the importance of this plant is that it is one of the world's most nutritious leafy vegetables because of its high protein content. The leaves are tender and sweet and can be served raw or steamed (leaf for life).

It can produce up to 60 tons of leaves per hectare using a multiple harvest system.

There can be big differences in leaf shape, colour and production and flavour but leaves are usually palmate, about 10 cm (4 inches) across. In Papua New Guinea, plants are propagated from stem cuttings with harvesting commencing 2 - 3 months after planting and continuing for 1-2 years, and the young leaves are picked and are cooked in coconut cream or water. They are also fried. Leaves may be added to soups. Blanched leaves can be added to salads. The leaves make a tasty addition to an omelette.

### *Abelmoschus moschatus* – Ambrette

*Abelmoschus moschatus* (L.) Medic, Malvaceae (Syn. *Hibiscus abelmoschus* L.) is a tropical weedy shrub native to India valued for its scented seed. Ambrette is a close relative to Okra.In Hindi, it is popularly known as *mushkdana, kasturi bhendi* (kasturi = musk; *bhendi* = lady's finger). In other Indian languages it is known as *gukhia korai* (Assamese), *kasturi bhenda* (Telgu), *kattukasturi* (Malylam), *varttilai kasturi* (Tamil), *lalkasturika* (Sanskrit) (Krishnamurty 1993). The area under ambrette is presently low in India but is increasing rapidly (Oudhia and Tripathi 2000) with seed exports to France, Germany, Japan, Singapore, Spain for its use as an aromatic oil. Indian drug manufacturers are introducing new herbal drugs containing ambrette for medicinal use.

*Abelmoschus moschatus* has four subspecies as listed below

- *Abelmoschus moschatus* Medikus. subsp. *biakensis* (Hochr.) Borss.
- *Abelmoschus moschatus* Medikus. subsp. *moschatus* var. *betulifolius* (Mast.) Hochr.
- *Abelmoschus moschatus* Medikus. subsp. *moschatus* var. *moschatus*
- *Abelmoschus moschatus* Medikus. subsp. *tuberosus* (Span.) Borss.

## Uses

Ambrette oil obtained from seeds possesses an odor similar to that of musk and its aromatic constituents have long been used in perfumery industry. Different grades of essential, or aromatic absolute, are marked in Europe as high-grade perfumes (Singh *et al.* 1996 ) The seeds are valued for the volatile oil present in the seed coat. Seed analysis report 11.1% moisture, 31.5% crude fiber; 14.5% lipids, 13.4% starch, 2.3% protein, volatile oil (0.2-0.6% ) and ca/ 5% resin (Srivastava 1995).

Analysis of volatiles report myricetin-3-glucoside and a glycoside of cyanidin in flowers, an aromatic constituent in seeds, beta-sitosteral and its beta-D-glucoside, myricetin and its glucoside in leaves and petals and beta-sitosterol from dry fruit husk (Rastogi and Mehrotra 1991a,b).

In India, roots, leaves (rarely), and seeds of ambrette are considered valuable traditional medicines. The bitter, sweet, acrid, aromatic seeds are used as a tonic and are considered "cooling, aphrodisiac, opthalmic, cardiotonic, digestive, stomachic, constipating, carminative, pectoral, diuretic, stimulant, antispasmodic, deodorant, and effective against "*kapha*" and "*vata*," intestinal complaints, stomatitis; and diseases of the heart, allays thirst and checks vomiting. According to *Unani* system of medicine seeds allay thirst, cure stomatitis, dyspepsia, urinary discharge, gonorrhea, leucoderma and itch. Roots and leaves are cures for gonorrhea (Agharkar 1991). Even use against venomous reptiles has been reported (Lindley 1985).

*A.manihot - with edible leaf*

Abelmoschus caillei

Abelmoschus esculentus

### *A.caillei*

Among the *Abelmoschus* species, very little attention has been devoted to this species. Its genetic diversity is clearly shown by the wide range of morphological characteristics displayed by this species. This may well point to the considerable genetic resources inherent in Abelmoschus *caillei*, which is endemic to the region and which is widely used in subsistence farming. The tender pods of this species is used as a vegetable.

Although a great degree of similarity exists between *A.esculentus* and *A. caillei* there were marked morphological differences between members of the two groups. This reflects the difference in cultivation zones: *Abelmoschus caillei* is commonly grown in the high-rainfall zone of West Africa and mainly in subsistence systems. In contrast, *A.esculentus* is commonly grown under irrigation in areas with less rainfall (Schipper 1998). However, a worrying finding is the small amount of *A.caillei* germplasm present in institutional reserves compared with the more common *A.esculentus*. Nevertheless, the prevalence of *A.caillei* in homesteads and cultivated fields within the region is indicative of its richness as a genetic resource and the preference of local farmers.

*A.caillei* has been a choice material for the plant breeders as it is a source of resistance to yellow vein mosaic virus (YVMV) and shoot and fruit borer. Hence it can be used for interspecific hybridization programmmes.

The other species A. *ficulneus*, *A. crinitus A.tuberculatus* and *A. angulosus* exist purely as wild forms. They can be utilized for distant hybridization programmes.

## References

Agharkar, S.P. 1991 Medicinal plants of Bombay presidency. Scientific Publ. Jodhpur, India. First reprint p. 1-2

Bates DM. 1968 Notes on the cultivated Malvaceae. 2. *Abelmoschus*. Baileya 16:99–112.

Chavalier A. 1940 L'origine, la culture et les usages de cinq *Hibiscus* de la section *Abelmoschus. Revue Botanique Appliquée* 20:319–328.

Duke, J.A. and Wain, K.K. 1981 Medicinal plants of the world. Computer index with more than 85,000 entries. 3 vols.

Hochreutiner BPG. 1924 Genres nouveaux et genres discutes de la familes des Malvaceae. Candollea 2:79–90.

Linnaeus C. 1773 Hortus Cliffortianus. Amsterdam.

Lindley, J. 1985 Flora medica.. Ajay Book Service, New Delhi.

Medikus FK. 1787 Ueber einige kunstliche Geschlechter aus der Malven-Familie 45–46.

Rastogi, R.P. and B.N. Mehrotra. 1991 Compendium of Indian Medicinal Plants. Vol. II. (1970-1979). Central Drug Research Institute, Lucknow and Publ. and Information Directorate, New Delhi.

Reed, C.F. 1976 Information summaries on 1000 economic plants. Typescripts submitted to the USDA.

Schumann K. 1980 Malvaceae. In: Engler A, Prantl K, editors. Die natürlichen Pflanzenfamilien, Vol. 3. Engelmann, Leipzig, Germany. pp. 30–35.

Singh, U., A.M. Wadhwani, and B.M. Johri. 1996 Dictionary of economic plants in India. Indian Council of Agricultural Research, New Delhi.

Srivastava, U.C. 1995 Ambrette seed. p. 887-897. In: K.L. Chadha and Rajendra Gupta (eds.), Advances in Horticulture Vol. 11-Medicinal and Aromatic Plants (1995). Malhotra Publ. House, New Delhi.

Stevels JMC. 1988 Une nouvelle combination dans *Abelmoschus* Medik. (Malvaceae), un gombo d'Afrique de l'ouest et centrale. Bulletin du Museum National d'Histoire Naturelle de Paris. 4e Serie. Section B. Adansonia. Botanique Phytochimie 2: 137–144.

Van Borssum-Waalkes J. 1966 Malaysian Malvaceae revised. Blumea 14(1):1–231.

Whitely, E.L. 1981 Kenaf, *Hibiscus cannabinus*. p. 259–268. In: McClure, T.A. and Lipinsky, E.S. (eds.), CRC handbook of biosolar resources. vol. II. Resource materials. CRC Press, Inc., Boca Raton, FL.

## Internet References

NCAUR New Crops Database - Oil Seeds, circa 1999–2000. www.ncaur.usda.gov/nc/ncdb/search.html-ssi

Plants for a Future: Database Search Results. www.scs.leeds.ac.uk/cgi-bin/pfaf/find_

LAT=malvaceae & COM=&FAM=. April 2002.

Dr. Duke's Phytochemical and Ethnobotanical Databases. www.ars-grin.gov/duke/ethnobot.html. April 2002

The Langdon Research Extension Center. www.ag.ndsu.nodak.edu/langdon/research

# Underutilized Root and Tuber Crops

**Underutilized and Underexploited Horticultural Crops, Vol. 2, 2007**
*K.V. Peter (ed.), pp. 219-253*
New India Publishing Agency, New Delhi (India)
E-mail : newindiapublishingagency@gmail.com
Web: www.bookfactoryindia.com

# 13 Taxonomy of Temperate Underutilized Root and Tuber Crops

**RANA J C., PANDEY, S.K* AND JAIGOPAL****
* Central Potato Research Institute, Shimla - 171001 (HP)
** National Bureau of Plant Genetic Resources Regional Station Shimla - 171 004 (HP)

The Mother Nature bestowed humanity with about 250,000 known plant species and perhaps the same or may be higher number of unknown plant species. Out of this treasure, 80,000 plant species are edible but human beings have probably exploited about 3,000 of these species for food and was cultivating about 150 on a commercial scale. However, largely human beings consumed only 20 different kinds of crops the world over and four of which viz. rice, wheat, maize and potato are supplying about 60 % of total calories. Under the circumstances, the food crops base even for well-known crops are shrinking at an alarming rate and in this process, many species have extincted from man's food bowl. While doing so, we have considerably reduced the range of organic compounds that we now consume, thus making ourselves vulnerable to various stresses. Realising importance of diverse food base in our life, thinking is going on among people of all lifestyles that more and more crops particularly those, which are known to be edible but have remained as "under-utilised crops", should be brought into cultivation. This will ensure both food and nutritional security of people. Among these, tuber crops constitute an important crops group, which was sustaining food security much before agriculture came into existence. Men right from hunting stage have been eating tuber crops. In fact, roots and tubers belong to the class of foods that basically provide energy in human diet in the form of carbohydrates. The terms refer to any growing plant that stores edible material in subterranean root, corm or tuber. These crops have saved humanity at famine times because tuber crops are last to go under adverse calamities. A further reason for the spread was the fact that during tribal warfare and invasions, the invader could not destroy or remove the food reserve, which could be kept conveniently under the ground, thus giving added food security to the population. However, while considering such crops for edible purposes, proper taxonomic identity of the species must be ensured because most of the plant species contain poisonous substances, which may be harmful, if eaten. The present chapter describes taxonomically and ethno-botanically, many underutilised tuber crops occurring in Himalayan region.

## Amorphophallus

It is essentially a genus of tropics and distributed from West Africa to the Pacific Islands but a few species also are grown in temperate Himalayan region and largely used for tubers. They mostly grow in secondary forests or disturbed spots in primary forests and forest margins. It is also cultivated in homestead gardens and locally known as *suran, zimikand, hausa, barari* etc. It is a monoecious herb with an underground storage organ usually a tuber. Leaf solitary, rarely paired, emerging from top of tuber or rhizome, lasting one growing-season or rarely long-lasting. One single leaf emerges from the tuber, consisting of a vertical petiole (stalk) and a horizontal leaf-blade. The latter is dissected into a few or numerous small leaflets. Once plants are mature, an inflorescence may develop. The inflorescence may replace the leaf in one season, or develop alongside it. Inflorescence consists of a foliar organ, the spathe, which usually envelops a stalk-like organ, the spadix. Flowers are tiny and strongly reduced and found at base of spadix. Female flowers form the most basal zone and are each reduced to a mere pistil. Typical pollinator attracting organs (like a corolla) are absent. Above this zone is found a zone of highly reduced male flowers called 'staminodes'. They only exist or a group of stamens. Remainder of spadix usually is sterile, lacking flowers, and is called the appendix. This entire structure, the spadix and the spathe, seems to work as one large "flower" and obviously acts as a unitary structure to attract pollinators, a feature lost by the actual individual flowers. The pollinated flowers then develop a globose berry. These can be red, orange-red, white, white-yellow and blue. *Amorphophallus* belongs to family *Araceae* (aroid family) and is estimated to encompass about 170 species (Hetterscheid, 1994), of which 14 occur in India (Hooker, 1978). A few important species that are occurring in the temperate areas are given below:

### *A. bulbifer (Roxb.) Blume*

*A. bulbifer* (Roxb.) Blume = *A. aculatum* Hook.f, = *A. taccoides* Hook.f. = *A. bulbiferum* Roxb. It gets its name due to the bulbils that form on top of the leaves. The bulbils are like "little bulbs" and help in propagation. It dislikes low humidity and the leaflets may partially desiccate especially in low light (Vajravelu and Bhargavan, 1983). The foliage is also very striking, with attractively segmented leaves on a green-mottled petiole that should reach 90 cm or more in shaded humus rich soils. It is distributed in Khasia hills, Sikkim and mid hill areas of western Himalaya ascending to 1500 m. Leaf 30-45 cm dia, ultimately *bulbilliferous* at the base, forks and nerves above; leaflets 9-24 cm, obovate or lanceolate; petiole 90-120 cm and peduncle 24-30 cm, green and pink streaked. Spathe 15-24 cm, erect pale pink or yellowish clouded with pink within. Anthers short; stigma 2-lobed; spathe long- peduncled; tube broad, turgid; limb rather longer ovate, cymbiform, obtuse; spadix very stout shorter or a little longer than the spathe; appendage as long as inflorescence and broader than elongate, conoid or oblong, top rounded. Tuber globose.

### *A. paeoniifolius (Dennst.) Nicolson*

*A. paeoniifolius* (Dennst.) Nicolson = *A. campanulatus* (Roxb.) Blume = *A. giganteus* Blume. The geographical limits of this species are quite uncertain (owing to its being cultivated widely in the east for its tubers). However, in India, its distribution is restricted up to Himalayan region. Leaf 1-2; 30-90 cm long, broad, segments simple or forked; leaflets oblong, acute; petiole dark green with pale blotches. Peduncle very short; petiole rough; spathe 18-27 cm, campanulate; limb sub-erect or recurved, wavy and crenulated; spadix very stout, hardly longer than the spathe; spathe variable in colour, green, usually with white spots below, greenish purple above, rough and dark purple within towards the base. Appendage globosely conoid or amorphous sinuously lobed. Berries obovoid. Tuber a depressed bulbilliferous sphere, 24-30 cm dia.

## Arisaema

A genus of stemless, paradioecious or monoecious herbs with usually tuberous roots. Leaves digitately compound, sometimes pedate. Spathe finally deciduous, the spadix included or excluded. Male flowers with 2-5 stamens; anthers bilocular. Female flower with a unilocular ovary and two or more basal ovules; style short or absent. Berry few-seeded, red. Seeds globose to ovoid (Pradhan, 1997). Only few species are edible. Owing to their high content of calcium oxalate crystals, they give extremely unpleasant sensation similar to needles being stuck into the mouth and tongue if they are eaten fresh but they are easily neutralized by thoroughly drying or cooking the plant or by steeping it in water. The roots can be cut into very thin slices and allowed to dry for several months, after which they are eaten like potato chips, crumbled to make a cereal or ground into a cocoa-flavoured powder for making biscuits, cakes etc. They also carry very high medicinal value (Chopra, et al., 1986, Anon., 1952).

*Arisaema* belong to family Araceae. Its taxonmy is not very well understood and different taxonomic cultures sometimes clash and personal views permeate that are not always backed up by professional taxonomic analysis (Anon., 1952). It comprised about 300 species, native mostly to the mountainous region of Old World and few to North America. About 40 species occur in India and 15 species belong to temperate region (Sivadasan and Kumar, 1987, Hooker, 1978). Important species that are occurring in the Himalayan region are *Arisaema longecaudatum* Blatter, *A. curvatum, A. murrayi* (Grah.) Hook, *A. concinuum, A. intermedium* Bl., *A. propinquum* Schott., *A. flavum, A. speciosum, A. tortuosum, A. murrayi var. sonubeniae* and *A. costautum. A. pulchrum. A. wightii, A. jacquemontii* Blume and *A. leschenaultii* Blume. Locally *Arisaema* is known as *sarp chhalee, samo-ki-khumb, kiralu, kiri -ki –kuri* or cobra lilies (for all types) and *chamus* (for edible types).

## *A. triphyllum (L.) Torr.*

*A. triphyllum* (L.) Torr. = *A. acuminatum* Small = *A. triphyllum* L.,= *A. atrorubens* (Aiton) Blume. It is commonly refered as Indian wild turnip in which brownish purple hooded flowers may be seen in the shady depths of the woods in early summer. It is a perennial plant, reaches a height of from 30-90 cm. Corm is somewhat globular and shaped like a turnip (Collett, 1971). The lower part of the corm is flat and wrinkled, while the upper part is surrounded by coarse, wavy rootlets. Leaf 1-2, unfold with the flowers; borne on long, erect, sheathing stalks, consist of three smooth, oval leaflets, the latter are 9-18 cm long, and from 3-9 cm wide, net veined, and with one vein running parallel with the margins. The "flower" is curiously formed, somewhat like the calla lily. Spathe within which is inclosed the spadix. The spathe is an oval, leaflike part, the lower portion of which, in the flower under consideration, is rolled together so as to form a tube, while the upper, pointed part is usually bent forward, thus forming a flap or hood over the tube shaped part which contains the spadix. The corm when first dug is intensely acrid, but drying and heat diminish the acridity. It loses its acridity rapidly with age. The partially dried corm is used in medicine (Handa and Kaul, 1997).

## *A. tortuosum (Wall.) Schott*

*A. tortuosum* (Wall.) Schott = *A. tortuosum* Wall *A. tortuosum* Schott this plant grows in the highlands where there is heavy rainfall. It is distributed in temperate and sub-tropical Himalaya from Shimla to Bhuttan, Khasia hills and Manipur at higher elevation (3000 m). Leaf 2-3; petiole 30-90 cm, green or with the sheaths mottled with purple; leaflet 5-18; 24-54 cm long, distant or crowded, or lmost radiately disposed, sessile or petiolulate, ovate or linear-lanceolate, subcaudately acuminate. Peduncle 60-120 cm; spathe 12-18 cm, pale green or purplish; tube sub-cylindrical, gaping gradually dilating into the limb. Spadix uni-or bisexual; appendage like a rat's tail, quite smooth. Tuber spheroid, attaining 15 cm dia.

## *A. tortuosum var. curvatum (Roxb.) Engl.*

*A. tortuosum* var. *curvatum* (Roxb.) Engl. = *A. neglectum* Schott. = *A. tortuosum* var. *curvatum* (Roxb.). It flowers in June and July and distributed throughout sub-tropical Himalaya and locally known as *Kiriki kukri* (Collett, 1971). Usually monoecious but sometimes paradioecious. Leaf 1-2, pedate; leaflets 11-13, linear to lanceolate. Spathe longer than the foliage, green, ribbed; tube 13-6 cm long, cylindrical, gradually broadening into an ovate, caudate-acuminate limb, larger than the tube. The shape of the leaflets varies from linear-lanceolate to lanceolate. Spadix appendage 9-21 cm long, stout, at first curved and then prolonged upright beyond the limb, black when dry. In monoecious forms the male zone longer than the female. Stamens stalked; anther dehiscence porose, bright yellow to white. Ovary ovoid, c. 1.5 mm long; style short, stigma discoid, papillate. Berry 6-7 mm, broad. Seeds 4-5, angular-ovoid, testa rugose, coriaceous, and yellowish-brown.

## *A. jacquemontii Blume*

*A. jacquemontii* Blume= *A. cornutum* Schott, = *A. exile* Schott. Paradioecious. Bulb subglobose, 3-4.5 cm broad. Leaf 1-2; leaflets 5-7, elliptic-ovate or elliptic-lanceolate, caudate-subacuminate, subsessile, cuneate, subentire, wavy to crenulate; petiole up to 30 cm long. Spathe green, white-streaked, as long as or longer than the foliage; tube cylindrical, 6-9cm long; limb curved, oblong to oblong-ovate, slightly shorter than the tube, tail 4.5-9 cm. Spadix included in the spathe; appendage rather stout, curved, base dilated, truncate (in male) or cuneate (in female); in male, spadix 3-6 cm long (in female 7.5-12 cm at post anthesis), darker towards the tips. Male flowers: stamens stipitate, anthers subglobose, 1mm broad, dehisce by longitudinal slits or by an apical pore. Female flowers: ovary subglobose, 1.5-2 mm long, style subsessile, stigma disciform. Berries globose, 4-4.5 mm broad, red when mature. Seeds 3-8, subglobose, up to 2.5 mm broad, reddish brown, acutish at one end. It flowers in June and locally known as *chamus* in the hills and consumed by the people at its early growth stage when corms are not fully mature. The corm is boiled in water, made into paste and small balls are prepared which are generally galloped as such after dipping in green chutney.

Other edible species are *A. atrorubens* Blume. - The shredded roots and berries are eaten by hill tribes. The starch of the root is delicate and nutritious. It must, however, be obtained from the root by boiling in order that the heat may destroy the acrimonious principle. *A. costatum* Mart. - this is said to be a large aroid, called *ape* in N-E region, which is frequently planted in dry ground. It is considered inferior to taro. *A. curvatum* Kunth - the Lepchas of N-E region prepare a food called *tong* from the tuberous root. The roots are buried in masses until acetous fermentation sets in and are then dug, washed and cooked, by which means their poisonous properties are in part dispersed, but not entirely, as violent illness sometimes follows a hearty meal of *tong*.

## Asparagus

They range from herbs to somewhat woody climbers. They have flattened stems (phylloclades) that serve the function of leaves. In the Himalayn region, it is known as *sansfali, sansfai, satavar, satavaree, safed musli* etc. The best known member of the genus is the vegetable asparagus (*Asparagus officinalis*) which is extensively used for its tender shoots. There are a few other species like *A. racemosus, A. filicinus* and *A. adacendense,* praised for their tuberous roots in temperate regions of the world. Apart from its edible uses, tuberous roots of *Asparagus* have immense potential in pharma industry. Local herbal vendors use it to treat diseases related to function of reproductive organs, kidneys, stomach and lungs. It is popular as 'Female Rejuvenative', as it is helpful in cases of low milk production, low sex drive, menopause and infertility. It is an important ingredient of the drugs used to treat cough, dehydration, diarrhea, dysentery, fevers (chronic), hematemesis, herpes, hyperacidity, stomach ulcers etc. (Chopra, *et al.* 1986, Anon., 1952, Handa and Kaul, 1997).

*Asparagus* belongs to family Asparagaceae. It comprises of about 300 species, all from the Old World, introduced in many countries in both hemispheres and throughout temperate and tropical regions (Malcomber & Sebsebe, 1993, Fellingham & Meyer, 1995). Many species from Africa are now included in the genera *Protasparagus* and *Myrsiphyllum*. Three species (*Asparagus officinalis, A. schoberioides* and *A. cochinchinensis*) are dioecious species while the others are hermaphrodites (Jessop, 1966). About 18 species occur in India (Hooker, 1978).

### *A. racemosus Willd.*

*A. racemosus* Willd. = *A. floribunda 'Kunth'.* It generally occurs in broad-leaved forests along streams or valleys and essentially a temperate plant growing at mid to higher elevations (1,200 - 2,000 m). It is a hermaphroditic sub-shrubby weak creeper like thorny plant, greenish with numerous underground white tuberous roots (Collett, 1971). Stems climbing, branched, up to 2 m; branches usually distinctly striate-ridged, ridges cartilaginous denticulate. Cladodes in fascicles of 3 -6, linear, flat, midvein distinct. Leaf spurs spinescent; spine straight or sub-recurved, 1.5-4.5 cm on main stems, 5-10 mm on branches, woody, sharp. Inflorescences developing after cladodes, axillary, each a many-flowered raceme or panicle, 1.5-6 cm; bracts 1 mm. Pedicel 1.5-3 mm, slender, articulate at middle. Perianth campanulate, 2-3 mm. Stamens equal, 0.7 mm; anthers yellow, minute. Flower in September and red berries matures in October-November. Tubers can harvested in winters when plants enter into dormancy. Root preparations are diuretic and laxative. Folk medicinal use of asparagus root includes treatment of heart disease, hypertension, and rheumatism (Leung and Foster, 1996; Wren, 1988). Immune stimulant effects of roots may benefit cancer patients receiving chemotherapy (Thatte and Dahanukar, 1988).

### *A. filicinus Buch.-Ham. ex D.Don*

*A. filicinus* Buch.-Ham. ex D.Don = *Protasparagus filicinus* (Buch.-Ham. ex D.Don) Kamble *Asparagus filicinus* D. Don. Occurring in the forests, thickets, shady and moist places along valleys at an elevation of 1,200-3,000 m from Kashmir eastwards to Bhutan (Collett, 1971, Anon., 1952). The two botanical varieties viz. *A. filicinus* var. *giraldii* C. H. Wright; *A. filicinus* var. *megaphyllus* F. T. Wang & T. Tang are very common. Botanically, it is a dioecious herb having rootstock with a cluster of fusiform roots swollen at base. Stems much branched, 30-60 cm, subsmooth; branches generally striate-ridged, ridges sometimes slightly cartilaginous denticulate. Cladodes in fascicles of 5-8, linear, falcate, flat, midvein distinct. Leaf spur short, not spinescent. Inflorescences developing with cladodes or very soon afterward. Flowers of both sexes solitary or paired; pedicel 1-2 cm, articulate near middle. Perianth pale green or sometimes tinged with pale purple, campanulate, 2.5 mm; filaments free; anthers ovate, 8 mm. Berry deep green, 5- mm in dia., 2- or 3- seeded. Flower in the month of July-Aug and tubers mature in October-November.

### *A. longiflorus Franchet*

Occurring in forests, thickets and grassy slopes at elevation ranging from 2,400-3,300 m. It is a deciduous herb with slender tuberous roots. Stems suberect, 30-150 cm, slightly striate-ridged distally; branches distinctly striate-ridged, ridges usually cartilaginous denticulate, particularly on young branches. Cladodes in fascicles of 4-12, straight, subterete, slightly flattened, irregularly grooved, usually cartilaginous denticulate. Leaf spurs slightly spinescent; spine 1-5mm on main stems, short on branches. Inflorescences developing after cladodes. Flowers of both sexes paired; pedicel usually 6-13 mm. Perianth purplish, 6-7 mm; Berry red, 7-10 mm in dia., 4-seeded. Flower in April-May and fruiting in June-August.

Besides, there are other species, which are occurring in the temperate region but at slightly lower elevations 800-1,200 m valued for their tuberous roots. *A. capitatus* subsp. *gracilis* Browicz = *A. gracilis* Royle ex Baker =*A. royleanus* P. Daniel is a tall, much branched, lower leaves subspinescent, cladodes 3-6-nate or lower 8-nate acicular or setaceous strict, pedicels axillary or subterminal 2-4-nate, perianth segments spreading. *A. capitatus* (L.) Baker= *Protasparagus capitatus* (Baker) Kamble is slender, much branched, and subscandent; branches smooth, leaves subspinescent, cladodes solitary or 2-3-nate acicular or setaceous firm, flowers terminating short branchlets, pedicels very short, perianth segments spreading. *A. fysonii* J.F.Macbr., = *A. subulatus* Steud. ex Baker is endemic to Nilgiri hills. It is a small erect or tall and variable bush, densely branched; upper internodes very short. Cladodes ascending, spreading and curved, rather stout, variable in breadth. Pedicels jointed above the middle. Perianth segments oblong-lanceolate; anthers oblong. *A. curillus* Buch.-Ham. ex Roxb., =*A. nepalensis* Baker, subscandent, stem robust smooth, branches angled, not grooved, spines stout recurved, cladodes 3-6-nate, acicular spreading or ascending triquetrous acute, racemes short 1-3-nate sometimes branched below, perianth campanulate. *A. adscendens* Roxb. – is a most common species occurring in the subtropical parts ascending to as high as 1,800 m from Punjab eastwards to Kumaon. It has excessively branched stem, flexuous, and primary branches ashy white; internodes very short. Spines stout, straight. Cladodes densely crowded, soft. Racemes copious, 3-6 cm, often branched at the top; pedicels jointed above or below the middle; flowers 0.5 cm diam. Ovules many in each cell. Berry 0.75 - 1 cm dia., 1- seeded. This species is also valued in religious ceremonies and whole plant including roots is worshiped in the marriages in Shivalik hills.

## Bunium

It belongs to family Apiaceae and 30 species are known to be distributed in Mediterranean region, North Africa, South & Middle Europe, Middle East, Iran, Afghanistan, Central Asia, West Pakistan. About four species viz. *Bunium alpinum* Waldst. & Kit. (North Africa and Europe), *Bunium elegans* (Fenzl)

Freyn, Bunium *paucifolium* DC. (both in West Asia, Caucasus) and *Bunium persicum* (Boiss.) B. Fedtsch. (Hindu-Kush region) are used for their seed and tubers (Atkinson, 1980). It is a perennial, glabrous herb. Roots globose, brownish black papery outside, milky white inside. Leaf much divided, 2-3-pinnatisect; segments filiform to linear. Flowers white to pinkish. Bracts and bractlets linear to lanceolate or lacking. Calyx teeth obsolete. Stylopodium flattened or conical; styles short or long, reflexed. Fruit ovoid to oblong or cylindrical; ridges prominent; furrows 1-3-vittate. Inner seed face plane to slightly concave.

### *B. persicum (Boiss.) B. Fedtsch.*

*B. persicum* (Boiss.) B. Fedtsch. = *Carum bulbocastanum* L. = *Cuminum nigrum* L. Local names for that spice are *kala jeera*, meaning "black cumin") or *shahi jeera* (meaning "imperial cumin"), *himali jira* ("Himalayan cumin") or *kashmiri jeera* ("mountainous region of Kashmir") (Handa and Kaul, 1997). The plant grows wild in a wide range in specific habiats that are available in the cold arid region of India especialy in J&K and HP at an alttitude of 2,500-3,000 m (Singh and Kachroo, 1976). It reaches about 60 cm tall and 30 cm wide, bearing frilly leaves and hermaphroditic flowers, pollinated by insects and self-fertile. Root tuberous, Stem 18-90 cm, erect, branched. Leaves 2-3 pinnate, finely dissected, ultimate segment of all the leaves linear, bracteoles 0-8, linear; Rays 8-20, petals 1 mm long. Fruit pedicels thin, 5-10 mm long, fruit yellowish- brown, oblong not narrowed upwards. Carpels exactly oblong, hardly widened at the middle. This species is primarily prized for its highly aromatic seeds which are used as a culinary spice in the Hindu-Kush region. Tuberous roots are also eaten by the hill tribal in scarcity times. Tubers raw or cooked gives delicious taste, which is very much like sweet chestnuts when cooked, but the tubers are very small and fiddly to harvest. In Kinnaur region, they are eaten after roasting in the fire (buried for sometimes under the burning coals like potato/sweet potato) with salt and green chutney. Medicinally seeds are astringent (Anon., 1952; Chopra *et al.*, 1986). Leaves also used as a garnish and flavouring in much the same way as parsley.

## Campanula

This is a group of annual, biennial and perennial plants, which are found naturally in many countries of the Northern Hemisphere. Campanulas, commonly called Bellflowers, vary in size and habit of growth from dwarf arctic and alpine species under 6 cm high, to large temperate grassland and woodland species growing to 2 m tall (Atkinson, 1980; Long, Chun-Lin, 1990; White, 1991). The leaves are alternate, sessile, and often vary in shape on a single plant, with larger, broader leaves at the base of the stem and smaller, narrower leaves higher up; the leaf margin may be either entire or serrate (sometimes both on the same plant). Many species contain white latex in the leaves and stems. The flowers are produced in panicles (sometimes solitary), and have a bell-shaped, five-lobed corolla, typically large (6-12 cm), mostly blue to purple, sometimes white or pink. The fruit is a capsule containing numerous small seeds. The tuberous roots are cooked and recommended as famine food. It belongs to family Campanulaceae that comprised of about 300 speceis and several subspecies, distributed across the temperate regions of the world, with the highest diversity in the Mediterranean region (Shimizu, 1993). About 13 species are occurring in India (Hooker, 1978). Important speceis that are valued for its tuberous roots include *Campanula rotundifolia* L. *C. rapunculus* L., *C. persicifolia* L., *C. punctata* Lam., *C. alsinoides* Hook.f. & Thoms, *C. aristata* Wall., *C. cashmeriana* Royle, *C. pallida* Wall., *C. alpina* Jacq., *C. saxifraga* M.Bieb., *C. pyramidalis* L. and *C. urticaefolia* L.

### *C. alsinoides Hook.f. & Thoms*

It is distributed in the North- West Himalaya at 2,500-3,400 m. Stem 12-24 cm, slender, hairy, exceedingly numerous from a woody root. Stem leaves thin, herbaceous; leaves oblong-ovate subpetioled, obscurely toothed, pilose on both surfaces. Peduncles 0.75-1.5 cm, slender. Calyx teeth small narrowly lanceolate, entire or somewhat toothed. Capsule small subglobose.

### *C. aristata Wall*

It is distributed in the Alpine Himalaya from Kashmir to Sikkim at 3,500-4,000 m (Collett, 1971). Plant glabrous. Stem 24-72 cm, erect, simple. Radical leaves elliptic, long petioled, subentire; cauline leaves linear. Peduncle 6-30 cm; flowers solitary; calyx teeth linear; corolla long, broad, conic, lobed half way down, deep blue. Capsule narrowly oblong or sublinear, contracted near the summit. Seeds very small, ellipsoid.

### *C. rotundifolia L.*

Occurring at an elevation of 2,500-3,500 m is a flat, glabrous, slender, short to medium perennial with 3 cm blue flowers arising to 18 cm above the dark green foliage. Leaves are roundish and wither early. The leaves on the stem are linear, the upper ones being unstalked. The flowers are blue, 15 mm long and on long thin stalks and in loose clusters. The petal lobes are short. Harebells flower between July and October and are found on dry grassland and heaths in cold regions (Shimizu, 1993, Namhata and Mukherjee, 1992). It is also a well known and favourite wild flower, which can be easily spotted in short grassland and roadside verges in summer when it becomes covered in thimble-sized blue bells.

## Chaerophyllum

This is essentially a plant of temperate origin. It is herb, glabrous or hairy. Leaves pinnately decompound, or 2-pinnate with pinnatifid pinnules; umbels compound; bracts 1-2, linear, bracteoles 3-6, lanceolate, or linear; flowers white, usually polygamous, the central subsessile fertile, others pedicelled male or hermaphrodite; calyx-teeth obsolete and petals emarginated (Sdobnina and Pimenov, 1991). Fruit oblong, narrowed upwards, glabrous, laterally compressed, narrowed at the commissure; primary ridges broad, obtuse, occasionally prominent; carpels terete, inner face with a deep T-shaped groove; carpophore stout, undivided or shortly 2-fid; disc not prominent on the fruit. Seed terete, inner face with a deep narrow groove, widened into a hollow in the centre of the seed. Genus is widely known as wild carrot (*Jangli gajar*) in the hills. It belongs to family *Apiaceae* and comprised of about 40 species distributed in the North Temperate regions, chiefly in Europe and Asia. About six species are occurring in India and four are know to be edible for tuberous roots. The root is edible only at the end of fall when root reserves have undergone a biochemical change favored by cold temperature and eaten only after brief (15-20 min) boiling and after peeling.

### *C. villosum Wall. ex DC*

It is distributed in the Himalayan region expanding from Kashmir to Sikkim at an elevation of 1,800-3,500 m, mostly in moist shady places in the forests, road sides or open grassy places. Roots fusiform; stem 30-120 cm, erect, branched, always hispid below, and often hairy upwards. Leaves large, pinnate or pinnatifid, ultimate segments scarce 0.75 cm, with 2-4 lanceolate teeth, often with scattered hairs; segments of the lowest often, pinnatifid scarcely half-way down; upper sheaths usually inflated, hairy, with prominent white margins. Bracts absent; rays 3-6, 3-9 cm long, stout; bracteoles, linear, often white margined, soon reflexed; pedicels 10-15. Fruit 1 cm, equally broad its whole length, narrowed somewhat suddenly at the tip. Leaves 2-3-pinnate; segments finely divided. Flowering and fruiting occur in July–August.

### *C. aquilegifolium Rech. f. & Riedl.*

Plants slender. Stem glabrous. Root globose, 6-10 mm in diameter. Leaves petiolate, bipinnate; segments oval to ovate, up to 1.5 cm long, 5-8 mm broad, lobed; leaf bases sheathing, pubescent; upper leaves sparsely pilose. Involucral bracts lacking. Rays 5-6, up to 3 cm long. Involucel of greenish white, oval to ovate bractlets, 2 mm long. Pedicels 5-8 mm long. Ovary and fruit densely hairy; styles 1 mm long, white, sub-erect to horizontal.

### *C. reflexum Lindl.*

Plants up to 1 m tall. Stem glabrous to sparsely hairy; hairs deflexed, white. Root elongated. Leaves bipinnate, glabrous to pubescent; pinnae finely divided. Involucral bracts lacking. Rays 6-12. Involucel of 5-8 lanceolate bractlets. Pedicels of imperfect flowers reflexed. Fruit 1-3 in an umbellet; central fruit subsessile; fruit cylindric to ellipsoid, beaked, 7-12 mm long; styles short; ridges prominent; furrows 1-vittate. Inner seed face sulcate.

### *C. acuminatum Lindl.*

Plants up to 60 cm tall, pubescent. Roots fusiform. Leaves 1-2-pinnate; pinnae linear to lanceolate; margins regularly serrate. Involucral bracts lacking. Rays 7-10, unequal. Involucel of 5 lanceolate bractlets with long, white marginal hairs; apex acuminate petals white. Fruit cylindrical, 8 mm long; stylopodium conical; styles short; ridges conspicuous; furrows 1-vittate. Inner seed face deeply sulcate.

## Codonopsis

The most recent classification of the genus (Hong, 1983) recognized three subgenera: *Codonopsis*, *Obconicicapsula*, and *Pseudocodonopsis*. It is allied to *Campanumoea* and *Leptocodon*, and some authors suggest that *Codonopsis* should include these genera (Morris and Lammers, 1997). According to Yoo and Lee (1989) it is perennial plant with tuberous roots; scandent, twining, or less commonly erect or ascending herbaceous stems. Leaves alternate or subopposite, petioled. Peduncles axillary or terminal, 1-flowered. Flowers conspicuous, green, blue, yellowish or lurid, often purple veined, 5- rarely 4-6-merous. Calyx superior; lobes 5, long, foliaceous. Corolla campanulate, shortly 5-lobed. Stamens 5, free, inserted on the margin of an epigynous disc; filaments somewhat dilated at the base; anthers oblong, free. Ovary at first inferior, turbinate, with a short conic superior vertex, 3-celled; style cylindric, stigma of 3 ovate or oblong lobes. Capsule somewhat fleshy, ultimately dry; beak elongated in fruit, loculicidally 3-valved. Seeds very many, small, ellipsoid, slightly compressed, smooth, brown, solitary or paired. It belongs to family Campanulaceae. It comprises of 55 species most of which are endemic to eastern Asia to Afghanistan, Pakistan, the Himalayas, southern China, and Taiwan (Hong, 1983; Shimizu, 1993). *Codonopsis* is one of the most famous and widely used Chinese Tonic herbs. There are eight species used for their tuberous roots in the Indian Himalayn region. It can be dried and ground into flour. A famine food, used when all else fails. The roots and leaves are used to make a poultice for the treatment of bruises, ulcers and wounds.

### *C. viridis Wall*

It is distributed from Kumaon to Nepal at an elevation from 1,500-2,500 m. Roots tuberous with twining stem. Leaves opposite and alternate, base obtuse or slightly rounded but not cordate, subentire, minutely pilose above, at least when young; petiole 12 cm. Peduncles 6 cm. Calyx-teeth not (or very obscurely) widened upwards, pilose on both surfaces. Corolla lurid yellow. Capsule hemispheric, broad; beak long. Testa with very small and obscure reticulations.

### *C. griffithii Clarke*

Endemic to Khasi hills and occurs at 1,200-1,500 m. Roots tuberous, stem twining, leaves hairy, elliptic-oblong from a rhomboidal or truncate base densely pilose beneath, petioles have crisped or patent hairs; calyx-teeth distant linear broader upwards, peduncles axillary, patently pubescent, corolla in bud glaucous. Young corolla glabrous or occasionally with a few long white hairs near the top; the testa is much more coarsely and prominently reticulate, so that the seeds glisten under the pocket-lens.

### *C. affinis Hook. f. & Thompson*

Occurring in Sikkim Himalaya up to an elevation of 3,500 m. Roots thick and tuberous. Stem twining, leaves 2-4, opposite and alternate, subentire, sparsely pilose above, at least when young; petiole 3-6 cm Peduncles 3-15 cm, often dichotomous with a bractiform leaf, so that the inflorescence appears of few flowered racemes. Corolla long and broad, green with purple marks. Capsule hemispheric with long beak. Seeds not glistening under the lens, i.e. the reticulations of the testa are very small, not prominent.

### *C. rotundifolia Wall*

Widley distributed in Western Himalaya from Kashmir to Kumaon, at an elevation of 2,000-3,800 m. Roots tuberous; stem twining; leaves alternate, very variable in size; petiole ¾". Peduncles 3 -18 cm, axillary and terminal. Calyx -lobes often crenate or toothed. Corolla, purple or lurid grey-blue. Capsule hemispheric broad. Seeds minutely reticulate, not glistening, more than twice as large as those of *C. benthami* and its allies.

### *C. benthami Linn*

Occurring in Sikkim Himalaya up to an elevation ranging from 3,000-3,500 m. Stem rambling or subscandent, leaves nearly as of *C. rotundifolia*, but less rounded at the base. Calyx-teeth ovate-acute; corolla lurid purple-yellow. Capsule hemispheric broad; beak long. Seeds not glistening. Separable from C. *rotundifolia* by the more woody rambling, but not twining, stem, and seeds short ellipsoid scarcely 0.3 cm long.

### *C. subsimplex Hook. f. & Thompson*

Endemic to Alpine pastures of Sikkim Himalaya at an eleavtion of 3,500-3,800 m. Stem 30-60 cm., branched. Leaves 6 x 3 cm, alternate and opposite, closely resembling those of C. *rotundifolia*; petiole 1.5 cm. Peduncles 3-12 cm, terminal. Calyx-lobes 1.5x 0.75-1 cm, entire or slightly toothed. Corolla 2.25 cm long and nearly as broad, lurid pale-blue. Seeds not glistening, about 1/3$^{rd}$ the volume of those of C. *rotundifolia*, twice as large as those of C. *benthami*. This differs from C. *benthami* chiefly by the shorter much broader corolla.

### *C. foetens L.*

Distributed in the alpine pastures of Sikkim Himalaya at an eleavtion of 4,000-4,200 m. Closely resembling C. *thalictrifolia* except as to the corolla. Stems 18cm, flexuose. Leaves 0.75-1 cm dia. Peduncles 9-18 cm, terminal. Calyx-lobes 1 x 0.5 cm, approximate at base, minutely pilose. Corolla 3 x 2.5 cm. Capsule depressed obconic, 1 cm broad; beak nearly 1.5 cm long, more than double the length of the portion of the capsule below it. Seeds ellipsoid, not glistening. Separated doubtingly from C. *thalictrifolia*, but the corolla is very different, and the calyx-teeth much broader at the base. It scarcely differs from C. *ovata* by the smaller leaves, more globose corolla and smaller seeds.

### *C. ovata Benth.*

It is occurring in Western Himalaya from Kashmir to Garhwal at an elevation of 2,000-3,200 m. Root woody, fusiform, large. Stem 18-36 cm, decumbent, then erect. Leaves alternate and opposite, obtuse or acute. Petiole 0.25-0.50 cm. Peduncles 9-18 cm, terminal. Calyx-lobes approximate at base, minutely pilose. Corolla widened upwards, sky-blue. Capsule depressed obconic, broad. Seeds much longer than those of C. *foetens*, not glistening, narrowly ellipsoid, 0.2 cm long.

## Crambe

It is primarily a genus of leafy vegetable and oil seeds but one species C. *cordifolia* known for thick and swollen root stocks Usually stout, perennating herbs (rarely annual) with large thick tap root; stem erect, branched, glabrous or hairy (sparsely hispid) with simple hairs. Lower leaves often large, lyrate or simple, rounded, long stalked, irregularly toothed, glabrous or hairy; upper leaves usually small, linear, sometimes absent or few. Racemes often branched, many flowered, ebracteate. Flowers mediocre, usually white, often long pedicelled. Sepals oblong, open, subequal, inner slightly saccate at the base. Petals obovate, narrowed at the base into a short claw. Stamens 6; filaments of the inner stamens usually with an erect, spreading tooth-like appendage. Lateral nectar gland hardly conspicuous, middle short and semiglobose. Ovary 2-parted, lower part very short, sterile, upper part globose, bilocular with 1 ovule in each cell; stigma depressed capitate; gynophore very short but distinct, as broad as the stalk-like lower part of the fruit. Fruit biarticulated; upper part globose or ovoid, much larger than the lower part, indehiscent, 1-seeded, glabrous; seed subglobose, brown; cotyledons longitudinally folded, truncate. It belongs to family Brassicaceae and it comprises of about 28 species in world (White & Marie, 1978), out of which seven species available in India (Anon., 1952).

### *C. cordifolia Steven*

Largely distributed in the western Himalaya bordering western Tibet and occurring at an elevation ranging from 3,000-3,500 m. A tall stout bristly hairy perennial. Stem striated. Rootstock thick and swollen. Radical leaves on stout petioles, 3-6cm, petiole striated and dilated at the base, rounded and cordate, laciniate and variously toothed; cauline on shorter petioles, Ovate elliptic or rhomboid, toothed or lobed. Racemes loosely panicled. Pods globose, upper part pisiform, crustaceous. Tuberous roots are edible and used in adversities.

## Dioscorea

Yam is the common name for members of this pan-continental genus, found in Africa, India, Southeast Asia, Australia and tropical America (Jayasurya, 1984; Wilkin, 1998), with about 630 scientifically described taxa. Prain and Burkill (1936) reported the occurrence of about 50 different *Dioscorea* spp. in India, largely in the west, east and northeast regions. *D. deltoidea* is highly priced species of the temperate region due to its immense medicinal potential. Generally collected form the forests but with the increasing demand, some domestication attempts have also been made. Many of the *Dioscorea* species serve as a 'life saving' plant group to marginal farming and forest dwelling communities, during periods of food scarcity. They are used in a similar fashion to potatoes and sweet potatoes (Alexander and Coursey, 1969). There are hundreds of cultivars (varieties) avaialble among the cultivated species. The yam has a rough skin which is difficult to peel, but which softens after heating. The majority of the yam is composed of a much softer substance known as "meat". Yam tubers can be stored for four to six months without refrigeration, which makes them a valuable resource for the yearly period of food scarcity at the beginning of the wet season (Atkinson, 1980 and Watt, 1908. The stem-tubers or tuberous rhizomes of many species, known as "*Yams*", mostly contain starch and are therefore cultivated for food in tropical regions of the world, especially of America. They can be propagated easily either by the bulbils present in the axils of the leaves or by `Eyes' as in potatoes (Alexander and Coursey, 1969)

Dioscoreas are generally large perennial herbs with twining annual stems arising from tubers. Leaves alternate or opposite, simple or palmately compound, cordate-sagittate or hastate, multicostate with reticulate and divergent venation, often bearing bulbils in their axils; petioles long, twisted at the base. Flowers small, in spikes or racemes, unisexual, actinomorphic. Male flowers: perianth 6-cleft; segments short, spreading, in 2 whorls; stamens 6, all antheriferous or some times only 3 bearing anthers; filaments free or united at the base, anthers small, incurved, 2-celled, dehiscing longitudinally; pistillode thick and fleshy or absent. Female flowers: perianth of 6, small, free, persistent tepas; staminodes 3, occasionally absent; ovary inferior, 3-quetrous, trilocular, 2 superposed ovules suspended from axile placentas in each locule; stigmas reflexed. Fruit a 3-winged loculicidal capsule. Seeds flattened, often winged.

The taxonomy of quite a few species in this genus is considered problematic (Prain and Burkill 1936; Velayudhan et al., 1998), which is attributed to its highly continuous variability of morphological characters, especially of aerial parts, such as leaves and bulbils. This continuous variation makes it difficult for 'Linnaean taxonomists' to segregate distinctly the various taxa of this genus (Wilkin, 1999 and Jayasurya, 1984). The Linnaean system uses character classes pertaining to that of floral, fruit and seed (reproductive) characters as well as the direction of twining of the stem. The size, shape and nature of tubers, and sometimes of bulbils are also used as important taxonomic characters. It belongs to family Dioscoreaceae and comprised about 600 species, found in tropical, sub-tropical and few in temperate regions of the world. About 25 species are occurring in India (Hooker, 1978) and a few occurring in the temperate region are discussed as follows.

### *D. pentaphylla L.*

*D. pentaphylla* L. = *D. triphylla* L= *D. digitata,* Mill.: It is highly versatile species and occurring from sea level to as high as 1,500 m from Kumaon in the northwestern region, eastward to Burma and southward to Sri Lanka (Collett, 1971). Locally known as *tardee, husa, gazaria, kanta alu* etc. It is a most common species that is collected by the people from forest and sometimes from semi-domesticated habitats. Its tubers oblong, 150-180 cm. Stem slender, prickly at the base, rarely above, often bulbiferous. Leaves usually snow white beneath, obovate, acuminate or cuspidate. Male panicles and flowers glabrous hispidly pubescent or villous, spikes lax or dense-flowered; flowers sessile or pedicelled, 0.2-0.25 cm dia, fragrant; filaments and staminodes very short. Capsule 2-3 cm, rounded at both ends or base cordate and tip apiculate, glabrous or pubescent. Seeds 1-3 cm, wing broader than the nucleus.

### D. kumaonensis Kunth

Occurring in the temperate Himalaya at an elevation of 1,200-2,000 m from Kashmir to Sikkim. Morphologically, very near to *D. pentaphylla*. Stem glabrous, slender, twining clockwise. Tuber small, globose. Leaves alternate, 3-5-foliolate, glabrous; petiole 1.5-7.5 cm long, leaflets 4.5-19.5 cm long, 1-4.5 cm broad, narrowly lanceolate, finely acuminate; petiolule 1-5 mm long, bulbils black, coriaceous. Male racemes pubescent, 3-9 cm long, axillary, solitary or sometimes 2. Antheriferous stamens 3, alternating with 3 staminodes. Female spikes pubsecent, solitary, axillary, 3-6 long. Capsule 2.25 cm long, 0.75 cm broad, more or less truncate at the apex, rounded at the base. Seeds winged at the top.

### D. deltoidea Wall. ex Griseb.

*D. deltoidea* Wall. ex Griseb. = *D. nepalensis* Sweet ex Bernardi. It is distributed from Kashmir to Bhutan ranging from 2,000-3,000 m (Collett, 1971). Branches very slender, terete, unarmed. Leaves attaining 21x15 cm, membranous, reticulate beneath, very variable in breadth and depth of lobing, lobes rounded or subangular, sometimes dilated outward; petiole 6-15 cm, very slender. Male spikes very slender, rarely branched; flowers 0.25 cm diam., solitary or clustered; perianth-segments nearly flat; stamens very short; anthers didymous. Capsule very variable, orbicular, deltoid or obtusely quadrate, rarely broadly obcordate, base rounded truncate or broadly cuneate; Seeds very variable in form, winged unequally all round. The tubers contain saponin, acrid resin, diosgenin, starch and calcium oxalate. They are used for washing shawls and woollen cloth as well as a vermifuge and an anthelmintic for purging out intestinal worms. Diosgenin is said to be a basic material for hormone preparation. It is locally known as *singli-mingli, kniss, kriss, tar, kitra, kildri, krish* etc. (Anon., 1952).

### D. pubera Blume

*D. pubera* Blume = *D. anguina* Roxb.= *D. cornifolia* Kunth. Widely distributed in the sub-temperate areas from Kashmir to Bhutan at an elevation ranging from 800-1,200 m. Tubers small, sub-fusiform, stems glabrous, twining anticlockwise. Leaves ovate, rather thick, nerves strong beneath, nervules transverse; opposite and alternate, large, long-petioled, broadly ovate-orbicular, cordate 7-nerved margin cartilaginous. Male spikes short dense-flowered. Female spikes 6-15 cm long, solitary or 2 together, axillary. Capsule 3 cm long, 4.5 cm broad, obovate with a retuse apex or obcordate. Capsule cordate at the top and base. Seed orbicular, softly tomentose. The tubers are prized for culinary purposes and sold in bazaars under the name of *Turar*.

### D. glabra Roxb.,

*D. glabra* Roxb., = *D. hongkongensis* Uline ex R.Knuth= *D. glabra* var. *vera* Prain & Burkill. Occurring in sub-tropical or mid-Himalayan region at an elevation of 1,000-1,800 m from Shimla to Sikkim. Morphologically, stem is stout, somewhat flattened. Leaves extremely variable, caudate-acuminate, youngest acute at the base, older truncate, or deeply cordate, lobes sometimes 3 cm incurved, and overlapping; 7-9-nerved, margin not thickened or cartilaginous; petiole 3-12 cm. Male spikes 3 cm, rarely more; flowers scattered, rather large, globosely 8-lobed, often coarsely dotted. Capsule very variable in shape, subquadrate broadly obcuneate or obcordate, retuse at the tip and base. Seeds irregularly orbicular, quite glabrous.

### D. hamiltoni Hook.f.

It is also a species of sub-tropical Himalayan region and occurring at an elevation of 500-1,200 m and distributed in Sikkim, Assam and Chota Nagpur area. Its branches are slender, leaves opposite and alternate, undistinguishable from those of *D. glabra*. Petiole 3-6 cm. Male spikes whorled on the branches of very slender elongated axillary and terminal panicles. Flowers globose . Sepals broad; petals cuneate- ovate. Capsule 3-6 cm long.

### D. bulbifera Linn

*D. bulbifera* Linn= *D. sylvestris* De Wild.= *D. crispata* Roxb., A plant of the foothills ascending to about 1,000-2,000 m in the tropical and sub-tropical parts of Asia & Africa. Tubers mostly large, globose or pyriform. Stems glabrous, twining clockwise bearing numerous brown warted bulbils c. 2.5 cm in diameter. Leaves alternate, 4-5 cm long, 2.5-13.5 cm broad, ovate, more or less deeply cordate at the base, long acuminate, 9-nerved, petiole 2.5-12 cm long. Male spikes slender, drooping, 6.5-10 cm long, clustered in the axils or sometimes panicled; stamens 6, antheriferous. Female spikes simple, fascicled in the leaf axils, 10-25 cm long. Capsule c. 2 cm long, 1 cm broad, oblong. Seeds winged at the base. It is locally known as *Ratalu, suaralu, pitaalu*.

## Helianthus

It is tall, annual or perennial, growing to a height of 60-300 cm. The rough and hairy stem is branched in the upper part. The petiolate leaves are dentate and often sticky. The lower leaves are opposite, ovate or often heart-shaped. The upper leaves are alternate and narrower. They bear several to many wide, terminal capitula (flower heads), with bright yellow ray florets at the outside and maroon disc florets inside. It belongs to family Asteraceae. This genus comprises of 67 species, all native to North America (Chabbert *et al.* 1983, Duke, 1978). However, at least the two species *H. annuus* (sunflower) and *H. tuberosus* (Jerusalem artichoke) are cultivated in other temeparte region of the world particularly the Asia and Europe. The later is well known for its tubers, whereas former is an important oilseed crop.

### *H. tuberosus* L

It is native to North America, and long used by the American Indian for food (Duke, 1979, 1981). A perennial herb often cultivated as an annual with tubers produced on the ends and branches of underground stems or rootstocks as well as midway on the rootstocks. Tubers knobby, white, red or purple skinned, ranging in size from 9-15 cm long, 3-9 cm thick; stem erect 90-240 cm tall, hirsute; leaves opposite or the upper ones alternate, ovate to ovate-oblong, serrate-dentate, rough above, with winged petiole; heads 6-9 cm across, few to many, terminal on the branches; rays 12 to 20, light yellow, conspicuous, disk yellow, seeds pubescent. It is locally known as *Hatichuk, Brahrnokha, Hathipich, balyatee aaloo*. The tubers store the carbohydrate inulin instead of starch. For this reason, Jerusalem artichoke tuber is an important source of fructose for industry (Palz, and Chartier, 1980). Tubers can be eaten fresh or raw, cooked like potato, or pickled. Tubers are also used to fatten cattle, sheep and hogs. The alcohol fermented from the tubers is said to be of better quality than that from sugar beets. It is also reported to be aperient, aphrodisiac, cholagogue, diuretic, spermatogenic, stomachic, and tonic (Chopra *et al.*, 1986). Jerusalem artichoke is a folk remedy for diabetes and rheumatism (Duke and Wain, 1981).

## Inula

In the Himalayan ranges, this genus is known as *pushkar mool*. This genus comprises of about 60 species in the world, of which 20 species are found in India (Hooker, 1978). It belongs to ffamily Asteraceae. It is herb, rarely shrub. Leaves radical, alternate. Heads solitary, corymbose or panicled, heterogamous, radiate, rarely disciform; Ray flower; Male fertile; ligule long short or minute, yellow or white, 3-toothed; Disc flower: Female fertile, tubular, yellow, limb elongate 5-toothed. Anther bases sagittate, tails long simple or branched. Style-arms of female flower linear, broader upwards, obtuse. Achenes subterete, usually ribbed; pappus hairs rather short, 1-2 seriate, few or many, smooth scabrid or bearded.

### *Inula racemosa Hook. f.*

*Inula racemosa* Hook. f. = *I. royleana* C.B.Clarke. This species is primarily used for medicinal purposes but rhizome after thorough washings can be made edible especially is adversities. The most important species used for edible purposes is *Inula racemosa.* This is most recently domesticated plant, which is under cultivation at higher elevation particularly in Lahaul & Spiti in the western Himalayan region.It is occurring throughout the Himalayan region from 1,800-3,500 m and locally known as *puskar mool.* Tall stout herb, 30-150 cm., stem grooved. Leaves coriaceous, densely tomentose beneath, crenate, elliptic-lanceolate, radical 24-54 x 15-24 cm, narrowed into a petiole as long; cauline often deeply lobed at the base. Heads many; Achenes 0.5 cm, glabrous, slender; pappus 1cm, reddish on a stout erect woolly or hirsute peduncle. The rhizome is used in Tibetan medicine, it is said to have a sweet, bitter and acrid taste with a neutral potency. It is used in the treatment of contagious fevers that have not fully ripened and pain in the upper body, especially between the neck and the shoulders. Rhizome is also valued in cardiovascular system, angina and dyspnea (Singh *et al.* 1993, Tripathi *et al.,* 1984).

## Lilium

It is an erect leafy stemmed herbs. Majority of species form naked or tunic-less scaly underground bulbs from which they overwinter. Flowers have three petals along with three petal-like sepals, often fragrant, and come in a range of colours ranging through whites, yellows, oranges, pinks, reds, purples, bronze and even nearly black. Markings include spots, brush strokes and picotees. Flowers are formed at the top of a single erect stem, with leaves being borne at intervals up the stem. It belogs to family Liliaceae and comprised of about 100 species, of which 11 species occur in India (Hooker, 1978). Bulbs of most species are rich in starch and generally used as vegetable in similar ways to potatoes mostly in scarcity times. Some species such as *L. polyphyllum* are highly valued as medicine.

### *L. nepalensis* D. Don

It is distributed in Western Himalaya from E. Asia - Himalayas from Nepal to Assam and N.E. India at an altitude of 2,300-3,500 m (Collett, 1971). Plants are often found on exposed positions in limestone rocks and in bamboo thickets. Stem 60-90 cm., slender. Leaves 9-12 cm, 5-7 nerved, lanceolate; perianth 12-15 cm, broadly funnel-shaped from the base, recurved from beyond the middle; stamens extended; anthers 3 cm, yellow. Flowers solitary or few and racemed or subumbelled, drooping, sweet-scented; tube greenish outside with a stout rounded keel; segment orange yellow with purple dots or spots on the lower half of the recurved portion within, outer segments oblanceolate subacute, inner much broader, tips rounded; filaments bright red, rather spreading; style exserted.

### *L. polyphyllum D. Don*

*L. polyphyllum* D. Don = *L punctatum* Jacquem. ex Duch. It is distributed from Kumaon to Kashmir at an altitude of 1,600-3,000 m. Bulb narrow, very deep, milky white, long narrow subequal fleshy scales. Stem 90-120cm, slender. Leaves 12-15 cm, many nerved, lower sometimes whorled, linear or oblanceolate, margins papillose. Raceme raised on the naked top of the stem, 4-10 flowered. Flowers whorled; perianth broadly infundibular; segments revolute from the middle; stamens exserted; anthers 1.5 cm long; bracts whorled; pedicels elongate, cernuous; flowers pendulous, 6 cm dia., fragrant, dull yellowish or greenish outside, white within speckled with long purple streaks; segments oblanceolate. It is locally known as *kasheer kakolee* and is an important constituent of chavanprash. The plant has become rare now in its distribution.

### *L. wallichianum Schult. & Schult*

It is distributed in Nepal to Kumaon at an altitude of 1,000-1,250 m. Bulbs small, on a creeping rootstock; scales many, short, ovate, and acuminate. Stem 4-6′, base ascending few fold. Leaves 6-12, narrowly linear; nerves 3-5 faint. Flowers subsolitary, horizontal, sweet-scented, pedicel long; tube greenish outside; segments subequal, oblanceolate, 2″ broad. Perianth 18-30 cm, narrowly, tubular, infundibular; stamens much shorter than the perianth; anthers 3 cm, orange yellow; style recurved at the top; stigma conoidal. Bulb - cooked. Boil and roasted as a vegetable. It is usually dried first. The dried bulb scales are demulcent. They are used like salep (obtained from various species of orchids, especially *Orchis* spp.) in the treatment of pectoral complaints.

### *L. giganteum Wall.*

*L. giganteum* Wall.= *Cardiocrinum giganteum* (Wall.) Makino. It is distributed in the Temperate Himalaya from Garhwal to Sikkim at 1,800-3,000 m and Khasia Hills at 1,500-2,500 m. Very tall. Scales subequal; pedicels short. Stem 18-36 cm, 6 cm dia. At base, fistular. Leaves tufted, long, lancelate, 9-12 cm dia., penninerved; Racemes many fold; perianth white narrowly funnel shaped, 30-60 cm; bracts large. Flowers 15-21 cm, fragrant, drooping; tube purplish within; segments oblanceolate, tips rounded recurved. Stamens and style included, filaments decurved at the tip; anthers 1.5 cm, yellow. Capsule 6-9 cm, obtusely angled; carpels retuse; septa pectinate. Seeds broad.

## Polygonatum

It is commonly known as Solomon's seal, distributed in the temperate regions of the Northern hemisphere, mainly in the Himalayas. It is Herb, perennial, from long, knotty, creeping, occasionally branched rhizomes; leaf scars conspicuous; roots fibrous to fleshy. Stems simple, leafy, erect to arching, glabrous or pubescent. Leaves simple, alternate [opposite or verticillate], dispersed along stem, short-petiolate, subsessile, or clasping; blade ovate or lanceolate to linear, margins entire. Inflorescences racemose, with 1–10(–15) flowers on axillary peduncles, except for distal- and proximal-most axils; peduncle glabrous [pubescent]. Flowers nodding or pendulous; tepals 6, persistent, marcescent, connate basally into cylindrical tube, distinct, tips short, valvate; stamens 6, included, adnate to perianth tube; filament surfaces various; anthers basifixed [dorsifixed], dehiscence introrse; ovary superior, 3-locular, ovoid to globose; septal nectaries present; style slender, included; stigma capitate to obscurely 3-lobed; pedicel articulate, glabrous. Fruits baccate, dark blue to black [red], globose, glaucous, pulpy. Seeds yellowish tan to brownish olive, globose or subglobose, 3–4.5 mm (Abramova,1975, Gates, 1917). It belongs to family Ruscaceae and comprised about 60 species, of which 13 species occur in India (Anon., 1952; Hooker, 1978). Some species of this genus are purported to have medcinal properties, and some (in particular *P. sibiricum*) are used as an Asian in traditional Chinese medicine.

### *P. verticillatum (L.) All.,*

*P. verticillatum* (L.) All., = *P. macrophyllum* Sweet =. *P. minutiflorum* H.Lév., It is distributed from Kashmir (1,800-3,200 m) to Sikkim (2,000-3,500 m). Stem tall, 60-120 cm, very slender, or stout, angled, glabrous. Leaves 4-8-nately whorled, green above, glaucous beneath, often ciliolate on the margin and nerves, sessile, linear or linear-lanceolate, acuminate, membranous. Peduncles 2-3 fold, whorled, perianth constricted at the middle. Perianth very variable in size, greenish; tube slender. Stamens inserted above the middle of the tube, anthers subsessile. Style as long as ovary. Berry deep red, 0.75 cm dia., 2-3 seeded. Seeds globose, white. Locally, it is known as *meda* and cosntiueute and important ingadient of *chvanprash* and one of the important *Astavarga* group of plant. Tender leaves and young shoots are cooked as a vegetable. They can be used as an asparagus substitute. Roots are rich in starch and cooked like potato.

### *P. cirrhifolium (Wall.) Royle*

*P. cirrhifolium* (Wall.) Royle = . *P. fargesii* Hua = *P. cirrhifoliodes* D.M.Liu & W.Z.Zeng. It is distributed in the Temperate Himalaya from Shimla to Bhutan at 1,800-3,500 m. Stem 60-120 cm, very weak flexuous, often climbing by the tips of the leaves, terete, puberulous or glabrous. Leaves 3-6- nately, wholed, sessile, membranous with slender circinate tips. Peduncles 2-4 folds, whorled, perianth subcylindric. Flowers white green or purplish. Stamen inserted above the middle of the tube; filaments ciliolate. Style as long as the ovary. Berry deep red, vary in size, seed 2-3. Locally, it is known as *mahameda* and cosntiueute and important ingadient of *chvanprash*. Young shoots cooked and also used as an asparagus substitute. The roots are cardiotonic, sialagogue, stimulant and tonic (Chopra *et al.,* 1986). The roots are used in Tibetan medicine where they are said to have a sweet taste and a neutral potency. Antitussive, carminative and tonic, they promote bodily heat and dry up serous fluids. They are used in the treatment of loss of vigour, pain in the kidneys and hips, swelling and fullness in the abdominal region, accumulation of fluids in bone joints, skin eruptions and coughs.

## Pueraria

It is distributed in tropical Asia to Polynesia. A twining woody climber or trailer; roots sometimes tuberous. Leaf pinnately trifoliolate; leaflets entire or sinuately lobed, stipellate, stipules sometimes produced below the point of insertion. Inflorescence an axillary raceme or panicle. Bracts and bracteoles present. Calyx 5-lobed or appearing 4-lobed by the union of the upper pair of lobes. Vexillum with 2 inflexed auricles. Stamens mono or diadelphous, vexillary stamen usually fused with the tube, sometimes free, anthers uniform. Ovary multiovulate, style curved, stigma small, capitate. Fruit linear, many-seeded, pubescent. It belongs to family Fabaceae and there are about 20 species found in world, of which 8 species occur in India (Hooker, 1978). It is an ideal plant for a variety of situations where the soil is subject to erosion. It makes vigorous and rapid growth in places where most grasses and legumes do not come up. It yields large quantities of leaf litter, which adds to the fertility of soil. It is reported to suppress even weed growth.

### *Pueraria tuberosa (Roxb. ex Willd.) DC.*

*Pueraria tuberosa* (Roxb. ex Willd.) DC. = *Hedysarum tuberosum* Roxb.ex Willd. Generally occurs in the mid Himalayan region from Kashmir to Kumaon at an elevation of 1,200-1,800 m (Babu, 1977). Root very large, tuberous. Stems shrubby, the branches finely grey-downy. Stipules minute, deciduous, cordate-ovate; leaflets membranous, roundish, 15-30 cm , long, green, glabrescent above, densely clothed with whitish adpressed hairs beneath. Flowers in dense virgate leafless often panicled racemes, reaching 18-27 cm long; pedicels very short, densely fascicled. Calyx 0.5-0.75 cm long, densely silky, teeth subequal. Corolla bluish, not quite twice the calyx; limb of standard orbicular, distinctly spurred. Pod 6-9 cm long, membranous, flat, 3-6 seeded. Lamina of wings oblique- oblong. Pod clothed with weak bristles. It is locally known as *Sural, tirra, siali, bisal, badar, bedarikand, bilaikand, bodar, dedarikand. patalkand,*. The tubers are large, 30-60 cm. long and 25-30 cm. broad, weighing up to 3 kg. They are often found in strings connected with the main roots by thin roots. They taste like liquorice and are said to be eaten raw or boiled in scarcity time. They are often fed to horses and ponies. The tubers can be used for the extraction of starch. The root is used as a demulcent and refrigerant in the treatment of fevers. It is crushed and rubbed on the body in the treatment of fevers and rheumatism. It is also emetic, galactogogue and tonic (Chopra, *et al.* 1986).

## Typha

It is a marshy herb, typically 1-3 m tall (*T. minima* is smaller: 0.5-1 m), with spongy, strap-like leaves and starchy, creeping rhizome. Leaves alternate, mostly basal to a simple, jointless stem that eventually bears the flowers. Rhizomes spread horizontally beneath the surface of muddy ground to start new upright growth. Male flower spike develops at the top of the vertical stem, above the female flower spike. The male (staminate) flowers are reduced to a pair of stamens and hairs and wither once the pollen is shed, leaving a short, bare stem portion above the female inflorescence. The dense cluster of female flowers forms a cylindrical spike, 10 - 40 x 1 - 4 cm. Seeds minute (about 0.2 mm long), and attached to a thin hair or stalk, which effects wind dispersal.

*Typha* belongs to family *Typhaceae* and is a genus of about ten species of monocotyledonous flowering plants in the monogeneric family, *Typhaceae*. The genus is essentially a cosmopolitan but largely distributed in temeparte region. About 04 species are occurring in India (Hooker, 1978). *T. laxmanii* rhizomes are a palatable, nutritious and productive root vegetable (Rybricek, 1981). They are generally harvested in the fall and winter. The pollen is also sometimes used as a flour supplement. Roots are rich in starch and can be boiled and eaten like potatoes. The root can also be dried, ground into a powder and then used as a thickener in soups etc or added to cereal flours. The root contains a lot of fibre. The leaves are diuretic. The pollen is astringent, desiccant, diuretic, haemostatic and vulnerary. It is used in the treatment of nose bleeds, haematemesis, haematuria, uterine bleeding, dysmenorrhoea, postpartum abdominal pain and gastralgia, scrofula and abscesses (Chopra *et al.,* 1986).

### *T. elephantina Roxb.,*

*T. elephantina* Roxb., = *T. schimperi* Rohrb. It is distributed from Kashmir to Assam in the mid Himalayan ranges. Rhizomes at base of erect shoots, mostly horizontal, unbranched, starchy, firm, and scaly. Stem 180-360 cm. Leaves 3-6 cm broad, trigonous above the sheath, margins often undulate above the middle. Flowers mixed with clavate pistillodes, bracteolate; pollen 4-globate; stigma lanceolate. Male spike 24-36 cm; rachis clothed with short often forked hairs; bracts 3 or more; anthers 1-5, 0.3 cm long. Female spike much stouter, 18-30 x 1-3 cm diam.; bracteoles with fascteoles; tips much longer than the hairs, which are shorter than the stigmas. Locally known as *Mothitrina, bora, Boj, bori, gond, kundar, pitz* etc. The young shoots are edible, and taste like asparagus. The plant is a favourite fodder of elephants. The starchy rhizomes and the pollen are also eaten. The rhizomes are somewhat astringent and diuretic, and are reported to be employed in dysentery, gonorrhoea, and measles.

### *T. domingensis Pers.*

*T. domingensis* Pers.= *T. angustifolia* var. *domingensis* (Pers.) Griseb.,= *T. tenuifolia* Kunth. It is distributed in northern India from Kashmir to Manipur at elevation ranging from 1,500-2,500 m. It closely resembles the common European *T angustifolia* Linn, which is distinguished by the bracteoles shorter than the stigmas (Hayden, 1939). Erect shoots 150-300 cm, not glaucous; flowering shoots 5-12 mm thick in middle; stems 2-3mm thick near inflorescence. Leaves: sheath sides membranous, margin broadly clear, summit with membranous auricles which often disintegrate late in season; mucilage glands at sheath-blade transition brown, absent from blade and usually from sheath center near summit; widest blades on shoot 4-12 mm wide when fresh, 3-8mm when dry; distal blade usually markedly exceeding inflorescence. Bracteoles of female flower subspathulate, equaling the linear stigmas, both longer than the hairs. Female flowers mixed with clavate-tipped pistillodes; pollen simple.

### *T. laxmanni* Lepech

Occurring at much higher elevation (3,500 m) in western Himalaya. According to Rybricek (1981) it is a dwarf, 60-90 cm high; leaves slender semicylindric at the sheath. Male spikes 3-6 cm; female spikes 2-3 x 0.6-1 cm; female flowers ebracteolate, mixed with pistillodes; hairs very short much shorter than the subobtuse stigmas; pollen simple. At once distinguished from all the previous species by its small size and slender leaves, but very closely allied in the European and Oriental *T. minima*. The rhizomes furnish a meal, which is made into cakes; they are also used as a vegetable. The stamens are reported to possess astringent and styptic properties.

## References

Abramova, L. I. 1975 On the taxonomical structure of the genus *Polygonatum* Mill. Bot. Zhurn. (Moscow & Leningrad) 60: 490–497. Bush, B. F. 1927. The species of *Polygonatum. Amer. Midl. Naturalist* 10: 385–400.

Alexander, J. and D. G. Coursey, 1969 The origin of yam cultivation. (pp. 405-425). In: P. J. Ucko & G. W. Dimbleby (eds.), The domestication and exploitation of plants and animals. - Duckworth, London, 581 pp.

Anonymous, 1952 The Wealth of India. A dictionary of Indian raw materials & industrial products. (Raw Materials). - Council Scientific Industrial Research, New Delhi.

Atkinson, E. T. (ed.) 1980 The economic botany of the Himalayas. (Reprint from "The Himalayan districts of the North-West Provinces of India", 1882). - Cosmo Publ., New Delhi, 672 pp.

Babu, C. R. 1977 Herbaceous flora of Dehra Dun. (HF Dehra Dun).

Chabbert, N., P. Braun, J.P. Guiraud, M. Arnoux, and P. Galzy. 1983 Productivity and fermentability of Jerusulam artichoke according to harvesting date. *Biomass* 3: 209–224.

Chopra. R. N., S. L. Nayar. and I.C. Chopra. 1986 *Glossary of Indian Medicinal Plants (Including the Supplement).* Council of Scientific and Industrial Research, New Delhi.

Collett, H. 1971 Flora Simlensis- A handbook of the flowering plants of Simla. Bishen Singh Mahendra Pal Singh, Dehra Dun. 636p. (Reprint).

Duke, J.A. 1978 The quest for tolerant germplasm. (In) ASA Special Symposium 32, Crop tolerance to suboptimal land conditions. Am. Soc. Agron. Madison, WI. p. 1–61.

Duke, J.A. 1979 Ecosystematic data on economic plants. *Quart. J. Crude Drug Res.* 17(3–4):91–110.

Duke, J.A. and K.K. Wain. 1981 Medicinal plants of the world. Computer index with more than 85,000 entries. 3 vols.

Fellingham, A.C. and N.L. Meyer. 1995 New combinations and a complete list of *Asparagus* species in southern Africa (Asparagaceae). *Bothalia* 25: 205-209.

Gates, R. R. 1917 A revision of the genus *Polygonatum* in North America. Bull. Torrey Bot. Club 44: 117–126.

Handa SS and M.K. Kaul. 1997 Supplement to cultivation and utilization of aromatic plants. Jammu-Tawi : Regional Research Laboratory, Council of Scientific and Industrial Research xv; 569p.

Hayden, A. 1939 Notes on *Typha angustifolia* in Iowa. Iowa State Coll. *J. Sci.* 13: 341--351.

Hetterscheid, W.L.A.. 1994 Preliminary taxonomy and morphology of Amorphophallus Blume ex Decaisne (Araceae). In: M.M. Serebreyanyi (ed.), Proc. Moscow Aroid Conference 1992: 35-48.

Hong, D.Y. 1983 Flora Reipublicae Popularis Sinicae, vol. 73(2). Science Press, Beijing.

Hooker, J D 1978 Flora of British India (Vol. I-VII). International Book Distributors. Dehradun, India. (Reprin).

Jayasurya AHM. 1984 Systematic arrangement of the genus *Dioscorea* (Dioscoreaceae) in Indian Sub-continent, Revised hand book to the Flora of Ceylon IX. Royal Botanic Gardens, Kew . Richmond, UK.

Jessop, J. 1966 The genus *Asparagus* in southern Africa. *Bothalia* 9: 31-96.

Leung, A.Y. and S. Foster. 1996 Encyclopedia of Common Natural Ingredients Used in Food, Drugs and Cosmetics, 2nd ed. New York: John Wiley & Sons, Inc.

Long, Chun-Lin, 1990 Diversification of homegardens as a sustainable agroecosystem in Xishuangbanna, China. Suan V Symposium on rural-urban ecosystem interactions in development. (Bandung, May 21-24, 1990). - Inst. Ecol. Padjadjaran Univ., Bandung, 15 pp.

Malcomber, S.T. and D. Sebsebe. 1993 The status of *Protasparagus* and *Myrsiphyllum* in the Asparagaceae. *Kew Bulletin* 48: 63-78.

Morris K.E. and T.G. Lammers 1997 "Circumscription of *Codonopsis* and the allied genera *Campanumoea* and *Leptocodon* (Campanulaceae: Campanuloideae). I. Palynological data". *Bot. Bull. Acad. Sin.* 36: 277-284.

Namhata D, and A. Mukherjee. 1992 Some folklore medicines of Bankura district, West Bengal *J. Econ. Taxon. Bot., Addit. Ser.* 10. 265-266.

Palz, W. and P. Chartier. (eds.). 1980 Energy from biomass in Europe. Applied Science Publishers Ltd., London.

Pradhan UC. 1997 Himalayan cobra-lilies (*Arisaema*): their botany and culture. Kalimpong : Primulaceae Books 99p. Includes brief notes on uses.

Prain D, and I.H. Burkill. 1936 An account of the Genus *Dioscorea* in the East. Annals of the Royal Botanical Gardens, Culcutta. Longman, London , UK.

Rybricek K. 1981 *Typha laxmannii* Lepch. (*Typha laxmannii* Lepech. near Poprad.) Biologia, A, 36. (7): 549-551.

Sdobnina L. I. and M.G. Pimenov. 1991 Anatomical features of underground organs of some geophilous Umbelliferae. Bot. *Zh. SSSR.* 76. 1527-38.

Shimizu, T. 1993 Campanulaceae. In K. Iwatsuki, *et al.* (eds.), Flora of Japan, Vol. 3a. Kodansha, Tokyo, pp. 405-418.

Singh RP, R. Singh, P. Ram and P.G. Batliwala. 1993 Use of Pushkar-Guggul, an indigenous antiischemic combination, in the management of ischemic heart disease. *Int J Pharmacog;* 31:147-160.

Singh. G. and P. Kachroo. 1976 Forest Flora of Srinagar. Bishen Singh Mahendra Pal Singh, Dehradun.

Sivadasan M, and S. Kumar. 1987 A new species of *Arisaema* (Araceae) from India with a note on variation and evolution of staminate flowers. *Aroideana* 10. (4): 18-21.

Thatte, U.M. and S.A. Dahanukar. 1988 Comparative study of immunomodulating activity of Indian medicinal plants, lithium carbonate and glucan. *Methods Find Exp Clin Pharmacol* 10 (10):639–644.

Tripathi S.N., B.N. Upadhyaya V.K. Gupta. 1984 Beneficial effect of Inula racemosa (Pushkarmoola) in angina pectoris: a preliminary report. *Ind J Physiol Pharmac.* 28:73-75.

Vajravelu E, and P. Bhargavan. 1983 Notes on some rare plants from *South India. J. Econ. Taxon. Bot.* 3. (3): 969-973.

Velayudhan K.C, V.K. Muralidharan, V.A. Amalraj, and K.I Asha. 1998 Genetic resource of yams of Western Ghats. Indian Journal of Plant Genetic Resources of the Western Ghats 11(1): 69-80. National Bureau of Plant Genetic Resources, Regional Station, Vellanikkara, Thrissur , India

Watt, G., 1908 The commercial products of India: being an abridgment of "The dictionary of the economic production of India". - Murray, London, 1189 pp.

White D. 1991 A Himalayan Campanula mystery. Himalayan Pl. Ass. Newsl. no.3. 24-26.

White, G.A. and Solt, Marie. 1978 Chromosome numbers in *Crambe*, *Crambella*, and *Hemicrambe*. Crop Sci. 18:160-161.

Wilkin P. 1999 A morphometric study of *Dioscoreaquartiniana* A. Rich (Dioscoreaceae). *Kew Bulletin* 54:1–18.

Wren, R.C. 1988 *Potter's New Cyclopaedia of Botanical Drugs and Preparations.* Essex: The C.W. Daniel Company Ltd.

Yoo, K.O. and W.T. Lee. 1989 A taxonomic study of the genus *Codonopsis* in Korea. Korean *J. Pl. Taxon.* 19: 81-102.

*Plates Contd...*

*Amorphophallus paeoniifolius* Corn

*Angelica glauca*-plant and root

*Arisaema jacqemontii*-plant

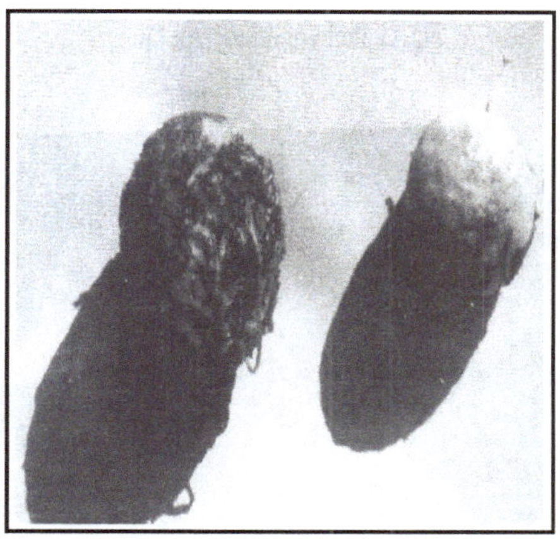

*Arisaema jacquemontii*-corns

*Arisaema tortuosum*-plant

*Arisaema triphyllum*-plant

*Asparagus adscendens*

*Asparagus filicinus* - plant and tubers

*Asparagus filicinus*-tubers

*Asparagus racemous*-plant

*Campanula alpina*-plant

*Campanula rotunidifolia*-plant

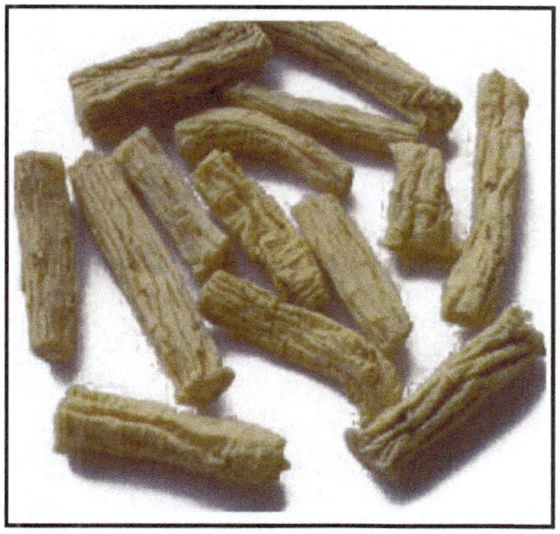

*Codonopsis rotunidifolia*-root

*Codonopsis rotunidifolia* -plant

*Codonopsis viridis*-plant

*Codonopsis ovata*- plant

*Dactylorhiza hatagirea*-plant

*Dioscorea bulbifera*- bulbils

*Discorea bulbifera*-corn

*Dioscorea deltoidea*-rhizome

*Dioscorea pentaphylla*- com

*Habenaria edgeworthi*-plant

*Habenaria intermedia*- plant and tubers

*Helianthus tuberosus*-plant

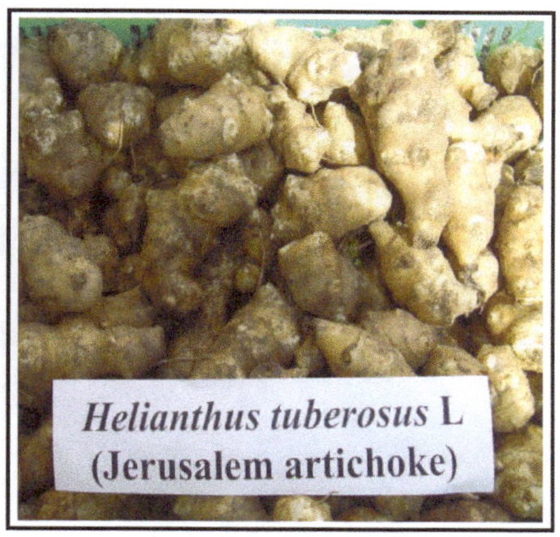


*Helianthus tuberosus* -tubers

*Clerodendron chinense*

*Clerodendron infortunatum*

*Clerodendron paniculatum*

*Clerodendron serratum*

*Clerodendron siphonanthus*

*Clerodendron speciosissimum*

*Clerodendron speciosum*

*Nymphaea micrantha*

*Nymphaea rubra*

# Underutilized Flowers

**Underutilized and Underexploited Horticultural Crops, Vol. 2, 2007**
*K.V. Peter (ed.), pp. 255-265*
New India Publishing Agency, New Delhi (India)
E-mail : newindiapublishingagency@gmail.com
Web: www.bookfactoryindia.com

# 14 Liliums in India — An Overview

**K. VALLIAPPAN**
President – Agro Division
Eureka Trading (P) Limited, Coimbatore - 641 012, Tamil Nadu

Lilies, both Asiatic lilies and Oriental lilies, along with their hybrids are cultivated in very small scale in India, but are classified under high value crops (based on per stem value). Lilies are also capital intensive crops, in terms of high cost of infrastructure required for protected cultivation and bulbs which are imported. This article deals with growing conditions, varieties and other cultural aspects, as well as, practical tips from field conditions, from areas where Liliums are commercially grown in India.

Lilies are grouped under "high value flowers" in commercial floriculture parlance and are generally regarded under minor crops in production terms (on quantity produced or on area under production or on per sq.meter yield per year basis). Lily is in fact a more productive and lucrative crop, specifically so in terms of income (on per sq. meter per year basis) than any other cut flower grown for commercial purposes in the open or protected cultivation in India.

## Lilies as Known World Over

Lilium is known for ages as evident from Shekels stamped by Palestinian coin makers in 143 BC (Bose *et al* 2003) and since, 1179 AD, lily is a part of the coat of arms of the kings of France. In America, *Lilium longiflorum* is a national symbol of pride and hope. Although white lilies are associated with funerals, generally lilies are perfect flowers for many different festive occasions, epitomizing innocence/purity /virginity. Lily, thus symbolizes regality, fertility, clarity, chastity and unity. Romans and Greeks crowned their brides with lilies. In Christianity, lily is a symbol of love that the Archangel Gabriel and the parents of Virgin Mary are often depicted holding a lily (Anon 2005).

World trade on cutflowers, estimated at US$ 847.3 million in 2003, has grown into a market of US$ 9.4 billion in 2005, with an estimated annual growth rate of 8% (Money 2006 and Anon 2006).

Lilies rank 10$^{th}$ among the top selling flowers in the world and lily bulb production in the Netherlands soared consistently during the second half of the 20$^{th}$ century, from about 100 ha in 1966 to over 3000 ha in 1993 (Pizano 2005).

## Lilies and Other Bulbous Crops as Cut Flowers or Pot Plants

World over, other bulbous crops, like Tulips, Amaryllis and Narcissus, are also popularly known as cut flowers. In India, these crops, along with other bulbous crops, still remain minor crops in terms of commercial value. There are many other bulbous crops like Achimenes, Canna , Cooperia, Crinum, Crocus, Dahlia, Eucharis, Eurycles, Gloriosa, Haemanthus, Hedychium, Hemerocallis, Hippeastrum, Hymenocallis, Nelumbo, Pancratium, Sinningia (Gloxinia), Zephyranthus (Maiti and Bose 1999) and these are popular only as garden / landscape components in India. Exceptions are bulbous crops like Tuberoses, Lilies, Calla lily, Heliconia, Red Ginger (Alpinia) and Alstroemeria, having commercial potential in India and are suitable for cultivation under Indian conditions (Srinivas, 2006).

Among these popular bulbous crops, which are cultivated in India for commercial cut flower purposes, tuberose and gladiolus are widely cultivated in different regions of India. Calla lily, on the other hand is area specific (high altitudes) and is grown for commercial purposes in select areas (Valliappan 2006). This article deals with various practical aspects of growing Lilium in Indian conditions.

## Types and Growing Areas

Lilies or Lilium, as commonly referred in commercial terms, are cut flowers which include mainly two types viz., Asiatic Lily (AS) and Oriental lilies (LO or OR) and recently, Lilium Asiatic hybrids have been introduced. The commercial cultivation of Lilium in India (Valliappan 2005) has been restricted to the following states:

*Tamil Nadu:* The Nilgiris (Coonoor, Kothagiri and Ooty), Kodaikanal, Shevroy Hills (Yercaud), Kalvarayan Hills (Karumanthurai), Hosur etc.

*Kerala:* Munnar and Peermedu

*Karnataka:* Bangalore

*Himachal Pradesh:* Kulu, Manali and Dehradun

*Uttar Pradesh:* Lucknow and

*Maharashtra:* Pune, Nashik, Sangli and Kolhapur

The area under Lilium in India is limited because of two reasons *viz.*, high cost of *planting material* (bulbs) and high investment required for *protected cultivation* (control of temperature, light and to protect from wind and frost injuries).

Lilies can be grown commercially in moderate and higher altitudes (sub-tropical and temperate climates). Lily is being grown in different altitudes in India, as given below:

*Plains:* up to 900 m: can be grown under poly roof and shade nets, with or without cooling (during Sep/Oct to Jan/Feb).

*Mid-Hill regions:* From 900 - 1200 m; can be grown under poly house conditions throughout the year. Shading will be required in summer.

*Hilly regions:* Above 1200 m: The structure should have frost proof sides, for Oct-Jan months. Shading on adhoc basis will be required in summer.

In general, Liliums prefer a day temperature of 20-25°C and a night temperature of 10-15°C, although it has been grown successfully in lower temperatures, at both in cool temperature (Western countries) and in hot climates (Tropical countries). Under controlled conditions of temperature, relative humidity and light, Lilies are being successfully grown. Orientals prefer higher altitudes, while Asiatics can be grown at lower altitudes. But, in India, it is shown that in winter at lower altitudes and at summer in higher altitudes, reversal of such preference is possible. Colour expression in Lilium flowers also varies with higher / lower altitudes. As a thumb rule, a difference of 12°C, between maximum (day) and minimum (night) temperatures is considered ideal for commercial cultivation of lilies.

## Varieties

There are hundreds of varieties grown all over the world and each region has a particular preference for varietal choices, for colour expression for acceptability in markets, number of buds (for a given bulb size), duration of crop, crop height, light requirement / susceptibility etc. In general, a good mix of colour (orange, yellow and white in Asiatics, with a minor proportion of pinks and reds or pinks and whites in Orientals, with other shades like mild yellow / orange and peach in other hybrids) are recommended for planting together to satisfy the market thirst, for a wide colour range.

The popular varieties in *Asiatic Lily* (wide range of colours; no fragrance), *Oriental lily* (large flowers with wavy margins; limited colours; fragrant) and *LA hybrids* (mild colours, with mostly Asiatic parameters) are given below, in alphabetical order and these are grown and found suitable to Indian conditions, both from production and marketing angles:

Asiatic Lilies (LA)

| | |
|---|---|
| Orange | Avignon, Bariton, Beatrix, Brooklyn, Brunello, Elite (Gibraltor), Golf, Prato, Salsa, Tresor, Update |
| Yellow | Adelina, Bangalore, Conneticut King, Dreamland, Festone, Gironde, Golden Horn, Hilde, Iverna, Las Vegas, London, Madras, Mona, Nove Cento, Pisa, Polyanna, Romano, Solemio, Val Di Sole, Yellow Giant |
| White | Alaska, Navona, Snow Storm, Sorpressa, Umbria |
| Pink | Allegretto, Blue Eyes, Chianti, Chopin, Cote d' Azur, Manchestor, Marseille, Minstreel, Montreux, New Orleans, Pink Superior, Renoir, Sophia, Vermeer, Vivaldi |
| Red | Amarone, America, Black Out, Black Jack, Camry, Diabora, Gran Paradiso, Ilonca, Marrakech, Nello, Red Bull, Saidja |

Oriental Lilies (LO or OR)

| | |
|---|---|
| Pink | Acapulco, Action, Aktiva, All Seasons, Ambon, Bernini, Brisbane, Brittania, Canberra, Chinon (Espresso), Creation, Crystal Star, Kiss Proof, La Mancha, Limasol May Tay, Medusa, Melanie, Mero Star, Miami, Mondeo, Monte Christo, Olympic Star (Olymphische Ster), Passage, Pesaro, Sakura (Peru), San Pedro, Sartre, Sissi, Sorbonne, Star Fighter, Star Gazer, Tiber, Tropical |
| White | All In, Alma Ata, Baluga, Canada, Carillon, Casa Blanca, Cassandra, Cordoba, Helvetia, Louvre, Marco Polo (Vedea), Melbourne, Royal Wedding, Sapporo, Siberia, Simplon, Snow Pearl, Twilight time, Universe, White Cup |

Lilium Asiatic (LA) Hybrids

| | |
|---|---|
| Yellow | Ceb Dazzle (Aladdin's Dazzle), Club House, Daytona, Indian Diamond, Madrid, Menorca, Milagro, Mother's Day, Mount Bachelor, Rovigno, Royal Delight, Royal Justice, Royal Perfume, Royal Trinity, Salmon Pride, Tennesee, Twilight Diamond, Yellow Thunder |
| Orange | Manhattan, Maranello, Royal Masquerade, Tropical Dream |
| White | Auckland, Bach, Courier, Donatello, Iceberg, Golden Torch |
| Pink | Algarve, Brindisi, Corelli (Zanlarelli), Dolce (Zanlalce), Money Maker, Pink Glow, Rosy Baron, Rodeo, Royal Club, Royal Parade, Samur, Top Gun, Turandot |
| Peach | Ball Room, Royal Rain, Salmon Queen, Stretto |
| Red | California, Eliath, Fangio, Glossy Wings, Medoc, Mount Rainier, Narbonne, Red Alert, Rodela, Royal Sea, Scarlet Star |

There also OT or OA hybrids and Longiflorum hybrids which are becoming popular in other parts of the world and are yet to be tried in India.

## Propagation

Though fine protocols are available for propagation of Lilies by seed and by tissue culture, commercial production is done mainly using micro or mini tubers / scales. European countries, especially Netherlands and of late, Spain and Chile, have emerged as major players in Lily bulb production.

Though there are efforts in India to produce bulbs in Kashmir Valley (Kulu, Manali, Nainital and Dehradun), with private or Government collaboration with Dutch breeders and bulb producing companies, the production of Indian bulbs is very much in nascent stages.

## Crop Duration and Bulb Sizes

Generally Asiatic lilies come to yield in 90-110 days and Orientals take 10-20 days more.

Again, *altitude and temperature* (in a particular growing season *viz.*, winter or summer), which are naturally inter-related, play a vital role in the duration of crop and quality of the produce (stem length, stem thickness and number of buds for a given bulb size). Same variety in a given bulb size can take longer cropping duration at higher altitudes (or in winter) than at lower altitudes (or in summer) or vice-versa.

For Asiatic lilies, *bulb sizes* of 10/12 and 12/14 are preferred, while for Orientals, 14/16 and 16/18 are ideal. The high cost of Orientals, prohibit Indian growers to look for sizes beyond 16/18., Indian growers often settle for smaller sizes in Orientals (as in LA hybrids) with 12/14 size, although Italian and Spanish growers prefer larger sizes like 16/18 and 18/20 to grow a crop of Orientals.

Generally, bulb sizes are directly proportional to flower quality, contributing to stem length, stem thickness and more importantly, number of buds, which are the main quality parameters from the marketing angle. In Orientals, specifically, large bulb size has a direct correlation to (later) leaf senescence (after harvest) and longer vase life. Simple logic is that Lilium being a bulbous crop, coming up with its own supply of nutrients, in the form of bulb, depends mainly on internal resources, rather than growing new roots, in the form of *stem roots* (directly related to *planting depth* of 6-10 cm, depending on planting season) and new roots, for absorption and translocation of the nutrients being supplemented, through fertilizers, by way of soil application, fertigation and foliar nutrition.

## Suitable Soil Types and Manuring

### *Soil Types*

Lily prefers well drained and light soils. Generally, clayey soils, with high pH and EC should be modified with addition of equal part of organic manure / red soil / sand as to that of the native soil. Amendments will be also required to correct pH level of soils (alkali soils with gypsum and acid soils with lime). Generally, on a per ha basis, a *fertilizer dosage* of 45 kg N and 60 kg P2O5 as basal dose, 45 kg N and 40 Kg K20 as top at 25-30 days after planting and 45 kg N, 60 kg P2O5 and 40 Kg K20 at 75-80 days after planting are recommended. But, this depends on the area (soil nutrient level) and type of fertilizer application i.e., by conventional soil application, through drip with or without combination of foliar feeding. In fertigation, the dosers can be programmed to deliver the above NPK evenly through out the life period on a two or single daily dosage *via* drip irrigation system.

*Organic manure,* vascular arbuscular mycorrhizal (VAM) fungae have been shown to increase the quality of Lilium. Similarly, soil cover with *organic mulches* (saw dust and rice husk) can also help in increasing nutrient status, as well as reduce soil temperature in tropical and sub-tropical conditions.

### *Population*

Generally, 1 m. wide crop beds of height 30-45cm are recommended for lily cultivation and a population of 45-55 bulbs per sq m is advised, although the thumb rule is to give less population in winter (to avoid mutual shading and eventual bud drop) and more in summer. This also depends on bulb sizes, type of growing (in containers or beds or in artificial media) and growing elevation.

While there are academic research papers available in plenty on nutrition, plant growth regulators etc. (Bose *et al* 2003), every grower has to perfect a protocol for the given locality (altitude and day/night temperature), media composition, polyhouse type, type of feeding, growing season (summer/ winter), type of lily (AS or OR or LA) and its variety. This is easier said than done because, lily growing (like every other crop in commercial floriculture) depends on learning from the field performance and perfecting the guidelines available in the form of advise or literature, academic or otherwise.

## Plant Protection

It is to be generally noted that the cost of plant protection and fertilizers is very minimal in Lilium, when compared to the cost of bulbs and infrastructure, required to grow Lilium. Similarly, when compared to damages by insects, diseases and physiological disorders are major problems with lilies. Hence, right from *soil fumigation* (methyl bromide @ 20g per sq.m or its formaldehyde equivalent), *pre-plant dip of bulbs* in any broad spectrum fungicide (as well as post-planting drench with fungicides) is recommended.

The *Fungal Diseases* of great significance in lily are: *Rhizoctonia solani, Rhizoctonia tuliparum, Pythium sp., Colletotrichum liliacearum, Fusarium oxysporum, Botrytis liliorum, Botrytis elliptica, Phytophthora parasitica* etc. Clean bulbs (mother bulb crop is to be inspected), Soil fumigation, pre-plant dip and post-plant schedule of fungicides, in rotation, can be adopted. Crop rotation, after every second crop of lily and shifting the polyhouse to a fresh area, every 3rd year will be good precautionary measure.

Similarly, reports in lily also include occurrence of *Bacterial diseases* (*Pectobacterium carotovorum* and *Bacillus lilii*) and viral *diseases* (*Lily Symptomless (Carla) Virus* or LSV, *Cucumber Mosaic Virus* or CMV, *Tulip Breaking Polyvirus* or TBV and *Lily X Potyvirus* or LXV. Preventive steps include treatment with bacteriocides, chemotherapy at harvest / storage stage of bulbs and control of vectors for viral diseases. Certain mycoplasmal diseases are also found in literature, but are rare in clean cultivation practices.

Among *insects*, aphids, mites, lily beetle (*Lilioceris lilii*), thrips, with occasional occurrence of stem borer and stem weevil are also reported. *Nematodes* can also be a problem for lily in red soils and in higher altitudes (acid soils).

When *selecting a field* for cultivation of Lilium, it is to be ensured that no tuberous crops have been grown in past one to two years in the same field or in the field from which media components would be incorporated. This will ensure a proper sanitation requirement with respect to residual pathogens and nematodes. Similarly, when nearby crops are vegetables, especially potatoes, crops belonging to Cruciferae or Cucurbitaceae or Solanaceae, extra care should be taken with respect to sides of insect proof nets, on all vents in the sides as well as the roof.

*Micronutrient deficiency* (especially, yellowing due to lack of Iron) should be monitored regularly, in conjunction with soil and/or leaf analysis and corrective measures, with chelated forms of microelements are to be taken up. Leaf scorch can also be due to Calcium deficiency (or physiological disorders like lack of light) and so, a good micronutrient mix should be kept handy for foliar feeding. Soil pH, in excess of 7.5 or less than 6..5 can lead to micronutrient deficiency and specifically for Iron, Fe-EDTA, Fe-EDDHA or Fe-EDDHMA or Fe-DTPA are helpful in such adverse pH situations. Similarly, at active growing stages and bud formation stage (1/3 and 2/3 period of the lily life cycle), lily craves for high potassium and so, foliar feeding of KNO3 can be handy to avoid potassium deficiency, as well as limiting uptake of other nutrients, especially Iron.

In analysing soil for EC, pH, macro and micro nutrients, both the media in the top layer of bed as well as, in the 30-45 cm region of the bed should be analysed, to get a fair idea of pH at stem root zone and rooting depth, to enable identification of micronutrients that may be lacking or those that will be in demand at vigorous growth stages of Lilium.

Physiological disorders like *leaf scorch* (mainly, lack of light), *bud scorch / bud blast* (dry climate in greenhouse and a need to increase humidity, shortage of water, or high levels of nitrate in soil), *bud drop* (too close a spacing, lack of stem roots or insufficient stem roots, which relate to depth of planting the bulb in various soils or climates) and *frost injury* (stunted plants, chlorosis and puffy leaves) cannot be generalized for a specific recommendation to particular area / region or for a specific season, but have to be expected and remedial measures kept ready.

## Harvest and Post-Harvest

Harvesting of stems is generally advised in early hours of day and at a time when the lowest bud shows colour. This again depends on market to be reached. The vase life of lily in tropical conditions is 10-12 days, if harvested in *"tight bud" stage* and that starts from the day of harvest. If the use of flower is immediate or if the market is nearby, then harvest of stems can be done a little later (when the lowest bud opens with a pin hole or later when it starts to open outwards). Too open stems are difficult to pack as they get damaged in transit or do not travel well at all.

*Post harvest storage* of flowers should be done at 2-4°C. The bulbs of size 12/14 are often left in the field for a month (of harvest of stems and stopping of water), then harvested and sent to cold store (at -1.5 to -2 degree C) in India, where a *second cycle* can be grown, without spending again on bulb cost. This however, is dependent on specific varieties, bulb size used to begin with and how well the first crop was taken care of in the field (spacing, fertilizer and plant protection) to conserve enough of the stored energy in the original bulbs, to spend again in its second lease of life from cold storing.

Similarly, long cold storage can also lead to blind shoots or black shoots, especially in oriental lilies (Henk and Kok, 2006). Long cold storage can also lead to fungal disorders like scale rot and basal rot and so, cold storage beyond 10-11 months is generally not advised for lily bulbs.

## Thrust Areas for the Future

Consolidation of growers and grower associations in respect of Lilium, to share the acquired skills in Lilium cultivation, to evolve a *regionwise package of practices* (type of poly houses and their configurations, suitable varieties, media composition, production planning with 1$^{st}$ and 2$^{nd}$ cycle bulbs out of cold storing, fertigation schedules and plant protection measures, both on regular and adhoc basis) is needed.

Common infrastructure to store bulbs (at – 1.5 to 2°C) and flowers (at + 2-4°C), with *regionwise cold store access* is to be developed to help small and marginal growers.

Research on forcing of bulbs is to be accelerated to *production of lily bulbs in India*, to meet the present need of nearly four million bulbs per annum (2005-06) and to replace import of the same with export of bulbs, to South East Asian markets to start with. Since, most varieties are out of the purview of Breeders' Right, it would be prudent to develop newer types and varieties which are more adapted to Indian conditions, by intensive plant breeding programs. This would enable Indian growers to cut down on their project costs, with nearly 50% being spent on cost of bulbs.

Encouraging research will be helpful in *popularizing native species* of lily, like *Lilium niligiriensis*, which is locally adapted, fragrant (like Oriental lilies) and bright yellowish orange in colour. Similar germplasm, native to India can be preserved, to develop local hybrids, multiplied by rapid means like Plant Tissue Culture and efforts are to be made to earn Royalties by patenting such species and varieties.

*Awareness* is to be spread among small growers and large planters alike, that competition for Indian grown lilies lies *outside* India and therefore, an united stand will benefit Indian growers at large. This would enable Indian growers, spread over different climatic zones across the country, to take advantage of lilies, in their respective regions with ideal climate.

Lilies are fast emerging as the *largest income contributor* (on a comparative scale of return/ sq.m/ year basis, when compared with any other cut flower or plantation crop in India) today and the benefit should reach small and marginal growers, with land holdings as small as 0.25-1.0ha, either on their own or in long term lease of say, ten years, in ideal climatic zones for growing lilies.

Increasing *market avenues* (to plan and plant the right mix of colours and combination of Asiatic / Oriental / hybrid types and in conjunction with 1$^{st}$ and 2$^{nd}$ cycle bulbs), both at domestic level and in the International market (especially, South East, Far East and West Asian markets), will avoid unusual "dumping" at festive occasions when everybody doubles or triples their per-day-production level, leading to unusual and eventual price fluctuations, which in turn, affect small and marginal growers, who are primarily dependent on domestic trading markets in Bangalore, Mumbai and New Delhi.

## References

Anon. 2005 Know more about lilies. *Floriculture Today* 10 (2): 14-17.

Anon. 2006 Germany, a big market. *Floriculture Today* 10 (12): 36-38.

Bose T K, Yadav L P, Pal P, Parthasarathy V A and Das P. 2003 Lilium. (In) *Commercial Flowers-II.* Naya Udhyog, Calcutta.pp.113-162.

Gude, Henk and Kok, Hans. 2006 Black shoots in Oriental Lilium. *Floraculture International.*16 (2): 44.

Maiti R G and Bose T K.1999 Other bulbous ornamental plants. (In) *Horticulture and Lanscaping,*pp.565-590.Bose T K, Maiti R G, Dhua R S and Das P (Eds).Naya Udhyog, Calcutta.

Money K.S. 2006 Floriculture-Advantage India. *Floriculture Today* 10 (2): 8-12.

Pizano, Marta.2005 The Changing lily scenes. *Floraculture International* 15 (9): 10.

Srinivas, Meenakshi. 2006.Bulbous Flowers becons Indian growers. *Floriculture Today* 11 (4): 54-56.

Valliappan K. 2005.Tamil Nadu-Flower bowl of India. *Floriculture Today* 10 (2):36-37.

Valliappan K. 2006. Calla Lily-a wonderful cut flower. *Floriculture Today* 10 (12):47-48.

**Underutilized and Underexploited Horticultural Crops, Vol. 2, 2007**
*K.V. Peter (ed.), pp. 267-281*
New India Publishing Agency, New Delhi (India)
E-mail : newindiapublishingagency@gmail.com
Web: www.bookfactoryindia.com

# 15 Underexploited Flowers of Our Surroundings

**U. SREELATHA*, JYOTHI BHASKAR** AND SOBHANA, A****
* Krishi Vigyan Kendra, Thrissur
** AICFIP, College of Horticulture, Kerala Agricultural University Vellanikkara, KAU P.O., Thrissur - 680 656

While setting up a garden or landscaping an area, the natural floristic wealth is ignored very often. But many of those flora are much adapted to prevailing soil and climatic conditions of the area. Plants with attractive flowers or foliage could be retained to maintain the natural appearance as well as to reduce cost of landscaping.Very familiar plants of our surroundings having good ornamental value are discussed in this chapter.

## *Goanese Ipecacuanha*

*Naregamia alata* commonly known as Goanese Ipecacuanha is a member of family Meliaceae. It is known as Tinparni in Hindi, Triparnika in Sanskrit, Nilanarakam in Malayalam, Nilanreagam in Tamil and Nelanaringa in Kannada. Common name was derived from its use of roots in treatment of malaria. Vernacular name, Nilanarakam originated from its morphological resemblance to members of Rutaceae.

### *Habitat*

Plant is distributed through out South India up to 900 m. It is an endemic plant to Western Ghats and is distributed as an under shrub in deciduous forests and also in lateritic zones in plains.

*Botany*

*Naregamia* is a woody herb, growing upto 30 cm high, with much spreading branches. Leaves are digitately trifoliate, elliptic to obovate, obtuse with cuneate base. Petioles are 2-3 cm long and winged. Flowers are white, solitary, axillary and pedicillate. There are five sepals and petals. Stamens are 10 with a white, club shaped staminal tube with yellow anthers. Fruit is a capsule, three valved and clavate. Cells are two seeded and seeds pendulous. The plant perrenates from its roots.

*Propagation and Cultivation*

The plant is propagated from roots, cuttings, and seeds. Micro propagation through tissue culture is reported by John, *et. al.,* (1997). Site selected for planting should preferably sea shady place. Care should be taken to avoid water logging. It does not come well in coastal sandy soils. Enriching soil with organic manures enhances growth and flowering. Watering is essential to keep the plant alive during summer season. Otherwise it will naturally dry out to sprout again during next rainy season.

*Uses*

*Naregamia* is a well known medicinal plant used for various purposes and quoted in many literatures. Roots are emetic, expectorant, useful in acute dysentery. Leaves and stem in decoction are used for biliousness (Warrier, *et.al.,* 1995). Plant is also used in rheumatism and itch. Roots contain alkaloid - naregamin.

Ornamental value of the plant is not yet fully understood. Being an under shrub with low, spreading branches and dark green shining leaves, the plant is a potential ground cover especially in shade. *Naregamia* flowers profusely during rainy season and can be introduced in large landscapes and parks for its attractive flowers. Pure white flowers in contrast to dark green leaves are a great attraction. Yellow anthers at the tip of the staminal tube add to the beauty of flower.

### *Clerodendrons*

*Clerodendrons* are a group of ornamentals of family Verbenaceae. Popularly they are known as "Witches Tongues". Majority of them originated from African and Asian countries. Some like *C. bungei* is native to Mexico and South America. Wide variability exists among members of Verbenaceae. Habitwise some of them are trees, some shrubs and yet others are climbers. Great variability is also observed in morphological characters like leaf shape, flower form, colour and size. Some clerodendrons produce fragrant flowers.

*Botany*

Habit wise clerodendrons are trees, shrubs or climbers. Leaves opposite, rarely verticillate, simple, entire or toothed or lobed. Flowers are small, borne in axillary cymes or panicles. Calyx campanulate, rarely tubular, truncate, five lobed, persistent or accrescent, often colored in fruit. Corolla salver or funnel shaped, tube usually long and slender, limb more or less oblique with five spreading lobes. Petals are arranged in butterfly like fashion in many species. Stamens four didynamous, long, exerted, filaments filiform, involute in bud, anthers ovate or oblong. Ovary imperfectly 4- celled, single ovuled. Style filiform, extended and shortly bifid. Fruit is a globose drupe, succulent rarely dry, four grooved separating into four pyrenes.

Of the species reported, there are many synonyms even for a single species. Hence the actual number of species may be less than already reported. Some of the important species are described here;

## *Cleordendron paniculatum*

### Pagoda Flower, Orange Tower Flower

Originated in South East Asia. This is an erect semi woody shrub with large evergreen leaves. Flowers are produced in terminal panicles of orange red/ scarlet colour. Though the species is of tropical origin, it has considerable cold tolerance. Usually the plant grows to a height of 90-150cm with single stem. Lower leaves are lobed while upper ones are entire. Flowers within a pyramidal shaped cluster are seen in tiers like a Japanese pagoda. It displays its showy flowers through out rainy season. The plant has long flowering period of 3-4 months. When pruned after season, the plant puts forth new growth and flower again provided there is adequate soil moisture.

## *Clerodendron speciosissimum (C. fallax, C. japonicum)*

The species originated in tropical Africa and Pacific Islands. It is a beautiful ornamental in warm humid tropical climate. When compared with *C. paniculatum*, this species shows elegance in mild climate. Flowering starts in rainy season and through out the rains, plant will be in flowers. Panicles are bigger than *C. paniculatum*. Flowers are bright red and more attractive. Leaves are round, entire and velvety. Seed setting is also observed in the species.

It is an important medicinal plant in South West China. Roots and flowers are used to treat arthritis, liver problems, eye dysfunction, hemorrhoids, hernia and insomnia.

### *Clerodendron thomsonae*

#### Bleeding Heart

This species also originated in Africa. It is an attractive vine. Leaves are dark green and glossy. Flowers have white sepals and dark red petals and more suitable for arches. Field life of sepals is less when compared to other clerodendrons. The plant flowers during rainy season and prefers sunny situation.

### *Clerodendron splendens*

#### Flaming Glory Bower

Native to West Africa. As the name indicates, the plant produces large clusters of dark red flowers. Unlike many clerodendrons, this species flower during winter season. It prefers open sunlight for better flower production. Flower clusters are bigger in places with cool nights. The plant is suitable for arches.

### *Clerodendron speciosum (C. delectum)*

#### Bleeding Heart

It a hybrid between *C. thomsonae* and *C. splendens*. By habit, it is a vine. The plant flowers freely and it comes well in shade also. It is suitable for standards, trellises and arbors. Petals are dark red and have lavender colored sepals. After closing of the corolla, sepals assume dark lavender color and last for a few weeks. It can be tried for dry flower arrangements.

### *Clerodendron fragrans (C. chinense)*

#### Chinese Glory Bower

Originated in tropical Africa and Asia. The plant produces highly fragrant flowers in round terminal clusters. Flowers are double and often tinged with pink.

### *Clerodendron marosiphon (C. incisum)*

Native to East Africa, an elegant erect shrub it never grows beyond 1m. Pure white and bud like flowers are borne in terminal clusters during summer and rainy seasons. Flowering is in several flushes but flowers are short lived. Leaves are smaller compared to other species.

### *Clerodendron siphonanthus*
#### Tube Flower

Originated in Eastern India, an erect growing shrub it grows about 3m tall. It is common in waste lands and road sides. Leaves are narrowly lanceolate. Long white flowers hang from upright branches. Sepals are long lasting and plants bear red berries. The species comes in coastal sandy soil also.

### *Clerodendron infortunatum (C, indicum, C. viscosum)*
#### Champagne Clerodendrum

Origin is Malay Archipelago. It is a semi woody shrub, 2-2.5m tall and branched. It is seen as under growth in forests upto 1500m. Flowers are white tinged with pink and are sweet scented. *C .infortunatum* is used in traditional medicine to treat cough, skin diseases, and tumours and as a vermifuge. It has anti-inflammatory action also. The plant is used in IPM in coconut. Leaves are used to control breeding of rhinoceros beetle which is a major pest attacking coconut (Chandrika Mohan, *et. al.,* 2001).

### *Clerodendron serratum*

Distributed through out India, in forests up to 1500m. It is woody shrub, somewhat straggling with bluntly quadrangular stem and branches. Leaves are usually three at a node and serrated. Flowers blue in color produced in cylindrical tiers. The plant flowers during August to December. It is suitable as a flowering shrub in shady land scapes. When pruned after each flowering, it bears a pretty appearance. The plant has well known medicinal uses in Ayurveda. Roots are bitter, acrid, anti inflammatory, digestive, carminative, expectorant and also used against skin diseases (Warrier, *et.al.,* 1995)

### *Clerodendron inerme*
#### Forest Jasmine

Native to India it is a straggling shrub, which forms a compact and clean hedge. The shrub can be pruned very low and it does not produce any leafless woody stem, and remains evergreen. White fragrant flowers appear during rains.It is easily propagated from cuttings. It is widely used for topiary. The leaf is used as insecticide especially against mites. The bio efficacy of the aqueous plant extract of *C. inerme* on tea mosquito bug is also proved. This species is used for management of root knot nematode in okra. It has fungicidal value and has inhibitory effect on TMV infection and controls transmission of the virus. Indian traditional practitioners use leaf extract of this species to treat oral cancer. The anti cancerous property was recently studied by Manoharan, *et.al.,* 2006.They reported that leaf extracts of *C. inerme* is very effective for treatment of oral cancers.

## *Clerodendron trichotomum*

### Harlequin Glory Bower

It is a small deciduous shrub having a height as well as spread of 2-3.5m, with multiple trunks creating appearance of a large shrub. A small tree can be produced by staking the plant until a sturdy trunk develops. The crown has round upright shape with moderate density and an irregular outline. It has simple oval dark green opposite leaves (15cm long) and delightfully fragrant beautiful white tubular flowers. White flowers emerge from the tubular red calyces in racemes of 18-27cm length. Flowering is usually seen in late summer in full sun. Fruits are blue in colour arising from red star shaped calyx and have metallic look and are fleshy. Flowers and fruits are sometimes found on the plant at the same time.

It requires a full sun to light shade and prefers moist well drained loamy soil. This species is moderately drought tolerant and usually not affected by pests and diseases. This shrub can grow more like a vine if it is not trained or pruned. The branches droop as tree grows and requires pruning to develop a strong structure. It can be trained to a standard or specimen plant in a container and used near patio or deck as a flowering ornamental or can be displayed at centre of a prominent location.

It sprouts and produces suckers regularly at the base of plant. *C. trichotomum* 'Carnival' is a new selection with bold cream variegated foliages. *C. trichotonium* fargesii is a late flowering shrub, with white flowers.

The plant has high medicinal value and leaves are anti-inflammatory as well as antioxidant. Stems are used as antihypertensive in Far East Asia, Japan and China Korea. Bark has anti proliferative action against cancerous growth. Active ingredient of *C. trichotonium* is identified as clerodendrins.

## *Clerodendron ugandense*

It is an evergreen climber native to tropical Africa. It bears blue and white flowers in panicles. It prefers fertile, moist and well drained soil and full sun. Pruning is recommended after plant has finished flowering. Flowering lasts to early summer.

## *Clerodendron wallichii (C. nutans)*

Commonly known as bridal veil, it originated in India and Japan. This graceful shrub grows up to 2.5m and has pointed narrow leaf of 12-27cm length. The white fragrant flowers are held in loose panicles which are more or less pendulous and around which the cascade of flowers appears. The plants should not be exposed to late afternoon sun.

### *Clerodendron kaempferi (C. squamatum)*

Originated in China it is a showy shrub with 1.5 to 2 m height having large round leaves. The scarlet flowers are borne in large clusters and are long lasting. It prefers partial shade. It is a good pot plant, propagated by tip cuttings.

### *Clerodendron bungei*

Native to Mexico and South America, it is otherwise called peanut butter plant because the leaves when crushed smell like bread spread. Flowers are large corymbs of rose red (bright pink) colour .The plant flowers from August to October. The plant is medium sized and semi herbaceous in habit, with bold and aromatic foliage. Flowers are also fragrant. The leaves have anti proliferative action against cancerous growth.

## Propagation and Cultivation

Majority of clerodendrons are shrubs and hence prefer shade especially in afternoon. Climbers like *C. splendens and C. thomsonae* flower profusely in full sun. Clerodendrons come well in humus rich soil with good drainage. They prefer slightly acidic soil. Most of the clerodendrons flower profusely during rainy season and hence have great potential as garden plants in regions with heavy rainfall. Propagation is by root suckers, cuttings and seeds. Since suckers are produced even at a distantance of 1m, this group of plants can be highly invasive. Care should be taken to restrict growth by severe and regular pruning and desuckering. In species like *C. paniculatum and C. speciosissimum,* lower stem of plants appears unattractive after flowering. Such species may be planted behind plants that maintain a low profile.

## Water Lily

Water lilies are aquatic perennials with beautiful flowers. There are about 50 species of water lilies in the world, plus numerous hybrids and varieties. They come in a wide variety of colours, including some varieties which bloom at night. A well grownup plant has several flowers open everyday and each flower lasts for about four to five days. This water plant can be grown in ponds or basins/containers which can hold 4 to 5 litres of water.

Water lilies are natives to Northern America, while a few belong to South Africa and Australia. It was introduced as an ornamental plant in many parts of the world and now it emerged as a weed in Canada and North America. Fragrant water lily and its hybrids are extremely popular water garden plants and can be readily obtained from nurseries. Water- lily is the National flower of Sri Lanka.

### *Botany*

Water lily leaves are perennial herbs belonging to family Nymphyaeceae. Leaves floating, usually peltate, with a thick waxy covering and 24-52cm wide. Mature leaves which have cleft at base with pointed leaf lobes are attached to flexible underwater hollow stalks rising from thick fleshy rhizomes in mud. The stem brings gases from the water's surface to the rhizomes and roots, and also conveys waste gases to the surface. The horizontal creeping and branching rhizomes (2-3cm in diameter) are attached by adventitious roots arising in groups below the leaf bases. The petioles leave crescent shaped scars on the rhizome when shed.

There are four sepals and numerous petals, gradually passing to stamens. Stamens are with petaloid filaments and small introrse anthers. Carpels sunk in the fleshy torus and with it forming many celled ovary crowned by radiating stigmas, with numerous ovules. Fruit is a spongy many seeded berry, ripening below the surface. Seeds minute, embedded in the pulp enclosed in a sac like aril.

## Types of Water Lilies

Water lilies are mainly of two types, tropical and hardy.

### *Tropical Water Lilies*

Tropical water lilies usually hold their flowers above the surface of water. There are day-blooming and night-blooming varieties. The flowers are open for 3or 4 days each. They need a water temperature of 21° C, though established plants can survive at a still lower temperature. The spread of Tropical water lilies ranges from 90 to 360cm.

Day-blooming varieties open in morning and close in evening whereas; night-blooming varieties open at about dusk and close by noon the next day.

### *Hardy Water Lilies*

Hardy water lilies can be grown in cooler climates. As long as the rhizomes do not freeze, they survive the winter. The spread of Hardy water lilies ranges from 90 to 240cm. They are available in different hues and some are changeable, they change colour over their 3 to 4 day-blooming period. Hardy water lilies do not bloom as abundantly as Tropical water lilies and they do not hold their flowers high over the water. Since these water lilies are hybrids, most are sterile and need to be reproduced through vegetative means. These varieties will not thrive in warm climates, as high temperatures cause them to burn and wilt.

### *Nympaea alba*
European White Water Lily

Naturally seen in temperate areas, they are robust hardy plants producing highly fragrant white flowers open through out the day.

### *Nympaea candida*

It is also a temperate species similar to *N.alba* but with smaller, round leaves cleft to leaf stalk. Flowers are pure white with yellow stamens.

### *Nympaea caerulea*
Egyptian Blue Water Lily

It is an African species. Lower surface of leaves have purple spots.It is a day blooming water lily with large leathery leaves and faintly scented light blue flowers with narrow petals.

### *Nympaea capensis*
Cape Blue Water Lily

A sub tropical species with day blooming flowers., the plant has robust habit with sinuate leaves with large sky blue flowers with pale center.

### *Nympaea gigantea*
Giant Water Lily

Plants with big leathery, glossy green leaves, about 50cm diameter, flowers light blue with dark blue tips and carried well above water surface. Petals are broad. It is a day blooming species with fragrant flowers.

### *Nympaea odorata*
Fragrant Water Lily

It is a temperate species. Leaves round, very thick with a split to the center. Flowers are white sweet scented, open only during morning hours.

### *Nympaea micrantha*

A viviparous plant. Buds in the sinus region get activated and grow into new plants. Leaves glabrous green on upper surface and light purple with brown spots on the lower. Flowers are violet in color with bright yellow anthers.

### *Nympaea omrana*

Plant with pink flowers. Stamens dark purple without any appendages. Leaves dark green on upper surface with slight purple color and green on the lower surface.

## *Nympaea rubra*

### Indian Red Water Lily

A beautiful tropical night blooming species, remains open until mid day. Flowers are deep carmine red. Leaves are red bronze in color on both surfaces.

#### ***Medicinal value***

Species named *Nymphaea caerulea* contains an anti-spasmotic called Nuciferin, and likely contains aporphine It is much like Cannabis, codeine or propoxyphene; maybe a little hallucinatory (at higher doses)- but mainly hypnotic like Cannabis/opiods. Actually, it is a pretty good sedative with a mild, opiod like feeling. Nuciferine, the drug, is known as an antispasmodic and a common constituent of plants in the *Nymphaea* genus. Called "Viagra" of the Pharoahs, Blue Water-lily was used by ancient Egyptians to enhance sex drive. Egyptian medicinal practitioners used this flower to stimulate blood flow, and as an anti aging treatment. The ancients worshipped Blue Water-lily as a visionary plant and it was the symbol for origins of life. When this flower was soaked in water or wine, and then ingested, it acted as an intoxicant. The Indians and Buddhist highly respected it for its visionary properties.

Recent biochemical studies showed a flavonoid in Blue Water-lily, similar to Ginkgo. It can be used by men and women as an aphrodisiac and as a general remedy for all illness. Blue Water-lily is a pain reliever; it stimulates circulation, relieves muscle spasms, and enhances sex drive. It is possible that Blue Water-lily could be effective in treating Alzheimer'sdisease and migraine.

#### *Propagation and Cultivation*

Water lilies prefer good sun light for producing good blooms and blooming heavily. It grows rooted in muddy or silty sediments in water up to 180–210cm deep. They grow in water as shallow as 18cm and as deep as 450cm. It prefers quiet waters such as ponds, lake margins and slow streams and grows in acid or alkaline waters. Beaver, moose, muskrat, porcupine and deer eat water lily leaves and roots and water fowl eats the seeds.

Plants are propagated through seeds and sprouting rhizomes or bulbs. The fruit which develops underwater is a spongy round berry, which when ripe releases up to 2000 seeds. *Nymphaea* rhizomes can be cut into 10 cm or larger pieces for propagation. A planted rhizome covers about a 450cm diameter circle in five years. The saplings and rhizomes can be planted within 18-24 cm deep from water surface and barely covered with soil. When several new leaves appear, more water can be added or moved to deeper water in the ponds.

*Seed propagation*

When a Water Lily blossom is fertilized, the flower retreats beneath the water's surface in a coiling spiral after normal blooming period of 3 or 4 days. Unfertilized flowers just float downward in water, their stems remaining straight. Seeds can be collected by placing muslin bags over the blossom as seeds ripen or by gathering a fertilized pod about 10 days after its submersion to allow it to ripen in a container of distilled water. Once the seeds are released from their pods, they may be planted. Shallow seed pans or margarine tubs can be filled with good garden soil. Spread the jellied seeds or individual seeds evenly over the surface of soil and lightly sprinkle sand to cover them; they are then gently watered. Place the containers in an aquarium or bowl of water with not more than 6or 9 cm of water covering soil's surface. They need to be set in a warm, well-lighted spot. Strong, direct light, however, can damage or kill the little plants. Once the first leaves reach the surface of water, plants may be picked out in clumps, carefully washed clean of soil and pulled apart from each other with toothpicks. Plant them separately in small pots or margarine tubs with the soil used above and cover them with 6 to 12 cm of water over the soil's surface. Bone meal and aquatic plant fertilizer should be incorporated in bottom half of container; monthly feedings aren't necessary. It will probably take 6 months for plants to grow large enough to plant outdoors. As the seedlings grow, care should be taken to avoid algal growth. Carefully pick away the algae by twirling it around a toothpick.

*Vegetative propagation*

Water Lilies have four kinds of roots. They all produce growing "eyes" or growing points on the tubers. The mexicana rhizomes produce tubers like corms or bulbs at the base or even embedded within the mother plant.

## Planting

Only one or two water lilies should be planted in a pot, container or small pool. The planting of rhizomes can be done at any time of the year. While planting water lily rhizomes, trim off the roots, leaving only a few, new roots to anchor and establish it, and cut leaf stems close to the rhizome. Remove dead and hollow pieces of rhizome and trim rhizomes so that they are only 6 or 9cm long. Any roots and foliage that are left attached should be covered with wet newspaper or damp towels to prevent them from drying out before they are planted.

Tuberosa and odorata -type of rhizomes should be planted in large and strong containers. Fill $3/4^{th}$ of pot with good topsoil enriched with bone meal. Mound the dampened soil and place the cut end of the rhizome against the wall of the pot. Tilt the tubers at 45° angle so that the growing tip will be at soil level. If there are roots, spread them across the soil. Fill rest of space with soil, patting it to decrease air pockets and do not cover the growing tip. Pot can be carefully lowered in water. Mexicana-type rhizomes and tropical water lilies should be planted in center of container because they usually grow vertically rather than horizontally.

Most containers used for potting are chosen based upon size of water garden and the purpose. The size of the pool or pond dictates size of container and varieties that will be grown. For the most part, containers that are about 45 cm in diameter and 30 cm deep are suitable for hardy water lilies. For planting tropical water lilies, containers that are 60cm in diameter and 30cm deep are suitable. The choice will be made based on the plant and the fact that larger containers produce larger plants and smaller containers tend to keep plants smaller in size.

Planting can be done in two ways; one method is by planting rhizome directly in potting media of the pool and the other method is by planting rhizome in the potting media filled in a pot or small container which is then placed in the pool. In the case of an empty or newly constructed pool, planting is done directly in potting media of the pool. A mixture of soil and compost or well-rotten cow dung manure is spread at the bottom of the pool, at least 15-25 cm deep. After planting, water is filled in pool slowly so that the crown is just above water surface. After 7-8 days, water is again let in slowly and this is continued till pond is completely filled. In about 6 to 8 weeks, the pond will be almost full. The second method of planting is adopted when planting is to be undertaken in an already existing pool which is filled or in case of changing a variety and introducing a new one. In such cases the planting is done in a pot or a container containing the potting media and then placing the pot in the pool in such a way that the crown of the plant is just above the water surface. The pot can be kept on bricks in the pool for about 7 to 8 days and can be lowered gradually when new growth appears on the plant.

### *How to Plant Water Lilies*

Soil used for planting should be damp. Fill containers about two-thirds full of soil. For hardy water lilies, the rhizome should be placed so that the growing point is directed toward the center and at a slight angle. Look for buds or "eyes," similar to what would be found on a potato to determine which end is the growing point. This is done so that it can grow across center of the pot. After the rhizome is in place, cover it with soil so that the growing point is just barely above soil level.

Tropical water lilies are planted much like hardy water lilies with one exception; they are planted in center of the pot. Potting emergent and submerged plants is similar to planting other aquatics. Place some soil in the pot, center the plant and add soil to cover the roots. Another method that works well is to fill container with soil, wet it thoroughly so it becomes muddy, and gently press the roots of the plant into the soil. Once planted, the soil should be covered with a 1.5 to 2cm layer of pea gravel. This prevents soil particles from clouding the water and discourages fish from digging into soil. After planting, water thoroughly and carefully lower the containers into the pond at the proper depth.

### *Placing Plants in the Pond*

Hardy water lilies grow well at water depths of 36-54cm over top of the pot. The plant can be placed in the pot at an initial depth of 18 cm and then lowered to its final placement or 36-54cm as the plant grows. If the pond is in full sun, the minimum amount of water over the top of the pot should be from 18-24 cm for the plant to grow well. If the pond is in shade (less than six hours of sun a day), then 18 cm of water is sufficient for growing tip to receive enough light. Tropical water lilies prefer to grow at water depths of 36cm but will tolerate 18-24 cm of water. In order to conveniently adjust water depth over the pots, bricks or inverted pots can be used as props under the plants to position them. As plants grow, the depth can be easily adjusted by adding or taking away supports.

### *Timing Placement of Plants in Pond*

Hardy water lilies and emergent plants are tolerant of cooler water temperatures and can be placed in the pond when temperatures reach about 10°C. Tropical water lilies and some floaters cannot tolerate cold temperatures and should not be placed in the pond until the water temperature reaches at least 21°C. Placing them out too early causes dormancy, injury and restrict potential growth of the plant.

### *Fertilizer Application*

Water lilies require good fertilization for getting continuous flowering and large sized blooms. Bone meal, groundnut cake, cow dung, goat manure and vermi compost can be applied at fortnight intervals.

### *After Care*

One or two oxygenators or aerating plants should also be grown along with water lilies as they complement the fishes which are generally grown in the pool. These oxygenators or submerged plants help to maintain natural balance which is very important for natural existence of plant life and fish. They release oxygen which is taken in by the fishes and consume carbon dioxide exhaled. About two dozen oxygenator plants are required for a two square metre pool. A few fish, preferably Gold fish are grown in the pool to check the growth of mosquitoes in the pool as they eat the eggs and larvae. A few scavengers like snails are also recommended but these are not so useful in a small pool.

Overgrowth of algae is problematic and to control, grow submerged plants like Anacharis. To keep the pond looking good, remove yellowing leaves and spent flowers every week. If an individual lily spreads too much, outer ring of leaves can be removed to reduce size of plant without affecting flowering. If aphids appear, hose them into water each morning to provide food for fish. Fish must be grown in ponds to prevent growth of mosquito larvae.

## Conclusion

Plants like Naregamia, clerodendrons and water lilies can be grouped as under exploited because either their ornamental, medicinal or other uses are not being utilized on a commercial scale. Naregamia an excellent flowering ground cover during rainy season can be utilized for landscapes in shaded conditions. Clerodendron flowers are nectar rich and attract butterflies and hummingbirds adding to aesthetic value of the area planted with these plants. Role of these plants in honey production is yet to be studied. Water lilies, though naturally seen in backwaters, canals and ponds, are yet to be utilized in ecotourism, for bio energy production and for treatment of polluted water from industry as these plants absorb many heavy metals.

## References

Chandrika Mohan, Nair C.P.R and Mohan, C.2001 A shrub to combat the rhinoceros beetle, *Indian Coconut Journal*, 31(11): 2-3.

John, S., Soniya, E.V., Valsala,K and Nair, G.M. 1997 Adventitious shoot formation from mature leaves and leaf derived calli of *Naregamia alata* W& A. *Indian Journal of Experimental Biology*, 35(11): 1249-1251.

Manoharan, S., Kavitha, K., Senthil, N and Renju, G.L. 2006 Evaluation of anticarcinogenic effects of *Clerodendron inerme* on 7, 12 – dimethylbenzanthracene – induced hamster buccal pouch carcinogenesis. *Singapore Medical Journal*, 47(12): 1038.

Warrier, P.K., Nambiar, V.P.K.and Ramnkutty, C. 1995 Indian Medicinal Plants - A compendium of 500 species, Vol.2. p.121-6, Orient Longman, New Delhi.

Warrier, P.K., Nambiar, V.P.K.and Ramnkutty, C. 1995 Indian Medicinal Plants - A compendium of 500 species, Vol.4.p.108, Orient Longman, New Delhi.

# Underutilized Trees

**Underutilized and Underexploited Horticultural Crops, Vol. 2, 2007**
*K.V. Peter (ed.), pp. 283-294*
New India Publishing Agency, New Delhi (India)
E-mail : newindiapublishingagency@gmail.com
Web: www.bookfactoryindia.com

# 16 Mahua

**SANJAY SINGH**
Central Horticultural Experiment Station
Vejalpur (CIAH- ICAR), Panchamahal, Gujarat - 389 340

Mahua botanically known as *Bassia latifolia* Roxb, (Hindi name: Mahua, Family: Sapotaceae) is an economical and multipurpose tree. Its flowers, fruits and oil obtained from seeds are eaten in various ways. Corolla commonly called mahua flowers is a rich source of sugar, containing appreciable amount of vitamins and minerals. Flowers are used for preparation of distilled liquors and potable spirits. Flowers and spent flowers (after fermentation) are also used as feed for livestock. Fruits are eaten as raw or cooked. Fruit pulp can be used as a source of sugar for alcoholic fermentation, whereas dry fruit husk can be used for preparation of absolute alcohol. Seed is a good source of oil. Amount of oil obtained from seeds is higher than many oil seed crops and oil seed-bearing trees. Oil obtained from kernel is used for edible purpose and used for preparation of vegetable oil. In Mahua oil, linoleic and unsaturated fatty acids are found which are useful for cardiac patients because it reduces cholesterol content in blood serum. Mahua oil is used in manufacture of soap, lubricating grease, fatty alcohols and candles. Mahua fat is a rich source of palmitic acid used in cosmetic and pharmaceutical industries. Mahua seeds can also be used for preparation of defatted flour, which has great potential in bakery products. Saponin obtained after extraction has industrial and commercial applications. Cake obtained after extraction of oil is used as manure and has insecticidal properties. The trees provide quality timber wood for various uses. Every part of Mahua yields an economic product of great potential value; hence it is a very useful tree.

## Origin and Distribution

A characteristic tree of dry region, it is found in North and Central India. It grows in Eastern Uttar Pradesh, Madhya Pradesh, Chattisgarh, Maharashtra, Bihar, Jharkhand, Orissa, Andhra Pradesh and Gujarat. In Rajasthan mahua grows on wastelands particularly in Southern part of state. Tree is very well known to rural folk since ages in India.

## Area and Production

Information regarding area and production in India are not available. Trees are common in villages in many states. They can be seen growing in parks, on roadsides, avenues and as windbreak. Production in India is concentrated in drier states and produce is collected by villagers and sold in local market. Its cultivation spreads to arid and semi- arid areas, resource-poor areas and wastelands where other crops cannot be grown.

## Soil and Climate

Mahua is very hardy and thrives well on rocky, gravelly red soils and also on saline and sodic soils. It grows even in pockets of soil between crevices of barren rock. For its better growth and productivity, well-drained deep loam soil is ideal.

Mahua prefers tropical and subtropical climate. It withstands drought admirably. Trees of *Bassia latifolia and B.longifolia* grow up to an altitude of 1,200 m. The *B. malabarica* grows in Western Ghats from Kannur to Travancore and also in the Himalayas. Trees of *B.butyracea* grow in Himalayan regions upto an altitude of 4,500m.

## Improvement

### *Reproductive Biology*

Singh *et al*, (2006) recorded that peak period of flowering and fruit set was March - April in different genotypes. Number of flowers (42.00) and fruit set/fascicle (9.50) were the highest in MH 2, while it was the least in MH 15. Variable (%) of anthesis was registered in different genotypes. Peak period of anthesis was recorded from mid night till morning in all genotypes. Anthers dehisced 2-4 days before time of anthesis. Pedicel length, pedicel thickness, bud length and breadth at flower opening varied from 4.20-5.50 cm, 2.00-3.00 mm, 1.30-1.80 cm and 1.00-1.20 cm respectively in different genotypes. Total number of sepals was four. Ovary and style length varied from 4.50-6.25 mm and 25.00-35.00 mm respectively. Pollen viability and pollen germination ranged from 90.00-98.50% and 26.50-40.20% respectively. On the basis of different reproductive attributes studied, fruit set/ panicle may be considered as one of positive traits while screening the elite genotypes of *Mahua*.

## *Genetic Diversity*

It is a highly heterozygous and a cross-pollinated fruit crop. As such, seedlings exhibit a wide range of variation, which aids in selection of superior desirable genotypes. Due to cross pollination and predomination of seed propagation over a longer period of time, it gives immense opportunity to locate elite trees having horticultural traits. Wide variations were observed in sweetness, acidity, size, shape and bearing habits in Mahua under Uttar Pradesh conditions (Singh *et al*, 1999).

Variability in physical characters of mahua fruits (Singh *et. al.* 1999)

| Maturity group | Flower wt. (g) | Juice (%) | Pomace (%) | Stamens (%) |
|---|---|---|---|---|
| Early | 1.5 - 2.8 | 66.4 - 69.6 | 28.3 – 30.8 | 0.8 – 1.4 |
| Mid season | 1.3 - 2.5 | 64.5 – 67.7 | 30.5 – 33.4 | 1.5 – 2.0 |
| Late | 1.2 - 2.6 | 62.8 – 64.1 | 34.1 – 35.0 | 1.3 – 1.8 |

Variability in chemical characters of fresh mahua flowers (Singh *et. al.* 1999)

| Maturity group | TSS (%) | Acidity (%) | Total sugar (%) | Vit. C (mg/100g) | Vit. A (IU) | Mineral (%) |
|---|---|---|---|---|---|---|
| Early | 25.0 – 26.7 | 0.09 – 1.0 | 23.8 – 25.5 | 42.9 – 60.9 | 637 – 650 | 1.6 – 1.6 |
| Mid season | 23.5 – 25.8 | 0.11 – 0.12 | 19.7 – 24.0 | 49.4 – 57.3 | 525 – 577 | 0.6 – 0.7 |
| Late | 21.0 – 26.0 | 0.10 – 0.12 | 19.9 – 24.0 | 48.8 – 49.2 | 528 - 696 | 0.7 – 0.8 |

Variability in physical characters of mahua fruits (Singh *et. al.* 1999)

| Maturity group | Fruit wt. (g) | Husk (%) | Seed (%) |
|---|---|---|---|
| Early | 16.8 – 35.3 | 58.0 – 71.7 | 28.3 – 42.0 |
| Mid season | 19.4 – 31.8 | 58.3 – 70.2 | 29.8 – 41.7 |
| Late | 18.5 – 20.5 | 70.8 – 71.9 | 28.1 – 29.2 |

Variability in chemical characters of mahua fruits (Singh *et. al.* 1999)

| Maturity group | Total sugar (%) | Vit. A (IU) | Vit. C (mg/100g) | Mineral (%) |
|---|---|---|---|---|
| Early | 13.0 – 13.8 | 586 – 667 | 47.5 – 59.2 | 1.6 –2.1 |
| Mid season | 11.5 – 13.2 | 823 – 860 | 48.3 – 72.5 | 1.8 – 2.3 |
| Late | 11.0 – 11.4 | 742 - 846 | 51.2 – 65.5 | 1.4 – 1.9 |

Variability in physical characters of mahua seeds (Singh *et. al.* 1999)

| Maturity group | Seed wt. (g) | Kernel (%) | Shell (%) |
|---|---|---|---|
| Early | 7.2 – 10.0 | 73.7 – 77.2 | 22.8 –26.3 |
| Mid season | 8.1 – 9.5 | 73.9 – 83.6 | 16.4 – 26.1 |
| Late | 5.2 – 6.0 | 76.8 – 78.8 | 21.2 – 23.2 |

Variability in chemical characters of mahua kernels (Singh *et. al.* 1999)

| Maturity group | Oil (%) | Free fatty acids (%) | Protein (%) | Minerals (%) |
|---|---|---|---|---|
| Early | 42 – 49 | 0.53 – 0.54 | 23.5 – 23.7 | 5.3 – 5.7 |
| Mid season | 42 – 45 | 0.59 – 0.62 | 22.6 – 22.7 | 3.9 – 6.4 |
| Late | 42 – 43 | 0.97 – 1.20 | 21.0 – 25.7 | 4.4 – 4.8 |

Singh and Singh (2005) surveyed area of diversity in Gujarat to identify elite genotypes among its population. Early flowering was observed in 1$^{st}$ week of March in MH 1, MH 4 and MH 5, while it was recorded late (2$^{nd}$ week of April) in MH 6. Flowers (45.00) and fruits (8.00)/ fascicle were recorded the highest in MH 2. Dry flower yield varied from 30.00 to 46.00 kg / plant being he highest in MH 2. Variation in fruit yield among genotypes ranged from 35.00 to 82.00kg / plant and it was maximum in MH 2.The highest total soluble solids (27.80 %), total sugar (24.24%) and vitamin C content (64.00mg/ 100g) were recorded in flowers of MH 4. Maximum fruit weight (34.50g), seed weight (14.50g), kernel oil (46.50%), minerals (4.90%) and protein content (24.00 %) were found in MH 2. Promising genotypes are being further exploited as potential parents to develop high yielding stable genotypes having positive horticultural traits.

## *Varietal Wealth*

Generally, trees of *B. latifolia* in North and *B.longifolia* in South bear flowers containing a large amount of sugar and seeds rich in quality oil. There is no improved variety of mahua for an organized orcharding. A large variability exists in its fruits. Recently selections are made at N.D.University of Agriculture and Technology, Faizabad (Uttar Pradesh), which are discussed bellow.

### *NM – 2*

Flowering takes place in first week of April. Average number of flowers/fascicle is 66, average weight of fresh flower 2.8g; average yield of flower (dry) of 7-year-old grafted plant is 9.9 kg. Fresh flower recorded 25.50% total sugar, 19.60% reducing sugar, 5.6 % non- reducing sugar, 60.90 mg/ 100 g Vitamin C, 650.70 IU Vitamin A and 0.60 % mineral content.

Total sugar content of dry flower was 58.70% with reducing sugar 55.60% and non-reducing sugar 3.0 %. Mineral content of dry flower was 2.20%. Ripening of fruit took place by second week of June. Average fruit weight 16.80g; husk content 58.80g %, total sugar13.60g %, reducing sugar 7.8g %, non- reducing sugar 5.80g %, vitamin C 51.30 mg/ 100 g, vitamin A 586.70 IU and mineral content 1.90 % were recorded. Average seed weight and kernel weight were 7.2 g and 5.50 g respectively. Oil content (47.40%), free fatty acid (0.53 %), protein content (23.60 %) and mineral content (5.70 %) were recorded. Unsaturated fatty acid (53.80 %) with oleic acid (40.30%) and linoleic (13.50 %) and saturated fatty acid (45.40 %) with palmitic (24.30%), stearic (20.20%) and archidic (0.90 %) were recorded.

### *NM- 4*

Flowering is in the fourth week of April. Average number of flowers/ fascicle is 54.6, average weight of fresh flower 2.6g and average yield of flower (dry) of 7-year-old grafted plant is 8.5 kg. Fresh flower recorded 24.50 % total sugar, 19.60% reducing sugar, 4.80 % non- reducing sugar, 49.20 mg/ 100 g Vitamin C, 528.20 IU Vitamin A and 0.67 % mineral content.

Total sugar content of dry flower was 57.30 % with reducing sugar 53.80% and non-reducing sugar 3.30 %. Mineral content of dry flower was 2.30%. Ripening of fruit took place by third week of June. Average fruit weight is 19.50 g, husk content 65.30 %, total sugar 11.20 %, reducing sugar 8.20 %, non- reducing sugar 3.20 %, vitamin C 55.80 mg/ 100 g, vitamin A 860 IU and mineral content 2.1 % were recorded. Average seed weight and kernel weight are 6.00 g and 4.70 g respectively. Oil content 45.20%, free fatty acid 0.63 %, protein content 25.70 % and mineral content 4.40 % were recorded.

Unsaturated fatty acid (50.60 %) with oleic acid (38.00%) and linoleic (12.60 %) and saturated fatty acid (48.50 %) with palmitic (19.90%), stearic (27.00%) and archidic (1.60 %) was recorded.

### *NM-7*

Flowering takes place in second week of April. Average number of flowers /fascicle 64.30, average weight of fresh flower 2.50g and average yield of flower (dry) of 7-year-old grafted plant is 10.00 kg. Fresh flower recorded total sugar 22.70 %, reducing sugar 19.50%, non- reducing sugar 3.00 %, Vitamin C 48.50 mg/ 100 g, Vitamin A 564.50 IU and mineral content 0.82 %.

Total sugar content of dry flower was 58.40 % with reducing sugar 56.60% and non-reducing sugar 1.50 %. Mineral content of dry flower was 3.00%. Ripening of fruit took place by third week of June. Average fruit weight is 21.8 g, husk content (55.00 %), total sugar (10.10 %,) reducing sugar (8.00 %), non- reducing sugar (2.30 %), vitamin C (59.20 mg/ 100 g), vitamin A (620.70 IU) and mineral content (2.4 %) were recorded. Average seed weight and kernel weight were 9.50 g and 7.20 g respectively. Oil content (46.70%), free fatty acid (0.55 %), protein content (22.40 %) and mineral content (4.80%) were recorded. Unsaturated fatty acid (49.20%) with oleic acid (34.40 %) and linoleic (14.90 %) and saturated fatty acid (47.90 %) with palmitic (20.80%), stearic (26.90%) and archidic (1.20 %) were recorded.

### *NM- 9*

Flowering takes place in fourth week of April. Average number of flowers / fascicle 54.6, average weight of fresh flower 2.2g and average yield of flower (dry) of 5-year-old grafted plant is 6.80 kg.

Fresh flower recorded 21.80 % total sugar, 20.00% reducing sugar, 2.10 % non- reducing sugar, 30.20 mg/ 100 g Vitamin C, 802.00 IU Vitamin A and 1.40 % mineral content.

Total sugar content of dry flower was 55.70 % with reducing sugar 54.30% and non-reducing sugar 1.60 %. Mineral content of dry flower was 2.90%.

Ripening of fruit took place by third week of June. Average fruit weight 16.70 g, husk content 60.00 %, total sugar 15.40 %, reducing sugar 12.20 %, non- reducing sugar 3.70 %, vitamin C 72.50 mg/ 100 g, vitamin A 890.00 IU and mineral content 1.80 % were recorded.

Average seed weight and kernel weight were 5.20 g and 4.0 g respectively. Oil content 48.50%, free fatty acid 0.48 %, protein content 16.80% and mineral content 4.10% were recorded. Unsaturated fatty acid (51.60 %) with oleic acid (37.60%) and linoleic (14.00 %) and saturated fatty acid (47.60%) with palmitic (19.50%), stearic (27.50%) and archidic (0.60 %) were recorded.

## Plant Propagation

### *Seed Propagation*

Mahua plants are propagated generally by seed giving a long gestation period (15-20 years) and large variability. Seeds after 5 to 6 days of storage at room temperature showed 80 % germination. Germination (%) in freshly extracted seeds was poor.

### *Raising of Seedlings*

Seedlings are raised to be used as rootstock. Mahua seeds have no dormancy. Fresh seeds after storage of 5-6 days at room temperature show 80% germination. Seeds can be sown on raised beds or in polythene bags about 30 cm deep during July. Seeds germinate within 10-15 days. Seedlings become ready to use as rootstock after 1 year. Polyembryony is found in mahua seeds. Therefore, 15-20% seeds produce nucellar seedlings.

### *Vegetative Propagation*

#### Soft Wood Grafting

When one-year-old seedlings raised for rootstock commence putting on new growth and leaves are of light green colour, these are ready for soft wood grafting. Shoots of 3-4 months, which have prominent apical bud, are taken as scion material. Defoliation of such shoots is done 8-10 days before detaching from parent tree for grafting operation. At time of removal of these shoots, apical buds should remain intact. Naturally defoliated shoots (March-April) are used for better success. Union is tied with 200-gauge polythene strip. If selection of scion and rootstock are proper, 60-70 % success may be obtained during March, April, July and August.

#### Soft Wood Grafting *in situ*

This method was proved better for establishing *in situ* Mahua orchard in gravelly soils and drier tracts where mortality of nursery raised grafts is very high. Freshly extracted seeds are sown in polythene bags for germination. After attaining a height of 5-10 cm, it should be directly planted in field at desired distance. These vigorous seedlings, attaining pencil thickness after six months are soft wood grafted with scion sticks of suitable genotypes. Such plants grow very fast and attain stature in a few years.

### Veneer Grafting

Veneer grafting is a successful method for multiplication of Mahua on one-year-old seedlings as rootstock. Shoots are taken from spring flush and operation was done in July and August. July is the ideal time for veneer grafting. Plants become ready for planting in September, just 2 months after grafting.

### Air Layering

Shukla (2005) reported that one year old shoots on mahua trees were ringed and upper end of cuts treated with IBA (1000 ppm). Ringed portion after hormonal operation was wrapped with moist moss grass. Rooted shoots were removed from mother plant 60 days after treatment and planted in polythene bags having size of 15 cm x 30 cm and filled with soil and FYM mixture (1:1).

## Agrotechniques

### *Planting*

Seedling plants should be planted 10m apart, whereas grafted ones 8m apart. Pits of 1 x 1 x 1m size are dug and filled with a mixture of top soil + 25 kg farmyard manure up to a level of 60m from ground level. Water pits are taken to settle soil. Planting is done in center of pit during July-September.

## Training and Pruning

Training is very essential to develop framework of mahua plants. Plants are allowed to grow straight with help of stakes. Framework should be developed by encouraging growth of 4-6 well-spaced branches on trunk at a height of 90-cm. from ground level. Pruning is not performed in mahua plants except removal of dead, diseased and crossing branches.

## Manuring and Fertilization

A dose of 10 kg farmyard manure, 100g N, 50g P and 75g K/plant should be given to one-year-old plant. It should be increased every year in same proportion up to 10 years. Fully-grown up trees require 100 kg farmyard manure, 1 kgN, 0.5kg P and 0.75 kg K Farmyard manure which should be applied during July-August. Half dose of N and full dose of P and K should be applied in July; remaining half dose of N should be applied by end of August under rainfed conditions. Manure and mixture of fertilizer should be spread under canopy of plants and incorporated in soil.

## Interculture, Weed Control and Intercropping

Light cultivation is required to keep plantation weed-free. Green manuring is useful if planted on poor soils. Intercropping with vegetables like bottle guard, ladys finger, cucumber and ridge guard may be followed in inter spaces upto 6- 8 years. Mahua based cropping system needs to be standardized.

## Irrigation

Irrigation is required during establishment of orchard. No irrigation is required for well grown up trees.

## Flowering and Fruit Set

In Mahua, flowers appear just after leaf fall (from February to April) in different agro climatic zones of country. Time taken for complete development of flower bud from its visible initiation to anthesis varies from 20 to 30 days. Average number of flowers / fascicle varies from 10 to 60. Young plantation of grafted Mahua (7- year old) shows heavy flowering with very poor fruit set (1.6-4.0 %). Dropping of young fruitlets was observed and only 8-13 % of set fruits reached maturity. Cause of poor fruit set may be due to poor pollination or self-incompatibility. Seedling trees of higher age have better fruit set than trees of lower age group.

## Maturity, Harvesting, Yield and Post Harvest Management

Singh *et al,* (2006) studied maturity standards in different genotypes of Mahua under Gujarat conditions and observed that fruit growth was faster initially and slowed down while reaching towards maturity. Total soluble solids, total and reducing sugars increased as fruits reached towards maturity. Titratable acidity increased during initial period of fruit development, then declined. Fruits of MH-1 and MH-5 are ready for harvest by 2$^{nd}$ week of May and that of MH-4 and MH-6 by 4$^{th}$ week of May. MH-2 and MH-3 may be harvested by 3$^{rd}$ week of May and 1$^{st}$ week of June respectively.

Mahua yields both flowers and seeds. Creamy coloured, fleshy corolla (flowers) falls in early hours of morning. They should be collected. Falling of corolla continues for 15-20 days. Ripe fruits are shed from trees during May - July. They are collected by hand for storage and in heaps for seed extraction. Yield of flowers (dry) varies from 100-150 kg and kernel 60-80kg/tree/year. Flowers before storage are dried. As soon as they are collected they should be dried on polythene sheet. Do not store them in bulk in moist condition. Flowers should be stored in thin layers. They should be turned over from time-to-time to bring lower layers to top. Dried flowers after packing in gunny bags are stored in a dry place up to 1 year under ordinary conditions. Seeds are extracted from ripe fruits. They are shelled within a week otherwise they germinate. Once seeds are germinated, they become unfit for oil extraction. Kernels obtained after shelling should be dried up to moisture content of 8% because seeds containing more than 7-8% moisture are liable to fungal attack. Dried kernels can be utilized for extraction of oil. Kernels can be packed in gunny bags and stored for a year. Both fresh and dried flowers are used for preparation of various nutritive food products, which form an article of diet. Mahua flowers are used largely in preparation of distilled liquors, alcohol and portable spirits. It is also used for preparation of cake, vinegar, jam, syrup and honey. Flowers and spent flowers (after fermentation) are utilized as feed for livestock. Fruit pulp can be used as a source of sugar for alcoholic fermentation. Dry fruit husk may be used for preparation of absolute alcohol (one tonne dry husk yields 130 liters of absolute alcohol). Seeds processed for oil are used to manufacture vanaspati, soap, greases and cosmetics. Seeds are also used for preparation of seed flour having great potential in bakery products. Saponin obtained after extraction has industrial and commercial applications. Mahua cake obtained is used as manure and has insecticidal properties. Every part of mahua yields economic products of great potential value; hence mahua tree constitutes most important raw materials for various industries.

## Marketing

About 75% of farmers sell their produce at farm level to the village merchants, retailers, big producers or to pre harvest contractors. They cannot afford to transport their produce to distant markets on account of non availability of transport facilities, expensive transport, and malpractices in market. Information regarding demand, supply, price, market outlook, knowledge of consumer's preference, marketing channels are important for marketing of produce.

## Plant Protection

### *Pest Management*

#### Bark Eating Caterpillar (*Inderbela spp.*)

It was observed as the most serious pest of Mahua. Larvae of this moth feed on bark of stem, interrupting translocation of sap and thus tree becomes weak and unproductive. Caterpillars 50-60 mm long have pale brown body, spines brown silken web on tree, which consists of their excreta and chewed wood particles. Pest is very active from February to March. To control the pest, remove webs from tree trunk and insert into holes cotton soaked in 0.06% dichlorvos (76 EC) solution. Close holes with mud paste. Monocrotophos (0.05 %) solution is also effective to control the pest.

### Mahua Leaf Roller (*Polychorosis cellifera*)

The larvae web leaves by folding tip downwards on both margins parallel to mid-rib and feed inside. In case of severe attack $1/4^{th}$ of lamina is eaten up. Regular clipping and burning of affected leaves can keep population under control. In case of severe attack, spraying with Chlorpyriphos 20 EC (2ml/l) or Endosulfan 35 EC (2ml/l) is recommended.

## Future Research Needs

There is considerable potential for expansion of Mahua cultivation in India. A few suggestions for future research priorities are given below.

The plant genetic resource (PGR) research needs to be undertaken on classification of genetic diversity through use of morphological, biochemical and molecular techniques. Efforts may be made in using molecular techniques for understanding genetic structure of a crop. Promising genotypes having tolerance to biotic and abiotic stress should be selected.

Model nurseries for local supply of quality plant material should be established.

Dwarfing rootstocks are the best because they induce dwarfness and enable to facilitate easy management of orchard. Such type of work may be initiated and be standardized for different agro climatic conditions

Information should be made available on Mahua based cropping system for different normal and problematic soils.

Agro techniques like integrated nutrient management, diversified farming system, high density planting system, weed management, canopy management and irrigation management should be standardized under different ecosystem of country.

Maturity, harvesting, grading, packaging and storage system should also be standardized.

There is need to develop new products from Mahua flowers and popularize them not only in domestic market but also in international market.

Research information on integrated pest management should also be made available.

## References

Singh I.S, Srivastava , A.K. and Singh, V. 1999. Improvement of some underutilized fruits through selection. *Journal of Applied Horticulture*, 1:34-37.

Shukla, S.K.2005. Propagating mahua by air layering. *Indian Horticulture*, April- June, p16.

Singh, Sanjay and Singh, A K 2005. Studies on genetic diversity in *Mahua(Bassia latifolia)* under semi- arid ecosystem of Gujarat". *Indian Journal of Agricultural Sciences*, 75: 519-523.

Singh, Sanjay, Singh, A K Bagle, B.G. and Dhandar, D.G. 2006. Developmental pattern and maturity indices in mahua (*bassia latifolia*) genotypes under semi- arid ecosystem of Gujarat. Paper presented in 8th Indian Agricultural Scientists and farmers congress, organized by B H U, Varanasi and Bioved Research and communication Centre Allahabad 21-22, February, 2006, p123.

Singh Sanjay, Joshi, H K, Singh, A.K., Bagle, B.G. and Dhandar, D.G. 2006. Reproductive biology of Mahua (*Bassia latifolia*). Paper presented in National symposium on under utilized horticultural crops, 8-9th June, 2006 at IIHR, Bangalore.p60.

**Underutilized and Underexploited Horticultural Crops, Vol. 2, 2007**
*K.V. Peter (ed.), pp. 295-309*
New India Publishing Agency, New Delhi (India)
E-mail : newindiapublishingagency@gmail.com
Web: www.bookfactoryindia.com

# 17 Chironji

**SANJAY SINGH**
Central Horticultural Experiment Station
Vejalpur (CIAH- ICAR), Panchamahal, Gujarat - 389 340

*Chironji or Charoli* (*Buchanania lanzan* Spreng Family: Anacardiaceae) is an excellent tree of agro forestry and social forestry. In wasteland development and Dryland horticulture, it assumes great significance due to multifarious uses and capacity to withstand adverse climatic conditions. At present, it is growing under forest condition as an under exploited fruit and gives monitory reward to tribal community of country and seems to be a boon for them. It is a valuable species found in dry deciduous forest throughout the country excluding eastern Himalayan forests (Singh, 1982). Flesh of ripe fruit is very palatable and is largely eaten raw or roasted and oily kernels are the most important part and are used in preparation of sweet meats and puddings (Singh, 1982). Mesocarp of fruit is edible and cherished by children (Munde *et al*, 2003). Kernel is highly nutritious and rich in protein (18.00 %) and yields sweet oil, which can be used to substitute olive and almond oil. Kernel contains 33.50 % oil, 1.90 % of which is unsaponifiable. Saponifiable part contained 20.00 % of linoleic acid. Chironji oil is non-repellant and non-toxic and is suitable for human consumption (Banerjee and Jain, 1988). Hemawathy and Prabhakar (1988) reported that total kernel lipid extracted from chironji was 65.60 % of dry kernel and consisted of 90.40 % neutral lipids, 3.4 % glycolipids and 6.2 % phospholipids. Kernel can be used for preparation of sweets. It provides quality timber wood for various uses.

## *Genetic Diversity*

No standard cultivars of chironji are available since a little work was done to exploit genetic resources of chironji. This crop is facing severe genetic erosion as a result of activities related to deforestation. A number of seedling strains with lot of variation in respect of fruit weight, total soluble solids, acidity, protein and earliness particularly in Bihar, Uttar Pradesh, Gujarat and Maharashtra provide good scope for selection of better varieties. A survey was undertaken in Uttar Pradesh to investigate nature and extent of variability present in chironji seedling progenies for morphological characters of tree and high degree of variability was observed with regard to fruit yield, seed size and chemical composition of fruit Rai, (1982). Singh and Chaturvedi, (1983) and Chadhar and Sharma, (1997) surveyed chironji growing area and recorded significant differences in morphomatrics and yield attributes. An extensive survey and exploration were carried out in Gujarat by Singh *et al,* (2006). They reported that the earliest flowering (first week of February) took place in 'CPT 1', 'CPT 5', 'CPT 10', 'CPT 15', 'CPT 17', 'CPT 19' and 'CPT 22', while the latest (4$^{th}$ week of February) occurred in 'CPT 13'. Maximum panicle length (35.11 cm) was found in 'CPT 1', while 'CPT 18' recorded the least panicle length (14.20 cm). Ripening time varied from third week of April to second week of May in different genotypes. The highest fruit yield / plant was found in 'CPT 7' (28.00 kg/ plant). With respect to all horticultural traits studied, the genotypes, 'CPT 1', 'CPT 2', 'CPT 5', 'CPT 6', 'CPT 7', 'CPT 8', 'CPT 12' and 'CPT 30' were promising. Vegetatively propagated promising genotypes are planted in field for their further evaluation.

Fruit yield and fruit quality attributes of different chironji genotypes (Singh *et al,* 2006)

| **Genotypes** | **Fruit yield/ plant (kg)** | **Fruit weight (g)** | **Pulp (%)** | **TSS (%)** | **Vitamin C (mg/100g)** |
|---|---|---|---|---|---|
| CPT 1 | 20.00 | 1.35 | 48.88 | 22.20 | 51.50 |
| CPT 2 | 16.50 | 1.31 | 58.02 | 21.21 | 46.50 |
| CPT 3 | 15.50 | 1.33 | 55.63 | 20.21 | 58.00 |
| CPT 4 | 11.00 | 1.33 | 50.37 | 20.58 | 57.94 |
| CPT 5 | 24.00 | 1.25 | 54.40 | 19.31 | 60.50 |
| CPT 6 | 26.80 | 1.12 | 61.60 | 21.84 | 42.21 |
| CPT 7 | 28.00 | 1.04 | 60.57 | 23.50 | 46.35 |
| CPT 8 | 18.00 | 0.98 | 61.22 | 23.00 | 47.00 |
| CPT 9 | 12.50 | 0.98 | 50.00 | 21.00 | 47.51 |
| CPT 10 | 14.00 | 0.97 | 49.48 | 19.00 | 55.35 |

*Contd...*

| | | | | | |
|---|---|---|---|---|---|
| CPT 11 | 16.90 | 1.09 | 44.04 | 21.00 | 61.31 |
| CPT 12 | 13.00 | 1.10 | 53.64 | 21.10 | 55.20 |
| CPT 13 | 15.00 | 1.20 | 48.33 | 20.00 | 57.00 |
| CPT 14 | 12.60 | 1.30 | 48.46 | 20.00 | 56.00 |
| CPT 15 | 10.20 | 1.07 | 53.27 | 19.50 | 64.00 |
| CPT 16 | 09.00 | 1.30 | 58.46 | 22.00 | 45.30 |
| CPT 17 | 13.25 | 1.34 | 61.19 | 20.00 | 47.20 |
| CPT 18 | 14.55 | 1.32 | 61.36 | 19.80 | 49.35 |
| CPT 19 | 15.00 | 1.20 | 61.66 | 19.90 | 42.26 |
| CPT 20 | 10.50 | 1.18 | 55.08 | 22.10 | 49.00 |
| CPT 21 | 14.00 | 1.19 | 57.14 | 21.00 | 56.35 |
| CPT 22 | 12.00 | 1.22 | 60.65 | 21.20 | 57.94 |
| CPT 23 | 10.00 | 0.98 | 62.24 | 21.50 | 52.32 |
| CPT 24 | 12.00 | 0.99 | 54.54 | 21.30 | 54.61 |
| CPT 25 | 13.20 | 1.13 | 62.83 | 21.40 | 58.37 |
| CPT 26 | 14.50 | 1.18 | 62.71 | 19.40 | 60.31 |
| CPT 27 | 15.80 | 1.18 | 56.77 | 19.80 | 52.62 |
| CPT 28 | 15.00 | 1.22 | 59.01 | 20.10 | 42.95 |
| CPT 29 | 16.00 | 1.24 | 57.25 | 22.12 | 42.36 |
| CPT 30 | 18.20 | 1.28 | 58.59 | 21.50 | 42.87 |
| CD (P= 0.05) | --- | 0.08 | 1.28 | 0.74 | 1.30 |

Stone and kernel quality of different chironji genotypes (Singh *et al*, 2006)

| **Genotypes** | **Stone weight (g)** | **Kernel weight (g)** | **Kernel (%)** | **Kernel protein (%)** |
|---|---|---|---|---|
| CPT 1 | 0.69 | 0.14 | 20.28 | 28.50 |
| CPT 2 | 0.55 | 0.13 | 23.63 | 30.70 |
| CPT 3 | 0.59 | 0.12 | 20.33 | 29.53 |
| CPT 4 | 0.66 | 0.10 | 15.15 | 26.34 |
| CPT 5 | 0.57 | 0.13 | 22.80 | 26.00 |
| CPT 6 | 0.43 | 0.12 | 27.90 | 28.14 |
| CPT 7 | 0.41 | 0.09 | 21.95 | 30.00 |

*Contd...*

| | | | | |
|---|---|---|---|---|
| CPT 8 | 0.38 | 0.09 | 23.68 | 26.77 |
| CPT 9 | 0.49 | 0.09 | 18.36 | 24.87 |
| CPT 10 | 0.49 | 0.11 | 22.44 | 27.00 |
| CPT 11 | 0.61 | 0.09 | 14.75 | 25.10 |
| CPT 12 | 0.51 | 0.08 | 15.68 | 24.00 |
| CPT 13 | 0.62 | 0.10 | 16.12 | 23.53 |
| CPT 14 | 0.67 | 0.12 | 17.91 | 25.90 |
| CPT 15 | 0.50 | 0.08 | 16.00 | 26.90 |
| CPT 16 | 0.54 | 0.13 | 24.07 | 26.32 |
| CPT 17 | 0.52 | 0.08 | 15.38 | 26.94 |
| CPT 18 | 0.51 | 0.10 | 19.61 | 28.31 |
| CPT 19 | 0.46 | 0.10 | 21.73 | 30.33 |
| CPT 20 | 0.53 | 0.11 | 20.75 | 30.28 |
| CPT 21 | 0.51 | 0.12 | 23.52 | 29.21 |
| CPT 22 | 0.48 | 0.11 | 22.91 | 28.37 |
| CPT 23 | 0.37 | 0.11 | 29.73 | 25.00 |
| CPT 24 | 0.45 | 0.11 | 24.44 | 24.94 |
| CPT 25 | 0.42 | 0.12 | 28.57 | 24.54 |
| CPT 26 | 0.44 | 0.13 | 29.55 | 25.24 |
| CPT 27 | 0.51 | 0.12 | 23.52 | 28.28 |
| CPT 28 | 0.50 | 0.10 | 20.00 | 29.29 |
| CPT 29 | 0.53 | 0.10 | 18.86 | 28.28 |
| CPT 30 | 0.53 | 0.11 | 20.75 | 24.94 |
| CD (P= 0.05) | 0.02 | 0.03 | 0.26 | 0.21 |

## *Botany*

Chironji or Charoli is highly heterozygous, cross-pollinated and as such seedlings exhibit a wide range of variation. It is a dicot woody plant of deciduous nature. Flowers are typical of family Anacardiaceae. Flowering starts January-February on well-developed panicles, which have hermaphrodite and pseudo hermaphrodite flowers, and they are located on annular prominent disc and fruit is drupe. Fruits become ready for harvest during April and May.

## *Reproductive Biology*

Reproductive biology of crop was studied at Central Horticultural Experiment Station, Vejalpur, Gujarat. It was observed that peak period of panicle emergence and flowering were recorded in January and February respectively. Peak period of fruit set was recorded in February in majority of genotypes. The highest panicle length (35.11cm) and fruit set / panicle (36.20) were noted in CPT 1. Time taken for complete development of flower ranged from 16-20 days. Variable % of anthesis / dehiscence was registered in different genotypes. Peak period of anthesis was recorded between 6-11 am in all genotypes. None of genotypes showed anthesis before 4 am and after 2 pm. Anther dehiscence commenced after opening of flowers i.e. at 7 am and continued till 3 pm. Peak period of dehiscence was recorded between 8 am - 12 noon in all genotypes. Flower diameter varied from 5.12-6.30 mm. Stamen and carpel length varied from 1.97 - 2.12 mm and 1.22 - 1.38 mm respectively being at top in CPT 13. Pollen viability ranged from 54.55-70.38 %. Pollen germination and pollen diameter ranged from 20.00 - 35.00 % and 49.10-63.18 micron respectively. Maximum stigma receptivity was recorded on day of anthesis. Fruit set / panicle was positively and significantly associated with panicle length and it may be observed while selecting elite genotypes.

## Origin and Distribution

A tree of dry region is found in North, West and Central India. Plants maybe seen in forest areas of Uttar Pradesh, Madhya Pradesh, Chattishgarh, Maharashtra, Bihar, Jharkhand, Orissa, Andhra Pradesh and Gujarat.

## Area and Production

Information regarding area and production of this fruit in India is not available because it is not grown on plantation scale. They can be seen growing in forest areas. Production in India is mainly concentrated in drier states and produce is collected by villagers and sold in local market. Its cultivation spreads to arid and semi- arid areas, resource-poor areas and wastelands where other crops cannot be grown.

## Soil and Climate

Chironji is very hardy plant and thrives well on rocky, gravelly red soils and also on saline and sodic soils. Though it is a very hardy tree, it does not survive under waterlogged conditions. It grows even in pockets of soil between crevices of barren rock. Trees are grown on degraded rocky area including salt-affected soils. For its better growth and productivity, well-drained deep loam soil is ideal. It prefers tropical and subtropical climate and withstands drought admirably.

## Plant Propagation

### *Seed Propagation and Raising of Seedlings*

Chironji plants are propagated by seed giving a long gestation period (15-20 years) and large variability. Germination (%) generally in freshly extracted seeds is poor because of hard seed coat. Seed germination of 83.00% could be achieved within 18 days with satisfactory seedling growth by mechanically damaging stony endocarp before sowing month of June, plant survival of 93-96 % with satisfactory plant growth was obtained when one year old seedlings were planted in 60x60x60 cm pit filled with red soil and 10 kg of well rotten farmyard manure coupled with proper mulching around plants during winter and summer season and with or without thatching (Shukla and Solanki, 2000). Seed treatment with sulphuric acid (5-7 %) was very suitable for promotion of seed germination in chironji. Seeds can be sown on raised beds or in polythene bags about 30 cm deep during June-July and it germinates within 25-35 days. Seedlings become ready for grafting after one year from date of sowing.

### *Vegetative Propagation*

#### Soft Wood Grafting

When one-year-old seedlings raised for rootstock commence putting on new growth and leaves are of light green colour, these are ready for soft wood grafting. Shoots of 3-4 months, which have prominent apical bud, are taken as scion material. Defoliation of such shoots is done 8-10 days before detaching them from parent tree for grafting operation. At time of removal of these shoots, apical buds should remain intact. Top of new growth of stock is cut and scion is fitted by cleft grafting or splice grafting. Union is tied with 200-gauge polythene strip. If selection of scion and rootstock are proper, 60-70 % success may be obtained during July-August.

#### Soft Wood Grafting *in situ*

This method was better for establishing *in situ* Chironji orchard in gravelly soils and drier tracts where mortality of nursery raised grafts is very high. Freshly extracted seeds are sown in polythene bags for germination. After attaining a height of 5-10 cm, it should be directly planted in field at desired distance. These vigorous seedlings, attaining pencil thickness after one year are soft wood grafted with scion sticks of suitable genotypes. Such plants grow very fast and attain stature in a few years. Main advantages of this method are better plant growth and no mortality.

### Chip Budding

Chip budding was reported as one of successful methods for vegetative propagation of Chironji (Tewari and Bajpai, 2001). In this method of budding, a healthy bud is selected in axil of 2 months old leaf emerging on new flush. Leaf blade is removed with help of sharp knife leaving petiole intact. Upper cut is given about 2.0 cm above bud which goes downwards, taking wood portion with it simultaneously, up to 1.0-1.5 cm below bud, then lower cut is given about 1.0 cm below bud so as to take out bud with wood. Same rectangle cut is made on rootstock and bud is placed at juncture. Bud is pressed by hand to remove open space if any and tied tightly with white polythene strip. In case, if cuts on rootstock are wider, at least one side bark of scion and rootstock must be matched properly. Rootstock is detopped about 10 cm above bud to facilitate bud sprouting. Bud sprouts within a month of budding indicating that bud establishes vascular connection with rootstock. After union, top of seedling is cut a little above bud union and polythene strips are carefully removed. If selection of scion and rootstock are proper, appreciable success may be obtained during of July-August.

### Root Cutting

Experimentally, this method was tried with varying degree of success. Due to poor viability of seeds and slow growth of seedlings, Buchnania species has not so far been raised in plantations despite high price of kernels in market. Rooting in root cuttings is vey difficult, however with use of auxins, appreciable success is obtained. Two- year old root cuttings having thickness of 3.6-5.5 cm, treated with IAA (1600 ppm) recorded 67.66 % rooting (Singh *et al.*, 2002). Root cuttings are successful, but the method has limited use as it results in serious injury to mother trees when large number of root cuttings are taken. Moreover, taproot is also absent in plants that affect longevity of plants particularly in degraded lands.

### Rootstocks and Use of Polycontainers

Rootstock selection for vegetative propagation of Chironji is important as it controls vigour and equilibrium between yield and quality. Dwarfing rootstocks are the best because they induce dwarfness and enable to facilitate easy management of orchard. Such type of work may be initiated and be standardized for different agro climatic conditions.

Raising of rootstock in nursery beds and lifting budded plants with earth ball in highly sandy soils are practically not feasible. Transportation of plants from long distance may also cause high mortality particularly under semi arid and arid environment. To reduce time for raising rootstock and to avoid damage during handling and transportation, polyethylene bags may be used on commercial scale. Generally polyethylene bags (25cm x 15cm) with small holes in bottom filled with porous rooting medium are used for raising rootstocks. Generally 1-2 seeds are sown in each poly bags and then placed in trench bed, so that it can be irrigated easily.

## Agrotechniques

### *Planting*

Seedling plants should be planted 10m apart, whereas grafted ones 8m apart. Pits of 1m x 1m x 1m size are dug and filled with a mixture of top soil + 25 kg farmyard manure up to a level of 30 cm from ground level. After onset of monsoon, soils are settled and planting is done in center of pit during July- August. Drenching of pits with insecticides are required if there is problem of termites.

## Training and Pruning

Training is very essential to develop framework of chironji plants. Plants are allowed to grow straight with help of stakes. Framework should be developed by encouraging growth of 4-6 well-spaced branches on trunk at a height of 90-cm. from ground level. Pruning is not performed in chironji plants except removal of dead, diseased and crossing branches.

## Manuring and Fertilization

A dose of 10 kg farmyard manure, 100g N, 50g P and 75g K/plant should be given to one-year-old plant. It should be increased every year in same proportion up to age of 10 years. Fully-grown up trees require 100 kg farmyard manure, 1 kgN, 0.5kg P and 0.75 kg K. Farmyard manure should be applied during July-August. Half dose of N and full dose of P and K should be applied during July and remaining dose of N may be applied by end of August under rainfed conditions. Manure and mixture of fertilizer should be spread under canopy of plants and incorporated in soil.

## Weed Management

Productivity of Indian orchards can be increased only when all aspects of production technology including weed management are given due consideration. Weeds injure crops very slowly in a subtle way. Weeds complete their life cycle in a shorter time compared to fruit trees and compete for light, water and mineral nutrients and reduce yields. In new and old orchards, hoeing, hand weeding and ploughing the land 2-3 times a year are done to suppress weed growth. Intercropping and mulching may also be followed to control weeds.

## Intercropping

Intercropping is intended to maximize land and space use efficiency to generate supplemental income particularly during initial unproductive phase of orchard to protect inter space from losses through weeds, erosion, impact of radiation, temperature, wind and water and enriching it by nitrogen fixing legume crops. Compatible crop combination is necessary with regard to species, cultivars, planting method and sequence. Peas, gram, lentil, black gram, cowpea, cluster been may be grown as inter crops in orchard. Experimental evidences are, however, not available.

## Irrigation

Plant survives under rainfed conditions. If irrigation facilities are available, basin system of irrigation should be used for young plantations. For fully-grown up trees, strip or bed system should be followed. Irrigation should be given during summer at interval of 15 days. Proper moisture after fruit set is important for retention and development of fruits.

## Flowering and Fruit Set

In chironji, flowers appear in January- February in different agro climatic zone of the country. Time taken for complete development of flower bud from its visible initiation to anthesis varies from 18 to 28 days. New plantation of grafted plants start flowering after 4$^{th}$ year. Cause of poor fruit set may be due to pollination problem or self-incompatibility. Seedling trees of higher age have better fruit set than tree of lower age group.

## Maturity, Harvesting and Yield

Singh *et al,* (2006)-studied developmental pattern in chironji genotypes and it was observed that fruit growth was faster initially and slowed down while reaching towards maturity and followed sigmoid growth curve. Deep purple colour appeared on fruit surface of different genotypes during peak period of ripening. Specific gravity showed increasing trend (more than one) in all genotypes during development. Total soluble solids, total sugar, reducing sugar and kernel protein increased as fruits reached towards maturity. Titratable acidity showed declining trend during development. Vitamin C increased during development and remained constant till harvest period. TSS and titratable acidity ranged from 19.50-22.00 % and 1.42- 1.50 % respectively during ripening. TSS/ acid ratio was recorded 15.71, 14.68, 13.47, 13.99, 13.73 and 14.69 in CPT 1, CPT 2, CPT 3, CPT 4, CPT 5 and CPT 6 respectively during ripening.

Fruits become ready for harvest during April and May. Singh et al, (2006) reported that ripening time varied from third week of April to second week of May in different genotypes and the highest fruit yield / plant was in 'CPT 7' (28.00 kg/ plant).

## Marketing

About 75% of farmers sell their produce at farm level to village merchants, retailers, big producers or to pre harvest contractors. They cannot afford to transport their produce to distant markets on account of e non availability of transport facilities, expensive transport and mal practices in market. Information regarding demand, supply, price, market outlook, knowledge of the consumer's preference, marketing channels are important for marketing of produce.

## Plant Protection

### *Pest Management*

#### Hoppers (Sucking Pest)

This is the most damaging insect during flowering. Both adults and nymphs damage the crop. They suck sap from tender shoots and panicles. Panicles wither away and fruit set is adversely affected. They excrete honeydew, which develops a sooty mould on leaves and panicles. It can be controlled by spraying with dimethoate (0.03%) or phosphomidon (0.05%), once at time of panicle emergence and then again at fruit set stage.

#### Mealy Bugs

These insects are characterized by a waxy coating over their bodies. Mealy bugs are found in large number on ventral surface of leaflets, base of leaf petioles, tender shoots and even fruits, and suck cell sap from different parts. Leaflets become chlorotic and fall down and in case of severe infestation, there may also be premature fruit fall. Spraying with dimethoate (0.03 %) or phosphomidon (0.05%) is effective to control pest.

### *Disease Management*

#### Phytophthora Gummosis

The fungus affects bark and penetrates in a limited way into wood. In advanced stages, bark cracks, shreds in lengthwise strips as it dries and gum starts exuding. In infection, if disease can be detected before bark covering more than half of circumference of a stem is damaged, scraping of affected portions with a little extra healthy tissue without injuring wood and application of Bordeux paste (1:2:16-20 water) on scraped and healthy portion around infected zone, may help in combating disease and recovery of plant.

### Powdery Mildew

During flowering season, in the beginning, appearance of grayish whitish powder on flower buds, fruitlets and rachis of panicles denotes the attack of fungus. In severe cases, whole panicle appears scorched. Within a few days of first visible symptom, all panicles are affected. To have effective control, 2-3 sprays of wetable sulphur at 15 days interval is the most economical method.

## Future Research Needs

There is considerable potential for expansion of chironji cultivation in India. A few suggestions for future research priorities are given below.

1. The plant genetic resource (PGR) research needs to be undertaken on classification of genetic diversity through use of morphological, biochemical and molecular techniques. Efforts may be made in using molecular techniques for understanding genetic structure of crop. Promising genotypes having tolerance to biotic and a biotic stress should be selected.
2. Model nurseries for local supply of quality plant materials should also be established.
3. Dwarfing rootstocks are the best because they induce dwarfness and enable facilitation of easy management of orchard. Such type of work may be initiated and be standardized for different agro climatic conditions
4. Information should be made available on chironji based cropping system for different normal and problematic soils.
5. Agro techniques like integrated nutrient management, diversified farming system, high density planting system, weed management, canopy management and irrigation management should be standardized under different ecosystems of country.
6. Maturity, harvesting, grading, packaging and storage system should be standardized.
7. There is need to develop new products from chironji kernels and popularize them not only in domestic market but also in international market.

8. Research information on integrated pest management should be made available.

## References

Anon. 2002 Annual report, Central Institute for Arid Horticlture, Bikaner, PP-3.

Banerjee, A. and Jain, M. 1988 Investigations on *Buchanania lanzan* Spreng seed oil. *Fitoterapia*, 59 : 406.

Chadhar, S.K. and Sharma, M.C. 1997 Trends in fruit production in *Buchanania lanzan* trees. *Vaniki Sandesh*, 21: 1-3.

Hemawathy, J. and Prabhakar, J.V. 1988 Lipid composition of chironji (*Buchanania lanzan* Spreng) kernel. *Journal of Food Composition Analysis*, 1 : 366-370.

Munde, V.M., Shinde, G.S., Sajindranath, A.K., Prabu, T. and Machewad, P.M. 2003 Correlation and path analysis studies in charoli (*Buchanania lanzan* Spreng). *South Indian Horticulture*, 50: 517-521.

Rai, Y.C. 1982 *Buchanania lanzan* Spreng. Studies on methods of propagation and estimation of fruit yield. *Indian Forester*, 108 : 501-511. 93-96.

Shukla, S.K. and Solanki, K.R.2000 Studies on seed germination, plant survival and growth of Chironji (*Buchanania lanzan* Spreng). *Journal of Tropical Forestry*, 6 : 44-49.

Singh, Jagdish, Banerjee, S.K. and Francis, A. 2002 Vegetative propagation of *Buchanania lanzan* Spreng root cuttings. *Indian Forester*, 128 : 700-704.

Singh, R.P. and Chaturvedi, O.P. 1983 Primary production of deciduous forest at Varanasi. *Indian Forester*, 109 : 255-260.

Singh, R.V. 1982 Fodder trees of India. Oxford and IBH Publication, New Delhi. PP-663.

Singh, Sanjay, Singh, A. K. and Apparao, V V. 2006 Genetic diversity in Chironji (*Buchanania lanzan* Spreng) under semi arid ecosystem of Gujarat. *Indian Journal of Agricultural Sciences* (Accepted).

Singh Sanjay, Singh A.K., Apparao V.V., Bagle, B.G. and Dhandar, D.G. 2006 Phsio-biochemical changes during fruit development in chironji (*Buchanania lanzan* Spreng). National symposium on under utilized horticultural crops 8-9th June,2006 at IIHR, Bangalore.pp 78.

Tewari R K and Bajpai, C K 2001 Propagation of chironji by chip budding. *Agroforestry News Letter*, 13: 1-2.

*Plates Contd...*

Chironji Seeds

Chironji Rootstocks

Soft Wood Grafting

Flowering in Chironji

Ripening Stage

Grafted Plant

MH- 10

MH- 14

Soft Wood Grafting

MH- 6

MH- 4

**Underutilized and Underexploited Horticultural Crops, Vol. 2, 2007**
*K.V. Peter (ed.), pp. 311-321*
New India Publishing Agency, New Delhi (India)
E-mail : newindiapublishingagency@gmail.com
Web: www.bookfactoryindia.com

# 18 Drumstick

**SALIKUTTY JOSEPH**
Assciate Professor, Kerala Agricultural University
P O KAU, Thrissur - 680 656

Moringa (*Moringa oleifera*) is one of the most popular perennial vegetables grown in South India. It is commonly grown as a unique vegetable. Its fruits, leaves and flowers are equally useful. They are cut into slices and used in the preparation of a number of food items. *Moringa oleifera* is the most widely cultivated species of a monogeneric family, the Moringaceae. The tree is indigenous to North West India. It is found wild in the sub Himalayan tract from the river Chenab eastward to the Sarda and in Tarai tract of Uttaranchal in India. Moringa is native to the sub-Himalayan tracts of India, Pakistan, Bangladesh and Afghanistan. This rapidly growing tree (also known as horseradish tree, drumstick tree, benzolive tree, kelor, marango, mlonge, moonga, mulangay, nébéday, saijhan, sajna or Ben oil tree, radish tree, West Indian ben), was utilized by ancient Romans, Greeks and Egyptians. It is now widely cultivated and has become naturalized in many locations in tropics. It is a perennial softwood tree with timber of low quality, which for centuries was advocated for traditional medicine and industrial uses. The name drumstick is derived from the pod shape resembling slender and curved stick used for beating the drum. Fruits also resemble the siliqua of radish and hence the name "radish tree".

The tree is widely distributed in India, Egypt, Philippines, Srilanka, Thailand, Malaysia, Burma, Pakistan, Jamaica and Nigeria. It is already an important crop in India, Ethiopia, the Philippines and the Sudan, and is being grown in West, East and South Africa, tropical Asia, Latin America, the

Caribbean, Florida and the Pacific Islands. All parts of the Moringa tree are edible. According to Fuglie (1999), the many uses for Moringa include: alley cropping (biomass production), animal forage (leaves and treated seed-cake), biogas (from leaves), domestic cleaning agent (crushed leaves), blue dye (wood), fencing (living trees), fertilizer (seed-cake), foliar nutrient (juice expressed from the leaves), green manure (from leaves), gum (from tree trunks), honey- and sugar cane juice-clarifier (powdered seeds), honey (flower nectar), medicine (all plant parts), ornamental plantings, biopesticide (soil incorporation of leaves to prevent seedling damping off), pulp (wood), rope (bark), tannin for tanning hides (bark and gum) and water purification (powdered seeds). Moringa seed oil (yield 30-40% by weight), also known as Ben oil, is sweet non-sticking and non-drying oil that resists rancidity. It has been used in salads, for fine machine lubrication, and in manufacture of perfume and hair care products (Tsaknis *et al*, 1999). In the West, one of the best known uses for Moringa is use of powdered seeds to flocculate contaminants and purify drinking water (Berger *et al*.1984., Gassenschmidt *et al*, 1995.and Olsen, 1987), but the seeds are also eaten green, roasted, powdered and steeped for tea or used in curries (Gassenschmidt. *et al*, 1995). This tree has in recent times been advocated as an outstanding indigenous source of highly digestible protein, Ca, Fe, Vitamin C, and carotenoids suitable for utilization in many of the so-called "developing" regions of the world where undernourishment is a major concern. The crop is valued mainly for the tender pods, which are cut into slices or pieces and used in culinary preparations. The leaves and flowers are also eaten as vegetables. The pods after harvest are boiled, dried, ground into powder and the resultant produce is exported to foreign countries to be used as "ready made powder" for culinary preparation. Drumsticks has highest demand in Singapore. Since the Drumsticks has antibiotic properties like Penicillin it has much more demand in countries like England, Japan, Canada, Bangkok, Doha, Dubai, Baharen, Muskat, Daman etc.

A paste of leaves is used as an external application for wounds. Flowers are used as tonic, diuretic and cholagogue (Chopra *et al*., 1949). Seeds are antipyretic, acrid and bitter. Seed oil is applied in rheumatism and gout. Fruits, leaves and flowers are used in culinary preparation. Immature fruits are cut into pieces and used in several culinary dishes either after scraping of tough skin or with skin intact. Roots of plant are used for seasoning pickles and in parts of South India for flavouring 'ghee' for which purpose; the leaves are put in melting butter just before butter turns to ghee. In some parts of Dutch East Indies, dried seeds are fried and eaten. It is also said that the seeds yield a valuable oil known as 'Oil of Ben' by watch makers for lubricating delicate machines (Esteves, 1962) and the oil is used in perfumery and illumination.

Fruits of moringa are rich in vitamin B and C and the trees are having medicinal properties and used in treatment of rheumatism and as cardiac circulatory stimulant.

### *Medicinal Use*

The drumstick is valued mainly for its tender pods, which are relished as vegetable but all its parts - bark, root, fruit, flowers, leaves, seeds and even gum - are of medicinal value.

1. They are used in the treatment of ascites, rheumatism and venomous bites as antiseptic and as cardiac and circulatory stimulants.
2. The root, in prescribed doses, is given in intermittent fevers, paralytic affections, epilepsy and hysteria and externally in palsy, chronic rheumatism, and enlargement of spleen, dyspepsia and also in bites by rabid animals.
3. The juice of the leaves or a compound spirit made of roots and orange peel with a little bruised nutmeg is useful in fainting fits, giddiness, nervous debility, spasmodic affections of the bowels, hysteria and flatulence.
4. A decoction or infusion of the root with addition of bruised mustard seeds is used in ascites caused by diseases of the liver and spleen.
5. This decoction or infusion is also useful as a gargle in soreness of mouth and throat, and pain in the gums due to dental caries.
6. Freshly extracted juice of root bark is used to relieve otalgia by pouring it into ears and also into hollow of the tooth in cases of dental caries.
7. Its seeds are acrid and stimulant. Oil of seeds is applied in gout and rheumatism. The leaves are rich in vitamins A and C and are considered useful in catarrhal affections.
8. The juice is mixed with honey and applied to the eyelids in cases of eye diseases.
9. A paste of the leaves is used as an external application for wounds. Poultice of the leaves is useful in reducing glandular swellings.
10. Its flowers are used as stimulant, tonic, diuretic and cholagogue. They are useful in increasing the flow of bile.

11. Its gum, mixed with sesamum oil, is dropped into the ears in otalgia. The gum, rubbed with milk and made into a paste, can also be applied to the temples in headache.
12. It is also applied to buboes and to painful bones in syphilis.
13. The pods made into a soup are prescribed as a diet in sub-acute cases of enlarged liver and spleen, articular pains, tetanus, debility of nerves, paralysis, pustules, patches and leprosy.
14. A curry made from the unripe pods acts as a preventive against intestinal worms.
15. Drumsticks other miraculous quality-its ability to purify water-has been used by households for centuries. But it has only recently been tested commercially. Powdered Moringa seeds, when added to murky, bacteria-laden water, act as a coagulant, binding to the bacteria and silt and falling to the bottom of the vessel. The clean water can then be poured out.
16. *M. oleifera* is also of interest because of its production of compounds with antibiotic activity such as the glucosinolate 4 alpha-L-rhamnosyloxy benzyl isothiocyanate.

### *Nutritional Importance*

Drumstick is a vitamin A rich, mineral packed, nutritious vegetable of the tropics and subtropics. Moringa trees are used to combat malnutrition, especially among infants and nursing mothers. It is advocated as "natural nutrition for the tropics." Leaves can be eaten fresh, cooked, or stored as dried powder for many months without refrigeration, and reportedly without loss of nutritional value. Moringa is especially promising as a food source in tropics because the tree is in full leaf at the end of the dry season when other foods are typically scarce.

A number of reports on nutritional qualities of Moringa now exist in both scientific and popular literature. Moringa leaves contain more Vitamin A than carrots, more calcium than milk, more iron than spinach, more Vitamin C than oranges, and more potassium than bananas. The protein quality of Moringa leaves rivals that of milk and eggs. Countless instances of lifesaving nutritional rescue are attributed to Moringa (Fuglie, 1999, Fuglie, 2000). Fruits contain crude fibre (4.8g/100g), phosphorus (110mg/100g), carotene (110µg/ 100/g), and vitamin 'C' (120 mg/100g) besides other nutrients. In fact, nutritional properties of Moringa are now so well known that there seems to be little doubt of substantial health benefit to be realized by consumption of

Moringa leaf powder in situations where starvation is imminent (Fahey 2005). The nutritive value of pods, leaves and leafpowder are given in Table below.

## Nutritinal Value of Leaves and Pods

Analysis of Moringa pods, fresh (raw) leaves and dried leaf powder showed them to contain following per 100 gram of edible portion:

| Components | Pods | Raw Leaves | Lead Powder |
|---|---|---|---|
| Moisture (%) | 86.9 | 75.0 | 7.5 |
| Calories | 26 | 92 | 205 |
| Protein (g) | 2.5 | 6.7 | 27.1 |
| Fat (g) | 0.1 | 1.7 | 2.3 |
| Carbohydrate (g) | 3.7 | 13.4 | 38.2 |
| Fiber (g) | 4.8 | 0.9 | 19.2 |
| Minerals (g) | 2.0 | 2.3 | - |
| Ca (mg) | 30 | 440 | 2003 |
| Mg (mg) | 24 | 24 | 368 |
| P (mg) | 110 | 70 | 204 |

## Botany

Drumstick belongs to family *Moringaceae*. There are two common species, *Moringa oleifera* and *M. concanensis*, former being the vegetable species. *M. oleifera* is distinguished by tripinnate leaves, leaflets 12-18 mm long with yellow or white petioles. *M.concanensis* is distinguished by bipinnate leaves and leaflet 15-30 cm long. Petals are with red streaks.

Drumstick is a perennial tree with a fragile stem and grows to a height of more than 10 meters. Recently, a few annual types have been developed which are high yielding and respond well to ratooning. Several types could be broadly classified based on morphological characters like branching habit of tree, pigmentation of floral parts and fruits (Irulappan, 1991).

The tree ranges in height from 5-15m with an open, umbrella shaped crown, straight trunk (10-30cm thick) with corky, whitish bark. The tree produces a tuberous tap root system, which helps in drought conditions. The evergreen or deciduous foliages (depending on climate) are spirally arranged, crowded at ends of branches, long petioled, incompletely tripinnate, on the rachises with glands between the pinnate and the leaflets (1-2 cm in diameter) which are stalked, ovate or obovate; base acute, obtuse or rounded, often oblique; apex obtuse, rounded or emarginate; entire, dull green on both sides, unequal, 0.7-1.4 cm long, 0.25-0.5 cm wide.

Flowers are pleasantly fragranced, white or cream coloured and are borne profusely in axillary drooping panicles 10-25 cm long. Petals are 5 unequal, yellowish, white with a greenish base, with thin veins, the two hindmost and the two lateral ones reflexed, ovate or obovate, obtuse, with a canaliculated base, inner side hairy at the base, otherwise glabrous, 1-1.17 cm long, 0.5-0.6 cm wide; the foremost petal erect, obovate, obtuse, glabrous in the inside, on the outside with two longitudinal rows of hairs, 1.4-1.6 cm long, 0.6-0.8 cm wide. Perfect stamens are 5 alternating with 5 subulate staminodes densely pilose at the base, the hindmost stamen the longest, 0.8-0.9 cm long, and the other ones much shorter. Ovary is stalked, densely clothed with rather long, appressed hairs. Ovules, style thin, curved, white, shortly pubescent, hollowed at the apex. The fruits (pods) are initially light green, slim and tender eventually becoming dark green and firm up to 120cm long depending on the variety. Fully mature, dried seeds are round or triangular shaped, the kernel surrounded by a lightly wooded shell with three papery wings.

## Varieties

There are a few named varieties. A variety, Jaffna is a perennial type grown in parts of South India, which bears long (60-90 cm) pods with a tasty soft flesh. The other types are " chavakacheri murunga", which bears fruits as long as 90 to 120 cm, "chemmuruuga" (red tipped fruits) which flowers throughout the year and wild "Kattumurunga" bearing small and inferior fruits. There are two distinct types in Thirunelveli district of Tamil Nadu, with the name "Palmurungai" and "Punamurungai" the former being refused for its thicker pulp and bitter taste. The "kodikkal murungai" of Trichy district mainly used as supports for the betel vine plants, produces many short pods of 15-20 cm length.

Besides these local types, different varieties and types of drumsticks are found in specific part of India. But no research has been done in different agricultural institutes and universities. The universities in south India are ahead in these researches. There are certain improved varieties developed by TNAU and Tamil Nadu state Department of Agriculture. Varieties of Drumsticks available are the following.

1. Kokan Drumsticks
2. Rohit-1
3. Coimbatore-2
4. P.K.M. -1
5. P.K.M. -2
6. Durga
7. Chem Moringa
8. G.K.V.K. -1
9. G.K.V.K. -2
10. G.K.V.K. -3
11. Jafna
12. K.M.1
13. Chavalavacheri
14. Yalpanam
15. Shabanam

### *KM-1 (Kuduminmalai -1)*

This variety was developed at Anna Pannai, Kuduminmalai of Pudukkottai. It is a selection from an annual type, propagated by seeds. The trees bear heavily producing 400-500 fruits /year. The plants are medium tall, so that the harvest becomes easy. The plants come to bearing 6 months after planting. Plants can be ratooned for 2-3 years by cutting the trunk at a height of 1 metre and can be replaced by fresh seedling after 3 years.

### *PKM-1*

This variety evolved at Periyakulam of TNAU, is a pure line selection from the population generated by continuous selfing of seed moringa type for six generations. The plants grow to a height of 4-6 m and flowers in 90-100 days after planting. As the seedlings become 75cm high, the shoot tips are to be nipped off to encourage the side branches and higher pod number. The first harvest starts 160-170 days after planting. Average yield is 33-kg/ year (200-225 fruits). The pods are 65-70 cm long with 6.3-cm girth and 150 g weight. This variety is very pulpy. Ratoon crops can be taken for 3 years. Every year after the harvest is completed; trees have to be cut back to about 1m from ground.

### *PKM – 2*

This variety is evolved at Periyakulam of TNAU.The pods are greenish in colour and tastes good. The length of each stick is 45 to 75cms and each plant can yield 300 to 400 sticks.This variety is a good yielder but requires more water.

### *Rohit-1*

This variety was developed by one farmer, Mr. Balasaheb Marale in Maharashtra after 7 years of continuous efforts. This variety has proved to be superior in production and quality.The fruits are dark green in colour with an average length of 45 to 60 cms with more pulp inside. The average weight of each stick is 60 to 70 g.Each plant yields about 400 to 600 pods per year which may go up to 1250 .Economic yield can be harvested for 10 years.

## Soil and Climate

Drumstick comes up very well in all soils, except stiff clays. In sandy loams rich in organic matter, it performs the best. This crop performs well under dry and arid tracts and yields profitably. A pH range of 6- 7.5 would be the most suitable. The crop puts forth luxuriant growth when temperature ranges 25-30$^0$C. Higher temperature of above 40$^0$C will result in heavy flower shedding. The crop is also lost by frost.

Originally considered as a tree of hot, semi-arid regions (annual rainfall 250 - 1500mm) it is well adapted to hot, humid, wet conditions with annual rainfall in excess of 3000mm. The tree was originally considered suitable only for lowland cultivation at altitudes less than 600m. The adaptability of the tree is demonstrated by discovery of natural stands at altitudes of 1200m in Mexico and in excess of 2000m in Zimbabwe. *M.oleifera* can also grow in a variety of soil conditions.It prefers well drained sandy or loamy soils, heavier clay soils will be tolerated but water logging should be avoided. The tree is tolerant to light frosts and can be established in slightly alkaline soils up to pH 9.

## Cultivation

Field is ploughed three to four times. Pits of 45cm$^3$ are taken at spacing of 6 m x 6m for perennial types and 2.5 x 2.5 m for annuals. Topsoil is mixed with 10 kg of FYM and pits are filled.

Perennial types are propagated by limb cuttings of 1 to 1.35 m length and 1 to 16 cm circumference. Limb cuttings are planted in situ during June-August. Sometimes the uneconomic trees are cut down leaving a stump, from which 1 to 2 shoots are allowed to grow. From the shoots, cuttings of 2 m length and 4-5 cm diameter are selected and used as planting material.

Annual types are propagated by seeds. Seeds can be either sown in situ in prepared pits or can be transplanted after raising seedlings in poly bags. About 625 g of seeds would be sufficient to raise the crop in one hectare. Seedlings can be planted one month after sowing. Additional 50-60 plants can be raised in polybags separately if direct sowing is resorted to, so that gap filling can be taken up one month after planting.

When seedlings attain 75 cm length, shoot tips are nipped off to encourage side branches. Three months after sowing, each pit is supplied with 100 g Urea, 100g Super phosphate and 50 g of Muriate of potash and irrigated. Again six months after sowing, the plants are top dressed with 100 g Urea.

There should not be any moisture stress till the establishment of cutting or completion of seed germination. Later, irrigation is done once in 10-15 days. If the soil is too dry or wet, there will be severe flower shedding, resulting in poor fruit set. In the young plantation, inter crops like cowpea and okra can be cultivated during initial years.

## Harvesting

The annual types come to harvest six months after sowing while the perennial types raised from cuttings take 8-9 months for bearing. The perennial types take even three years to bear, if propagated through seeds and hence not advised. During the initial two years, yield of plants raised from cutting will be low (80-90 fruits/ year). It gradually increases to about 500-600-fruits/ tree/ year in the fourth and fifth year and the pods are harvested mainly in March- June. Under South Indian condition, an additional second crop can be harvested during September- October.

The annual crops are seasonal in bearing and those crops sown during July- September alone will come to harvest in 6 months time. Fruits of sufficient length and girth are harvested before fibre development. The period of harvest extends for 2-3 months. After the completion of harvest, the trees are cut back to 1 m height, for ratooning. The ratooned plants develop new shoots and start bearing in 4-5 months. About three ratoons can be taken in a plant. At each and every ratoon, plants are supplied with 20-35 kg FYM and the recommended NPK.

The perennial types can be retained without ratooning for 15 years. Then they are removed and fresh planting is resorted to. The old trees can also be pollarded or cut back to about 0.3 - 0.45 m height from the ground level and manured with organic matter (25 Kg).

## Pest and Diseases

Drumstick is not affected by serious pests or diseases. Common infested caterpillar pests are *Eupterote mollifera* and *Noorda blitealis*. They infest the unopened flower buds, causing them to drop. Moreover, two species of ash weevil, *Myllocerus temiclavis* and *Millocerus viridanus* found to attack the drumstick trees. In Tamil Nadu, root- rot caused by *Diplodia sp*. was observed. The hairy caterpillar causes defoliation in drumstick, is controlled by spraying with fish oil resin soap. The caterpillar pest can also be controlled by burning with lighted torches and by rogor or 'fish oil resin soap' emulsion spray.

## Solar Dehydration of Moringa

Since it is a seasonal fruit, it can be utilized effectively by dehydration (Amutha and Krishnamoorthy, 1999). Selected moringas after washing are cut into pieces and exposed to blanching (steam blanched for 5 min and sulphited (steam blanched samples are dipped in 0.1% KMS solution for 30 min.). These are loaded uniformly in a tray and dried in a solar drier. Drying ratio is 1: 6.1.The dried Moringas can be reconstituted by cooking in boiled water for 20 minutes. Color and texture are better in blanched and sulphited pieces. Dried pieces can be packed in polyethylene bags and stored in airtight metal tins.

In solar drying, nutrient loss is minimal and dried products can be utilized throughout the year. It can be easily reconstituted and cooked well.

### References

Amutha, S. and Krishnamoorthy, P., 1999 Solar dehydration of Moringa, *Indian Horticulture,* Vol., p.19.

Berger, M.R., Habs.M., Jahn,S.A. and Schmahl.S., 1984 Toxicological assessment of seeds from *Moringa oleifera* and *Moringa stenopetala*, two highly efficient primary coagulants for domestic water treatment of tropical raw waters. *East African Medical Journal* Vol.61, pp. 712-716.

Chopra, R.N., Badhwar and Ghosh, S., 1949 Poisonous Plants of India, Delhi, pp.161-163.

Esteues, A.B., 1962 Hort *Abstracts*, Vol.32, p.5422.

Fuglie, L.J., 1999 The Miracle Tree: *Moringa oleifera*: Natural Nutrition for the Tropics. Church World Service, Dakar. pp. 68; revised in 2001 and published as The Miracle Tree: The Multiple Attributes of Moringa, pp.172

Fuglie, L.J., 2000 New Uses of Moringa Studied in Nicaragua. ECHO Development Notes #68.

Gassenschmidt, U., Jany K.D., Tauscher, B and Niebergall, H., 1995 Isolation and characterization of a flocculating protein from *Moringa oleifera* Lam. *Biochimica Biophysica Acta* 1243, pp.477-481.

Irulappan,I.,1991 (In), *Recent Advances in Tropical Vegetable Production*- Compendium of Lectures, Peter K V; Joseph, Salikutty and Indira P.(Eds).Kerala Agricultural University, Vellanikkara, pp.273-307.

Fahey, J. W., 2005 *Moringa oleifera:* A review of the Medical Evidence for its Nutrititional, Therapeutic and Prophylactic Properties, *Trees for Life Journal*, part I, 1:5.

Olsen, A., 1987 Low technology water purification by bentonite clay and *Moringa oleifera* seed flocculation as performed in Sudanese villages. Effects on *Schistosoma mansoni* cercariae. *Water Research,* 21(5): pp.517-522.

Tsaknis J., Lalas, S., Gergis, V., Douroglou, V. and Spiliotis, V 1999 Characterization of *Moringa oleifera* variety Mbololo seed oil of Kenya. *Journal of Agricultural and Food Chemistry* Vol.47, pp.4495-4499.

# Underutilized Spices

**Underutilized and Underexploited Horticultural Crops, Vol. 2, 2007**
*K.V. Peter (ed.), pp. 323-351*
New India Publishing Agency, New Delhi (India)
E-mail : newindiapublishingagency@gmail.com
Web: www.bookfactoryindia.com

# 19 Diversity in Turmeric (*Curcuma longa* L.)

**A. MANOHAR RAO**
Senior Scientist (Horticulture)
Acharya N.G.Ranga Agricultural University
Rajendranagar, Hyderabad - 500 030

Turmeric (*Curcuma longa* L.) is a tropical herbaceous plant extensively cultivated in India, Sri Lanka, parts of China and Pakistan. Country of origin is not known with certainity, it is either of Chineese or Indian origin. India ranks first in area (1.49 lakh ha.) and production (5.27 lakh tonnes.) in the world and it is grown in most of the states in the country. Major turmeric growing states are Andhra Pradesh (56,000 ha.), Orissa (23,000 ha.), Tamilnadu (17,000 ha.), Assam (12,000 ha.), West Bengal (11,000 ha.) Maharashtra (6,000 ha.) Karnataka (6,000 ha.) and Kerala (3,000 ha.). Andhra Pradesh ranks first in area and production in the country. Area under this crop is increasing in North Eastern States.

Turmeric belongs to family *Zingiberaceae* and genus *Curcuma*. Genus *Curcuma* has 42 species, of which *Curcuma longa* is extensively cultivated for culinary purposes, *Curcuma aromatica* for flavour and recently *Curcuma amada* (mango ginger) for preparation of chutneys. Turmeric is mainly cultivated for under ground rhizomes, rich in curcumin. Turmeric is traditionally used for medicinal, religious, culinary purposes and also as a cosmetic and dye (Shah 1997). Recent reports in controlling human cancer has given renewed interest on the crop. In Ayurveda, turmeric is regarded as aromatic, a stimulant tonic, carminative and anti helminthic. Essential oil of turmeric is antiseptic and is used in treating gall stones and gall complaints (Pruthi 1976). As this crop is grown in India since ancient times and also in different parts of country, a lot of diversity is reported for growth, yield and quality characters.

## Botany

Turmeric is an erect perennial herb, grown as an annual and in certain areas cultivated as a biennial crop. It possesses an underground stem or rhizome, thick and rounded, with short blunt fingers with varied plant height. Aiyer (1954) and Parry (1962) reported height of turmeric in between 50 to 120 cm. Similarly significant variation for plant height among varieties grown in different agro climatic conditions were observed by Pillai and Nambiar (1975), Rao *et al*. (1975), Shah *et al*. (1982), Kumar and Jain (1998) and Rao (2000). Leaves are borne in a tuft, alternate, obliquely erect or sub sessile, lanceolate, green acuminate with a long leaf stalk with varied number of leaves ranging from 8 to 9 leaves in Dindigam variety (Rao *et al*. 1975) to 33.2 leaves in Rajapuri by Cholke (1993). Leaf length varies from 30 to 40 cm and breadth 14 to 16 cm. Inflorescence is terminal and borne in between leaf sheaths. Flowers are pale yellow equalling the bracts. The underground rhizomes having commercial importance due to presence of curcumin consist of two distinct parts; the egg or spindle shaped mother rhizome with branched primary and secondary finger rhizomes with yellow to bright orange yellow colour inside the rhizome.

Detailed cytological investigations by Nambiar (1979) revealed that earlier reports of chromosome numbers of 2n=32, 62 and 64 for *C longa* and 42, 63 and 86 for *C. aromatica* were probably exceptional cases and correct chromosome numbers for these species are 2n=63 and 84 respectively.

### *Curcuma species and Genetic Resources*

The genus *Curcuma* has 40 to 50 species in Tropical Asia. Among 40 species, only two species viz., *Curcuma longa L.* and *Curcuma aromatica Salisb.* are commercially cultivated for production of turmeric. More than 50 turmeric types exist in India belonging to *Curcuma longa* and a smaller extent to *Curcuma aromatica* and most of these cultivars go by local names only (Nair *et al*. 1980). Other species, *Curcuma angustifolia, C. aerugnosa, C. caesia, C. mangga, C. purpurascens, C. xanthorhiza and C. zeodaria* are also cultivated in limited extent in different areas. Two closely related species *Curcuma perolata and Curucma roscoeana* are natives of Cambodia, grown for their decorative foliage and blossoms.

Important characteristics of cultivated species C. *longa,* C. *aromatica* and C. *amada* are as follows.

### *Curcuma longa: (Popularly Known as Longa Types)*

Most of the high yielding cultivars fall in this group and mature in between 8-9 months. Pseudo stem is stall, robust with oblong elliptic leaves narrowed at the base. Plant height reaches a height up to 1m with 8 to 10 leaves. Spike length ranges 10-18 cm.wiith pale yellow flowers (Bailey 1963). Starchy root tubers are not produced at end of fibrous roots. Rhizomes are bigger in size with more curcumin but moderate in volatile oil.

### *Curcuma aromatica (Popularly Known as Kasthuri Types)*

In this group cultivars mature early, within 6-7 months. Pseudo stem is short with eliptic/oblong leaves. Flowering is common in this species, tetraploid, fertile and set viable seed. (Nair *et al.* 1982). Plants produce peculiar root tubers at the end to fibrous roots. Rhizomes possess pleasant unique aroma due to rich volatile oil but moderate in curcumin content. Rhizome production is less as compared to *C. longa* types. It is used in preparation of dyes, cosmetics and drugs but not used as a spice or condiment.

### *Curcuma amada (mango ginger)*

Popularly known as mango ginger cultivated in India for its rhizomes, which have the odour of raw mangoes and used for preparation of pickles.

Popular cultivars of Turmeric grown in India

| Popular cultivars | Growing area/ Sustability | Yield (t $ha^{-1}$) | Reaction to pests & Diseases |
|---|---|---|---|
| *Andhra Pradesh*<br>*Popular Kasthuri types* | | | |
| Kasthuri Kothapeta | East and West Godavari Districts | 15-20 | Susceptible to leaf spot |
| Kasthuri Tanuku | West Godavari | 12-15 | Susceptible to leaf spot |
| Kasthrui Amalapuram | Central delta in East Godavari | 10-12 | Susceptible to leaf spot |
| Chaya Pasupu | Agency Area of Godavari,Vizag and Srikakulam | – | – |

*Contd...*

| | | | |
|---|---|---|---|
| *Popular Kesari types C. longa* | | | |
| Kesari Duvvur | Cuddapah | – | Susceptible of leaf blotch |
| Amruthapani Kothapeta | East Godavari | 25 | – |
| *Popular long duration types* | | | |
| Duggirala | Krishn and Guntur Districts | 32 blotch | Susceptible to leaf |
| Tekurpeta | Rayalaseema | – | Tolerant to leaf spot. |
| Mydukur | Cuddapah | 32 | Susceptible to rot and leaf spot |
| Armoor | Northern districts of Telangana | 25 | Susceptible to leaf spot, rot and fly |
| Sugandham | Cuddapah | 20-25 | Susceptible to blotch and rot |
| Vontimitta | Cuddapah | 20 | – |
| Avanigada | Kurnool | – | – |
| Nandyal | Krishna | 15-18 | – |
| *Tamilnadu* | | | |
| Erode | Tamil Nadu | 30-32 | – |
| Selam | Tamil Nadu | – | – |
| *Kerala* | | | |
| Alleppey | – | 25 | – |
| Mannutty Local | – | 24 | – |
| *Assam* | | | |
| Shillong | | 40 | Tolerant to leaf blotch and rot |
| Tallikarbi | Karbi region of Assam | 30-40 | Tolerant to leaf spot and Rot |
| *Maharashtra* | | | |
| Rajapuri | Maharashtra and Gujarat | 20 | Resistant to leaf spot and susceptible to blotch and rot |

*Contd...*

| | | | |
|---|---|---|---|
| Eavaigon | – | 45 | – |
| *Orissa* | | | |
| Duhgi | – | 10 | – |
| Jobedi | – | – | – |
| Katingia | – | 8 | – |
| *Uttar Pradesh* | | | |
| Gorakhpur | – | 15 | – |

*Source:* Rao and Rao (1994)

## Crop Improvement

Scheme for research on turmeric was first implemented in 1944 at G Udayagiri in Orissa state under the auspices of Imperial Council of Agricultural Research. Prior to the first five year plan period, very little attention was devoted to the scientific development of the crop. During the first five year plan, a scheme for undertaking research work on turmeric was initiated in 1955 at Peddapalem in Andhra Pradesh. Presently, crop improvement work in turmeric is under progress at Indian Institute of Spices research, Calicut, Kerala, ICAR and under aegies of All India Coordinated Spices Improvement Project, ICAR in different centers Pottangi (Orissa), Coimbatore (Tamilnadu), RARS. Jagital (A.P.) and Solan (H.P.) besides several Agricultural Universities in the country.

## Selections/Varietal Studies

In turmeric, so far only clonal selections were attempted, as hybridization could not be

Practiced due to sterility, since commercial types are sterile triploids of *Curcuma longa* (Ramachandran 1961).

Under irrigated conditions, Shankaracharya and Natarajan (1974) recorded fresh yield of 16.8 to 22.4 t ha$^{-1}$. In Andhra Pradesh, Rao *et al.* (1975) broadly classified turmeric accessions into three groups depending on their durations. They are (1) Short duration "Kasthuri" types belonging to *C. aromatica* maturing in 7 months referred as "Ca" types, (2) Medium duration "Kesari" types and referred to as intermediate or CLI types. (3) Long duration types belonging to C. *longa* maturing in 9 months and are referred as CLL types. Clonal selections were made among more popular cultivars in these duration groups (Rao *et al.* 1975, Subbarayudu *et al.* 1976) and resulted in identification of eight promising types in short duration, six types in medium duration and eight in long duration group.

Pillai and Nambiar (1975) recorded the highest yield in Karhadi local followed by Kasthuri and Tekuripeta. Philip *et al.* (1982) found Mannuthy local the best in respect of cured rhizome yield (8588 kg $ha^{-1}$).

In Maharashtra, Pujari *et al.* (1987) released a high yielding variety Krishna (72 q $ha^{-1}$) for general cultivation. In later studies, Jalgaonkar *et al.* (1988) recorded the highest yield in Krishna (53.86 t $ha^{-1)}$ of fresh rhizomes followed by waigon (48 t $ha^{-1}$) under Konkan conditions. Ten years of research work at the high altitude research station, Pottangi in Orissa indicated that PTS 10 and PTS 24 produced more cured yield due to their high dry recovery 31 and 26% respectively. Curcumin, essential oil and oleoresin contents were 9.3, 4.2 and 13.5% for PTS 10 and 9.3, 4.4 and 13.1% for PTS 24 respectively. Due to their orange yellow colour and comparatively less susceptibility to leaf spot (*Collectotrichum sp*), rot and scale they were released as Roma and Suroma and recommended for large scale cultivation to replace local varieties.

Nine high yielding accessions designated as PCT 1 to PCT 9 were selected (Ratnambal and Nair 1986) and gave for multi location testing. Among selections, PCT 8 had the maximum yield potential (31.2 t $ha^{-1}$) and high curcumin content (8.7%). Subsequently it was released as Survarna. Maurya (1990) recommended RH-10 for Bihar. Its yield potential is reported (419.77 q $ha^{-1}$) fresh rhizomes, dry rhizomes (75.34 q $ha^{-1}$) and curcumin content (8.4%). In Northern Telangana zone of Andhra Pradesh, in short duration varietal studies, Reddy *et al.* (1989) recorded the highest rhizome yield in PCT 13 and PCT 14 (28.04; 26.71 t $ha^{-1}$) fresh and cured yield (5.69 t $ha^{-1}$; 5.44 t $ha^{-1}$) respectively.

Nandi (1990) recorded the highest rhizome yield in PTS 25 (27.0 $ha^{-1}$) followed by CLS 9 (24.6 $ha^{-1}$). Indiresh *et al.* (1990) found the highest fresh yield in PCT 8 (32.29 t $ha^{-1}$) followed by waigon (31.64t $ha^{-1}$). The highest cured yield was recorded in PCT 9 (6.5t $ha^{-1}$). Ratnambal *et al.* (1992) reported fresh rhizome yield of 29 t $ha^{-1}$ and 28.82 t $ha^{-1}$ in PCT 13 and PCT 14 respectively. Hedge (1992) found the highest fresh rhizome yield in cultivar Cudapah (22.19 t $ha^{-1}$) followed by Bidar (21.69 t $ha^{-1}$). Cholke (1993) recorded the highest fresh rhizome yield in Cuddapah (33.37 t $ha^{-1}$), followed by Amalapuram and Cuddapah. Under rainfed conditions, Sheshagiri and Uthaiah (1994) reported that BSR 1 was the best in fresh rhizome yield (16.57 t $ha^{-1}$) followed by waigon (15.45 t $ha^{-1}$). Radha Krishna *et al.* (1995) recorded the highest yield in CO 1 (16.54 t $ha^{-1}$) of green turmeric followed by

BSR 1 (14.74 t $ha^{-1}$). Kurian and Valsla (1996) reported that VK 5 was superior in fresh rhizome yield (5.85 kg/m2). The cultivar Chayapasupu recorded the highest yield (43.73 t $ha^{-1}$) of fresh rhizome followed by VK 31 (42.42t $ha^{-1}$).

Yield of cured rhizome was the highest in VK 116 (5.63 t $ha^{-1}$) under 25-30 % shaded conditions. But in open conditions, VK 31 (35.06 t $ha^{-1}$) and VK 55 (34.66 t $ha^{-1}$) were the high yielders in terms of fresh rhizome yield (Latha *et al.* 1995). Ramakrishna *et al.* (1995) found PCT 13 was superior in yield (19.15 t $ha^{-1}$). Kurian and Nair (1996) reported the highest fresh rhizome yield in VK 121 (43.02t $ha^{-1}$) and cured yield in VK 116 (8.43 t $ha^{-1}$). Kumar *et al.* (1996) reported that cultivars RH 10 and Maurya produced the highest yields of fresh rhizomes (30.1 and 28.5 t $ha^{-1}$) respectively and cured product (6.1 and 5.6 t $ha^{-1}$) respectively. Curing percentage was the highest in G.L. puram (22.3).

## Mutation Breeding

Since turmeric is vegetatively propagated, efforts were made to induce variability by using mutagens. In Tamilnadu, two mutants CO-1 and BSR-1 identified from "Erode local" and released for large scale cultivation in the name of CO-1 and BSR-1. They have yield potential of 24.70 and 30.69 t $ha^{-1}$ respectively over 'Erode local' (20.2 t $ha^{-1}$) (Balashanmugam *et al.* 1986). These lines have moderate curing percentage (18.4 and 20.5) with curcumin content (4.2%). Recently, Chezhiyan and Shanmugasundaram (2000) reported a new variety BSR-2 derived by X ray mutagenesis from the variety Erode local. It produces 690-890 g fresh rhizome yield / plant, curcumin content 3.75% and curing percentage of 20.

In Anantharajupeta (A.P.), Colchicine, EMS and MNG each at 250, 500, 1000 ppm on Cv. Mydukar resulted normal germination in all treatments, number of leaves were more in EMS 1000 ppm treatment, but yield was poor with EMS 250 ppm (RFRS, Anantharajupeta annual report 1986).

## Hybridization

The crop improvement was concentrated on selection as hybridization could not be practiced due to sterility, since the commercial types are sterile triploids of C. *longa* (Ramachandran 1961). Nazeem *et al.* (1993) reported VK5 of *Curcuma longa* (2n=63) produced seeds when it was crossed with Kasthuri and Dindigam which belonged to C. *aromatica* (2n=84). They indicated possibility of crossing high yielding *longa* types with high curing(%) of selected *aromatica* types. Viable seed set in turmeric reported by Kumar *et al.* (1994) has induced for recombination breeding programme.

Studies on flowering behaviour, fruit set and germination in C. *longa and* C. *aromatica* revealed that in C. *aromatica,* flowering period was July-September, whereas in C. longa it was September-December. Opening of flowers took place between 6 and 6.30 am at Kasargod. Mature capsules were observed in October-November. Days for flowering in C. *longa* varied from 118 to 143 days, whereas in C.*aromatica* it varied from 95 to 104 days. Time taken for maturity of seeds from flowering ranged from 23 to 29 days in *aromatica.* Pollen stainability in acetocarmine varied from 45-48% in C. *longa* and 68-74% in C. *aromatica.* Two distinct types of seeds, dark heavy and light brown were extracted from mature capsules of C. *aromatica.* Seeds had white aril, smooth surface and an apical micripylar ring with a wavy outline. Germination (%) varied from cultivar to cultivar and even plant to plant and 70 to 90 was recorded in 8 to 20 days after sowing. No germination was observed after 20 days. All the progenies had somatic chromosome number 2n=84 like parental clones and seeds of individual plants had remarkable morphological similarities to parental clones. Seedling progenies produced from mainly roots, root tubers and rhizomes were very small during first year. Normal development occurred during second year. C. *aromatica* being a tetraploid semi wild species, it produces viable seeds and production seems to be mainly controlled by environmental factors. In C. *longa* two distinct types, flowering and non flowering was noticed. Seed production was not observed in any flowering group. They were complete sterile and had greater vegetative vigour and ability to spread by rhizome compared to C. *aromatica.* Hybridization of C.*aromatica* (2n=4x=84) with C.*longa* (2n=3x=63) could be attempted through polyploid and haploid breeding techniques to produce plants with differential chromosome types for selecting better types with short duration as in C. *aromatica* and the high rhizome yield and curcumin content as in C. *longa*. Nair (1980) cautioned that it would be undesirable to mix up *longa* and *aromatica* types even in crossing programmes as the latter cannot be used for any culinary purposes and that it will only reduce quality of *longa* types.

## Improved Varieties

Efforts were made through different methods to identify and select turmeric types having yield potential with high % of curcumin.

## Package of Practices

### *Time of Planting*

Planting time varied with region and cultivar and usual time for planting of turmeric in India is May and June. Yields were also greatly influenced by

Improved Turmeric Varieties

| Name | Duration (days) | Fresh yield (t/ha) | | Dry recovery (%) | Curcumin (%) | Oleroresin (%) | Essential Oil (%) |
|---|---|---|---|---|---|---|---|
| | | Average | Potential | | | | |
| CO-1 | 285 | 30.0 | 35.0 | 19.5 | 3.2 | 6.7 | 3.2 |
| Krishna | 240 | 9.2 | 11.8 | 16.4 | 2.8 | 3.8 | 2.0 |
| Sugandham | 210 | 15.0 | 20.0 | 23.3 | 3.1 | 11.0 | 2.7 |
| BSR-1 | 285 | 30.7 | 39.6 | 20.5 | 4.2 | 4.0 | 3.7 |
| Suvarna | 200 | 17.4 | 43.5 | 20.0 | 4.3 | 13.5 | 7.0 |
| Suguna | 190 | 29.3 | 60.3 | 12.0 | 7.3 | 13.5 | 6.0 |
| Sudharshana | 190 | 28.8 | 54.9 | 12.0 | 5.3 | 15.0 | 7.0 |
| Roma | 250 | 20.7 | 40.0 | 31.0 | 9.3 | 13.2 | 4.2 |
| Suroma | 256 | 20.0 | 44.9 | 26.0 | 9.3 | 13.1 | 4.2 |
| Rajendra sonia | 225 | 4.8 | - | 18.0 | 8.4 | - | 5.0 |
| Ranga | 250 | 29.0 | 37.1 | 24.8 | 6.3 | 13.5 | 4.4 |
| Rasmi | 240 | 31.3 | 37.5 | 23.0 | 6.4 | 13.4 | 4.4 |
| Prabha | 205 | 37.47 | - | 19.5 | 6.52 | 15.0 | 6.5 |
| Prathibha | 225 | 39.12 | - | 18.5 | 6.21 | 16.2 | 6.2 |

*Source:* Peter and Kandiannan, (1999)

time of planting. In Andhra Pradesh, it was observed that delayed plantings reduced vegetative growth and yield and increased incidence of leaf spot diseases. In short duration Kasthuri types, planting in first fortnight of June, while in long duration types planting from middle of June to middle of July was optimum (RFRS, Anantharajupeta annual report 1973). In Punjab, planting from April 10$^{th}$ to June 30$^{th}$ (Randhawa and Mishra 1974), while in Kerala, planting between 15$^{th}$ April to 15$^{th}$ May was optimum (Sivaraman Nair 1982).

### *Method of Planting and Spacing*

Spacing of turmeric varies with method of planting. Crop can be planted on flat beds or on ridges. Aiyadurai (1966) stated that flat beds were preferable in Orissa, while in A.P. the broad ridge method of planting was superior and more profitable than ridge and furrow method, as elevated beds provided better drainage. According to him, in Orissa, the spacing of 22.5 x 22.5cm gave the best, while in A.P. using the ridge and furrow method of planting, optimum spacing was 45 and 60 cm between the rows and 22.5 cm between plants. In heavy black soils of Andhra Pradesh, Rao *et al.* (1975) recommended a wider spacing of 46 x 23 cm. due to excessive vegetative growth. In Tamilnadu, Ponnuswamy and Muthuswamy (1981) recommended 45 x 20 cm for Cv. Erode local, while in Maharastra Rajput *et al.* (1980) advocated 45 x30 cm for local cultivars. A closer spacing of 22-25 cms with high plant densities was suggested in Kerala by Phillip (1985) and in Punjab (Radhawa and Misra, 1974). Phillip (1985) opined that further increase in population would decrease growth, yield and yield components due to severe competition for nutrients.

### *Planting Materials*

Turmeric is always propagated vegetatively by fingers or mothers with one or two buds. Seed requirement for turmeric is high i.e. 2 ½ tonnes of seed is required to plant one hectare. Different planting materials, mother rhizomes, cut mothers and primary fingers are used as seed and research results gave variable advice on form of rhizomes to plant. Aiyadurai (1966) stated that mother rhizomes were better than finger rhizomes in Orissa. In Andhra Pradesh, mother rhizomes when planted as whole or even as split halves gave rise to more vigorous sprouts than fingers, but fingers have better storage, remained more tolerant to wet soil conditions and involved lower seed rates. Though research results favour planting of whole mother rhizomes. (Umrani *et al.* 1980, Joseph Philip 1984) but planting of a primary finger is the common practice with major turmeric growing belts. It might be due to low quantity of requirement of seed materials of finger rhizomes. Primary fingers were also efficacious and cost of seed material was worked to be cheaper in Andhra Pradesh (Anjaneyulu and Krishnamurthy 1979).

### *Mulching and Intercropping*

Turmeric is cultivated as sole crop or intercropped with maize, castor and also between coconut trees. In Northern Telangana Zone of Andhra Pradesh, turmeric is invariably intercropped with maize, whereas in Rayalaseema Castor and in coastal Andhra Pradesh, it is grown as inter crop in coconut gardens. Due to slow germination (2 to 4 weeks for germination) and initial growth being slow, different types of mulches and short duration crops are tried to protect sprouts and also to conserve moisture besides reducing weed growth before turmeric makes good growth. In Anantharajupeta. Castor was the best among different intercrops, while in mulching dry mango leaf mulch was beneficial (RFRS, Anantharajupeta annual reports 1975, 76). Mulching reduced weed growth and enhanced germination. Studies of Rao and Reddy (1990) at RARS, Jagtial indicated that sowing of maize in every second and third inter row space of turmeric and maintaining 100 % population of both turmeric and maize and additional fertilizer to maize (recommended) resulted in higher rhizome yield and net returns as compared to farmers practice of sowing turmeric and maize in every inter row space without additional fertilizer to maize.

In Tamilnadu. Sekhar and Muthuswamy (1985) recommended double intercropping with femugreek and onion.

In Orissa, mulching with Daincha and Sun hemp leaves was useful (Aiyadurai 1966). In Bihar, Jha *et al.* (1983) reported increased yields in turmeric with Dalbergia leaves.

## Nutritional Requirements

Turmeric is a heavy feeder and more uptake of nutrients was reported by Nagarajan and Pillai (1979) and Sivaraman (1992). Manuring plays important role in management of turmeric and good response to nitrogen for boosting up vegetative growth (Umate *et al.* 1984), Phosphorous was necessary for root development and K was involved in transportation of carbohydrates to rhizomes (Muralidharan and Balakrishnan 1972). In addition, K may also improve curcumin content (Mohan Babu and Muthuswamy 1984, Ahmed Shah *et al.* 1988). Fertilizer recommendations varied from state to state depending on cultivars and soils. Aiyadurai (1966) reported that applications of 100 kg/ha ammonium sulphate doubled yield over that of un manured crop. Muralidharan and Balakrishnan (1972) and Nambiar (1972) recommended applications of 100 kg N, 100 kg P2O5 and 200 kg K2O/ha under Kerala conditions. In Punjab, Randhawa *et al.* (1973) advocated 150 Kg N, 50 Kg P2O5 and 50 Kg K2O/ha. In Maharastra, Umate *et al.* (1984) recommended 120 KgN, 40 Kg P2O5 and 40 Kg K2O/ha for black soils of Maharastra.

In Andhra Pradesh, studies on mineral nutrition in black soils did not give consistent results. In less fertile red loamy soils of Rayalaseema, trails with Cv. Mydukur indicated that the crop responds well to higher fertility levels and application of 312.5 kg N, 125 kg P2O5 and 187.5 kg K2O /ha was optimum (RFRS, Anantharajupeta annual report, 1973). Similar results were reported by Raghava Rao and Swamy (1978 and 1984) in less fertile loamy soils of Tirupati (A.P.). Rama Rao *et al.* (1975), and Rao and Reddy (1977) tried higher doses of NPK and observed increased plant height, leaf number, leaf area and rhizome production with higher levels of NPK but no interaction effect was noticed. Nitrogen 375 kg, P2O5 175 kg and K2O 237.5 kg/ha gave more yield. Higher levels of N, P and K showed adverse effect on curing (%) and curcumin content while low levels of P and K have a slight beneficial effect. Leaf concentration of N increased with increase in N and P levels of application while K application depressed it. Positive correlations were noted between levels of leaf N, P and K with yield of rhizomes. Nutrient uptake studies indicated that the uptake of nutrients was greater up to 3$^{rd}$ month for K, 4$^{th}$ month for N and 5$^{th}$ month for P with subsequent decrease in concentration of these nutrients. The crop attains maximum vegetative growth by 4$^{th}$ and 5$^{th}$ months suggesting early application N, P and K for increasing plant growth.

Turmeric responds well to organic matter and experimental evidences are available on beneficial effects of organic matter either alone or in combinations with inorganic fertilizers on growth and productivity of turmeric. Usually large quantities of organic manure in the form of FYM, oil cakes and green leaves (mulch) are applied in different turmeric growing areas. Applications of tank silt and sheep penning are also practiced in a few areas of Tamilnadu and Andhrapradesh for improving soil fertility (Rethinam *et al.* 1994).

In studies of Raghava Rao and Swamy (1978 and 1984) on cultivar Mydukur, increase in NPK content in inorganic form increased vegetative growth in terms of plant height, number of leaves, leaf area and yields. Organic manures were helpful in improvement of rhizome quality especially colour and curing (%). According Rao *et al.* (1975), 25 tonnes of cattle manure or compost and 63 kg N ha$^{-1}$ in the form of oil cake was optimum in Andhrapradesh.

In the studies of Saha (1988) under mid hill dry conditions of Mizoram, the highest yield and maximum profit were obtained with application of 50 t ha$^{-1}$ of cow dung manure alone, followed by application of 90 kg N, 60 P2O5 and 90 kg k2O ha$^{-1}$ in the form of Urea, Diammonium phosphate and Muriate of Potash. Balashanmugam *et al.* (1989) found increase of over 37 % in fresh rhizome yield over control (no FYM) by application of 25 tonnes FYM ha$^{-1}$.

Gopalakrishna *et al.* 1997 reported in turmeric that integrated use of N and FYM at all levels were better than inorganic N application alone. Application of N @ 250 kg ha$^{-1}$ + 5 FYM t ha$^{-1}$ gave maximum net profit and cost benefit ratio.

Sadanandan and Hamza (1998) found the highest rhizome yield (averaged over 4 cultivars) in NPK fertilizer plots 4884 kg ha$^{-1}$, followed by neem cake, groundnut cake and cotton cake treatments 4818, 4809 and 4623 kg ha$^{-1}$ respectively. Curcumin yield (average of 4 cultivars) was the highest in plots supplied with neem cake (287 kg ha$^{-1}$), followed by cotton cake, ground cake and NPK fertilizer (284,277 and 268 kg ha$^{-1}$ respectively). Application of organic fertilizers increased nutrient availability and improved soil physical conditions. Economic analysis showed that application of NPK fertilizer gave the highest benefit: cost ratio, followed by groundnut cake.

Rao (2000) noted significant differences for growth, yield attributes and yield due to use of organic and inorganic combinations over recommended fertilizer (inorganic) alone. The highest cured yield was in crop applied with Neem Cake 1.25 t ha$^{-1}$ + FYM 12.5 t ha$^{-1}$ + recommended fertilizer dose followed by FYM 25 t ha$^{-1}$ + recommended fertilizer dose over recommended fertilizer dose. Curcumin, essential oils and oleoresins were more in organic and inorganic combinations over recommended fertilizer alone.

### *Harvesting*

Turmeric is usually harvested 8-9 months after planting (Aiyadurai 1966). The harvesting time also depends on variety. Short duration varieties mature between 6 to 7 months, medium duration varieties between 7-8 months and long duration varieties between 8-9 months. Umrani *et al.* (1980) recorded maximum yield in Tekuripeta variety, when it was harvested between 8-81/ 2 months.

### Processing

Processing in turmeric consists of boiling, drying, polishing and colouring of raw turmeric rhizomes. Boiling is done in turmeric to kill the biological activity of cells and to get uniform colour throughout rhizome,

### Boiling

After digging, rhizomes are cleaned and finger and mother rhizomes are separated. Normally mother rhizomes take more time for cooking than fingers (Rao *et al.* 1975) and separate boiling is required to get uniform cooking.

Methods and equipments vary from tract to tract but principle is same. Raw rhizomes are cooked in water just sufficient to cover rhizomes in a boiling pan. In Duggirala zone, perforated troughs immersed in a boiling pan are used most economically and efficiently as there is less wastage of fuel and time (Rao 1965). Equipments for boiling consists of a boiling pan of dimensions 1.37x1.37x0.66 m, which accommodate four troughs of size of 0.6x0.6 m with holes at bottom and sides and fixed with looped handles. Each trough holds 175 kg of raw turmeric and pan can be efficiently used for continuous boiling and each boiling takes about half an hour and 700 kg of raw turmeric can be boiled at a time. Turmeric leaves are used as fuel. While boiling, over cooking renders the fried product brittle, resulting in a higher % of pieces during subsequent polishing (Govindarajan 1980). Cuing should be completed within 2 to 3 days of harvesting to get maximum colour,

The CFTRI, Mysore developed a simpler, hygienic and efficient technique of curing and colouring of turmeric. In this method, rhizomes are boiled in lime water or sodium carbonate. A water solution containing 20 g sodium bisulphate and 20 g hydrochloric acid / 45.3 kg of rhizomes is recommended to give them desired yellow colour. To export turmeric, any chemicals or mammalian excreta should not be added while boiling. The turmeric rhizomes should be boiled in clean and plain water.

### Drying

Cooking of rhizomes facilitates quicker drying and boiled turmeric is sun dried for 10-15 days to dry about 5 % moisture level, till it becomes quite hard and brittle. When mechanical driers are used for drying, temperature is maintained within 50-60$^0$ C as there is loss of volatile substances beyond this range (Natarajan and Shankaracharya 1974). Artificial drying is advantageous in sliced turmeric as it suffers surface bleaching in sun drying. Turmeric should be properly dried or other wise aflotoxins problems may arise due to more moisture content and may not be suitable for export.

### *Polishing*

Dried turmeric has rough and dull appearance with scales and root bits. It is subjected to polishing the outer surface by manual or mechanical drying (Shankaracharya and Natarajan 1973, Narayanan *et al.* 1978) to improve its colour and appearance. Loss in weight due to full polishing is 5-8% (Rao 1965, Shankaracharya and Natarajan 1973), whereas it is 2-3% in half drying. Improved methods is by using hand operated drums mounted on central axis and drum filled with turmeric rotated at 80-100 rpm to effect polishing by abrasion of surface against metal surface. There are power operated polishing machines also.

### *Colouring*

Rhizomes are made more attractive by resorting to artificial colouring. There are two methods of colouring viz., dry method and wet method. Turmeric powder is added in the drum in the last 10 minutes of polishing. The CFTRI, Mysore developed an improved and harmless method for colouring turmeric (Pruthi 1976). In this method fingers are taken in a wicker basket which is shaken continuously and simultaneously a prepared emulsion is poured in to it, when the fingers are uniformly coated with the emulsion and they are dried in sun. Composition of emulsion as recommended by CFTRI, Mysore for coating 100 kg of half polished turmeric is alum 0.04 kg, turmeric power 2.00 kg, castor seed 0.14 kg, sodium bisulphate 30 g, concentrated hydrochloric acid 30 ml. The wet process requires further drying before storage (Govindarajan 1980).

### *Quality Characters*

Turmeric, like other spices, is available as whole, ground form (powder) and as oleoresin. House wife in the west, as a consumer, buys fine ground turmeric, while those in growing countries buy mostly whole or splits and to a small extent in urban areas as powder. As it is used in food, quality plays an important role. Among fingers, round and split fingers had the best quality (Rosengarten 1969). Lewis (1973) found distinct differences in quality, quantity of oil and oleoresin in different types of turmerics grown in India. Turmeric grown in hills was of better quality than that of the crop grown in plains Chaurasia *et al.* (1974).

### *Curcumin Content*

Alleppey turmeric contained 6.5 % curcumin as against 3-4 % in other varieties (Lewis 1973), while Shankaracharya (1974) recorded 0.2 to 3.8 % of curcumin in dried rhizomes of turmeric. Among 38 types of turmeric, curcumin content varied from 3.0 to 8.1 % with the lowest in Nandyal and the highest in Vontimitta (Mathai 1974).

Chaurasia *et al.* (1974) obtained the highest curcumin content in Alleppey Fingers (5.2%) and Alleppy bulbs (4.8%). They found varieing curcumin content (1.15 to 3.94%) in different cultivars.

Cultivars of C. *longa* comprising medium and long duration types contained higher proportion of curcumin than C. *aromatica* types which recorded the lowest values (Rao *et al.* 1975). Murthy *et al.* (1976) found higher curcumin in *longa* types than aromatica types and observed variation of 8.9 to 14.5 % in curcumin content among 15 types of turmeric grown under Kasargod conditions.

Among 20 selected clones, Muralidharan and Kutty (1976) recorded the highest curcumin in Alleppey (6.2%) followed by Ethamukkala (5.63%), Kharadi local (4.35%) and Wynad local (4.17%). Singnificant variations among the 19 cultivars were observed for curcumin content and Mannuthy local was the best for curcumin content (Phillip *et al.* 1980).

Muthuswamy and Shah (1982) reported that the cultivar Selam recorded the higher curcumin (4.75%) than Erode type (3.90%). Different researchers, Ghosh and Govind (1982) found the highest curcumin content (7.40 to 7.90%) in cultivars No. 24 and CLL 323 (Avanigadda), Philip (1983) in Mannuthy local (7.38%) and Pujari *et al.* (1987) in Salem (4.8%).

Ratnambal *et al.* (1992) found that the PCT 14 was superior in curcumin content (7.9%). Hedge (1992) recorded the highest curcumin of 6.40 % in Moovattipuzha. The Duggirala cultivar was superior to Mydukur with respect to curcumin content and Suvarna (PCT 8) contained the highest curcumin content 4.95 % (Kumar *et al.*, 1992). In Dharwad conditions Cholke (1993) reported the highest curcumin content in PCT8 (8.16%) followed by PCT 14 (7.45%) and the least in Bidar-4 (2.45%).

Kumar *et al.* (1994) revealed that the many land races contain more than 6 % curcumin. Aieng turmeric of Manipur, Wynad local, Edepalayam, Thoduphuzha, Manathody, Pulpally types of Kerala, Aizwal type of Mizoram, Sugandham of Gujarat are rich in curcumin.

Venkatesha (1994) recorded the highest curcumin content in PCT 8 (8.20%) followed by PCT13 (4.50 %), BSR 1 (4.20 %) and the least in Mydukur (1.69%). In Meghalaya, RCTH 1 recorded a curcumin content of 6.8 %.

In Kerala, Kurian and Valsala (1996) found the highest curcumin content (7.82%) in VK 145, a selection from Kuruppampad closely followed by VK-

188 (7.69%). Patil *et al.* (1995) reported the highest curcumin content in Lakadong (5.44%) followed by Suvarna (4.65%) and CO-1 (3.79%).

Collections of Kerala representing Alleppy type VK-96 (7.85%) followed by VK-112 (7.23%) recorded the highest curcumin content (Kurian and Nair 1996).

## Essential Oil

Volatile oil, responsible for aroma, also varied with duration, groups and cultivars. Turmeric, dried and cured generally yield 1.5 to 5 % volatile oil, though a few selections of Agricultural Research Stations showed values up to 7%. There is confusion with C. *aromatica another* closely allied rhizome which is generally high in volatile oil content (4 to 8%) and low in curcuminoids (1.5%). Chemical and aroma characteristics of volatile oils of two species are different and C. *aromatica* can not be substituted for C. *longa* in its food use (Govindarajan, 1980), Studies of Rao *et al.* (1975) revealed that oils ranged from 2.2 - 6.8 %. Among three groups of turmeric, C. *aromatica* types and medium duration Kesari types recorded higher oil content while it was lower in long duration types. Subbarayudu *et al.* (1976) observed the oil content was higher in medium duration types (5.3 to 6.8%), low (2.2 to 4.2%) in long duration types, and moderate in short duration types (3.3 to 6.0%).

Ghosh and Govind (1982) recorded the highest oil content in culitvar Sugandam and minimum in Kasthuritanuku. Volatile oil content is more in C. *aromatica* (Reddy *et al.* 1990). Ratnambal *et al.* (1992) reported that PCT 14 and PCT 13 had volatile oil content of 7 and 6 %respectively. In IISR Prabha and Prathibha, Kumar *et al.* (1996) recorded an oil content of 6.5 and 6.2% respectively.

### *Oleoresin Content*

Use of oleoresins is becoming increasingly important in developed countries where processed food industry has made phenomenal progress. Oleoresins, as name implies, are generally mixture of compounds, volatile oils, other active ingredients like non-volatile fatty and resinous materials, extractable by solvents, are used singly, in sequence or in combinations.

Oleoresin content in turmeric varied from 4 to 7.5% (Murthy *et al.* 1972). Lewis 1973) reported 6 to 7 % oleoresins in six types of turmeric (24.3%). The bulb of Alleppey contained only 16.2 percent. Philip *et al.* (1980) observed significant variation among cultivars and Mannuthy local was found the best.

Ratnambal (1986) and Ratnambal *et al.* (1992) reported that the Konni (19.2%) and PCT 14 (15%) were superior for oleoresins content. In IISR Prabha and Prathibha, Kumar *et al.* (1996) recorded an oleoresin content of 15% and 16.2% respectively.

## Turmeric Value Added Products

### Turmeric Powder

Turmeric is valued for its aroma and colour. Colouring matter is stable to heat, hence "Cryogenic grinding' is not necessary. The powder is mostly consumed domestically for culinary purposes. Turmeric powder is an important constituent to prepare curry powders, various spice mixtures, pickles and soups.

### Oleoresins

Oleoresins is a highly viscous orange brown product containing 30-35% volatile oil and has a characteristic turmeric aroma. Turmeric oleoresin is extracted from good quality turmeric powder with solvents like acetone - alcohol or ethylene dichloride and subsequent desolventization. Since it is highly viscous, it is mixed with diluents like propylene glycol or vegetable oil to obtain a homogenous pourable product. Turmeric oleoresin finds application in meat and fish products pickles, dairy products and confectionary.

## Volatile Oils

Turmeric rhizomes containing 3 to 5% essential oil is usually obtained by steam distilltion of ground spice .Turmeric oil is light yellow in colour with harsh odour. Turmeric oil has not commercial importance at present. Major constituents of oil are Turmerones (58%) and Zigiberene (25%). Odour is mainly due to Turmerones. Zingiberene did not contribute much for aroma. Factors affecting yield and quality depend on variety, stage of maturity, method of curing and method of distillation.

### Pigments, Antioxidants and Medicinal Properties of Turmeric

Curcuminoids are brilliant orange pigments of turmeric and responsible for yellow colour. They belong to family of poly phenols which include all bio-flavonoids flavans and flavonols. All poly phenols are antioxidants with certain groups being stronger than others. Among them most powerfuls are curcuminoids.

Major Curcuminoids in turmeric are curcumin, demethoxy curcumin and bis demethoxy curcumin. These substances comprise 3 to 6% of turmeric composition. Curcumin makes 70 to 75% of curcuminoids, demethoxy curcumin 20% and bis demethoxy curcumin at about 3%. Curcumin is a well studied compound among curcuminoids. In pure form, it is an orange yellow, crystalline powder insoluble in water.

Curcuminoids have several antioxidant properties including scavanging of reactive oxygen species as super oxide anions and hydrogen peroxide, inhibition of lipid peroxidation and inhibition of oxidation of low density lipoprotein (LDL). Tetrahydro curcumin, a reduced derivative of curcumin has stronger antioxident activities.

Medicinal properties of turmeric are well documented in ancient Indian Medical literature. Bhavaprakasha, the ancient famous Ayurveda specialist mentioned four types of turmeric namely,

1. Haridra (Turmeric) - *Curcuma longa*
2. Vanaharida (wild turmeric) - *Curcuma aromatica*
3. Karpoora havidra (Camphor turmeric) - *Curcuma amada*
4. Dark Haridra (Tree turmeric) - *Berberis aristata*

Haridra and Vanaharidra are similar in action and two others have different uses. It is used in different ways in Indian traditional system as 1. Insect and ant repellant 2. Veterinary medicine 3. Heal wounds and ulcers of animals. 4. Application of turmeric powder on fresh wounds to act as antiseptic 5. Leach therapy 6. Purification of guggul 7. For cold and other allergies. 8. Joint inflammations. 9. For improving complexion (especially ladies). 10. Skin diseases. 11. Ringworm infestation.

Beside these traditional uses, other uses reported are 1. Analgesic 2. Anti diabetic 3. Prevention of gall stones or decreasing size of stones. 4. Osteo-arthritis and rheumatoid arthritis. 5. In reducing serum cholesterol levels. 6. Anti inflammatory. 7. In lever diseases. 8. Anti carcinogenic and anti tumor. 9. Leukemia. 10. Cardiac complaints. 11. Opthamalic. 12. Digestive system and Respiration. 14. Anti cholesterol action. 15. Immune protective action. 16. Anti helmentic. 17. Antiviral and anti fungal. 18. Inhibition of HIV-1 and HIV-2 proteases.

Possible anticarcinogenic activity of curcumin and other curcuminoids is accounted by a few mechanisms. These include inhibition of angiogenesis, upregulation of apoptosis, interference with certain signal transduction pathways, critical for cell growth and proliferation, inhibition of colonic mucosa cyclooxygenase (COX) and lipoxygenase (LOX) activities and inhibition of farnesyl protein trnasferase.

### *Diseases*

Crop suffers from foliar and rhizome diseases. Among foliar diseases, Leaf spot caused by *Colletotrichum capsici* and Leaf blotch caused by *Taphrina maculans* are important. In rhizome diseases, rot caused by *Pythium graminicolum* is the most serious malady of the crop.

## Leaf Spot

The disease caused by *Colletotrichum capsici* is found in all turmeric growing areas. The disease was studied extensively by Ramakrishnan (1954) Kandaswamy (1958), Reddy (1957), Dakshina murthy *et al.* (1966).

### *Disease Ecology*

Disease initiation and development depend not only on initial inoculum and weather conditions but also on variety, location, cropping pattern and soil fertility.

Disease generally appears in August-September when crop is about 4-5 months old (Ramakrishna 1954) and noted 50% yield losses in case of severe infection. The pathogen infects young leaves first and spread to other leaves. Younger leaves are more susceptible as compared to older leaves, attributed to loss of carbohydrates and phenols and more of total nitrogen in younger leaves compared to older leaves. Disease symptoms are elliptic to oblong spots of variable size, most evident on upper surface, enlarging to 4-5 cm long and 3 cm wide and frequently coalesce covering most of the leaf, which then dries up. Centers of spots are grayish white with a brown margin and an indefinite yellow halo. In very severe infection most of the leaves may dry up, presenting a scorched appearance of the field (Joshi and Sharma 1980).

### *Yield Losses*

Yield losses depend on time of initiation of disease. Reduction in rhizome weight by 62.7 % was observed by Nair and Ramakrishnan (1973) due to foliar infection by *Colletotrichum capsici.*

### *Disease Management*

In the past, several chemicals like Bordeaux mixture, Copper fungicides, Dithane Z-78, and Mancozeb Difolaton were used to control the disease. Dakshinamurthi *et al.* (1966) recommended four sprays of 0.2% Captan or Dithane Z-78 at monthly intervals for effective control of this disease. They found Copper fungicides were not effective as organic fungicides. Subbaraja (1981) observed application of nitrogen at 120 kg $ha^{-1}$ reduced disease incidence irrespective of split doses applied. Similarly, potash application at 70 and 120 kg $ha^{-1}$ reduced the disease, the incidence was 21.8 and 18.6 %, respectively as compared to 46.3 in control. Further, spraying of Bavistin (0.2%) was very effective in controlling disease as well as improved the yields. Rao and Rao (1988) delivered six sprays of 0.25% Dithane M-45 at fortnightly.This gave a benefit/cost ratio of 2.44. It was also found that Copper fungicides depress yields. They indicated that judicious application of fertilizer and maintenance of balanced soil fertility are essential to minimize yield losses due to this disease.

## Leaf Blotch

The disease is caused by *Taphrina maculens* and usually appears on lower leaves in October and November. The disease is characterized by appearance of small scattered oily looking translucent spots on both surfaces of leaves and more on upper surface. Infected leaves distort and have reddish brown appearance. In case of severe attack, hundreds of spots appear on leaves. Disease hastens defoliation process.

### *Disease Management*

Following crop rotations, selection of seed material from disease free areas and seed treatment with Mancozeb 3 g/litre or Carbendezim 1 g/litre of water reduces the disease. Spraying Dithane Z-78 (0.2%) followed by Dithane M-45, Blitox 50, Bavistin and Cuman were also effective as foliar sprays (Srivastava and Gupta 1977).

## Rhizome Rot

Rhizome rot is the most serious disease affecting turmeric and threatening its cultivation. In certain times, farmers are unable to realize even seed material also. The disease was first reported by Ramakrishnan and Soumini (1954) from coastal districts of Guntur and Krishna (AP). They found that it was caused by *Pythium graminicolum*.

The disease manifests initially in isolated patches in field. The infected plants showed progressive drying up beginning with older leaves. Diseased plants were stunted and stood out conspicuously in affected portion of field. In some areas, leaves of affected plants turn yellow and roll inward giving an appearance of wilting symptoms (Rao and Rao 1988).

Examination of diseased plants showed that both roots and rhizomes were damaged. In severe cases, all roots were dead and rhizome development was completely arrested. In partial infection, some of the roots were dead and tips of growing rhizomes showed brown discoloration. In advanced infections, fingers and mother rhizomes were shrunken and shriveled. Fingers were affected with soft rot giving foul smell. The disease gradually affected all fingers and mother rhizome and eventually plants died. In majority of diseased rhizomes, active maggots of *Mimegrella coerulifrons* were found (Rao and Rao 1988). This fly was primary cause of rhizome rot in Maharastra (Ajri *et al.* 1982), but in Kerala it was associated with rot but had no significant role in initiation of disease (Premkumar *et al.* 1980).

Planting of healthy rhizomes and seed treatment with Mancozeb (0.25%) after harvest and before planting, crop rotation, use of biological agents and good drainage help in reducing the disease. In Northern Telangana zone of Andhra Pradesh, PCT 13 (Suguna) and PCT 14 (Sudarshana) showed filed tolerance to this disease (Rao *et al.* 1994).

## Pests

### *Turmeric shoot borer (Dichocrocis punctiferalis)*

Caterpillar bores into shoots and cause dead hearts. Adult moth is orange yellow with black margins on both wings and lay eggs on leaves and other soft plant parts. Caterpillar is reddish brown with black blotches all over the body. Nature of damage and life history of shoot borer on turmeric were studied by Jacob (1981). At Kasargod (Kerala), under lab conditions (temperature range 30-33 degrees Celsius, relative humidity (60-90%) egg period lasted 3-4 days. There were five larval instars and they lasted for 3-4, 5, 3-7, 3-8 and 7-14 days respectively. Fully grown larvae were light brown and 16-26 mm in length. Pre pupal and pupal periods lasted for 3-4 and 9-10 days respectively. Adult females laid 30-60 eggs during its life span. In the field, 6-7 generations were completed duing a crop season.

### *Lace Wing Bug (Stephanicus typicus)*

Small, whitish lace wing bug is found in colonies on foliage causing a sickly and spotted appearance of plants.

Spraying (0.1%) Malathion recommended for control of pest. Thangavel *et al.* (1977) obtained the best control of pest with Dimethoate and Phosphomideon each at (0.05%).

### *Scale insect (Aspediotus hartii)*

It infects rhizomes both in field and in storage and cause considerable damage. In the field, under severe cases of infestation, plants wither and dry. In storage, pest infestation results in shriveling of buds and rhizomes. If infestation is severe, it adversely affects sprouting of rhizomes.

Nature of damage caused by scale insect was given by Jacob (1982). Adult females are minute, circular and light brown to grey. Females are Ovo-Viviparous and also reproduce panthenogenitically, about 100 eggs are laid by a single female.

Dipping of seed rhizome of turmeric in Quinalphos (0.1%) for 5 minutes after harvest and before planting was also effective in controlling pest infestation.

Survey of Kotikal and Kulkarni (2000) in Northern districts of Karnataka revealed incidence of major insect pests on turmeric such as rhizome fly, shoot borer, lacewing bug, leaf, thrips and scale insects in all the areas and minor pests like various species of leaf beetles, cater pillers and grasshoppers.

Other pests Thrips (*Panchaetothrips indicus* Bagn) leaf feeders (*Creatonotus gangis (L)*) and rhizome maggots (*Mimegrella coerulifrons*) were also reported on turmeric.

## References

Ahmed Shah H, Vedamuthu PGB, Abdul Khader Md., and Prakasam. V. 1988. Influence of different levels on potassium on yield and curcumin content of turmeric.pp 108-113. *Proceedings of National seminar on Chillies, Ginger and Turmeric* (edited by G. Satyanarayana *et al.,*)

Aiyadurai S.G. 1966. *A review of research on spices and cashewnut in India,* ICAR, New Delhi.

Aiyer YAK .1954. *Field crops of India.* pp 319-326. The Bangalore Printing and Publishing Company Limited, Banglore, 4th edition.

Ajri D.A. Ghorpade S.A. and Jadav S.S. 1982. *Research on rhizomefly Mimegrella coerulifrons on turmeric and ginger in Maharastra.*88 pp. ICAR report

Anjaneyulu V S R and Krishnamurthy D. 1979. The efficiency of Broad ridge method of planting turmeric. *Indian Cocoa Arecanut Spices Journal.* 2: 115-116.

Bailey L B. 1963. *Manual of cultivated plants.* pp 287-289.The Mac millan Company, New York.

Balashanmugam P V, Vangamudi K and Chamy A. 1989. Studies on the influence of Farm yard manure on the rhizome yield of turmeric. *Indian Cocoa Arecanut and Spices Journal* 12: 126.

Chaurasia L D, Kulakarni B M, Nair K N G and Mathew T V .1974. Curcumin content of Indian turmeric. pp- 55-56. *Proceedings of symposium on Development and Prospects of Spices Industry in India.*

Chezhiyan and Shan mugasundaram KA. 2000. BSR-2- A promising turmeric variety from Tamilnadu. *Indian Journal of Arecanut,Spices and Medicinal plants* 2 (1) :24-26.

Cholke S M . 1993. *Performance of turmeric (Curcuma longa L.) cultivars.* M.Sc. (Ag) thesis, University of Agricultural Sciences, Dharwad.

Dakshinamurthy V, Reddy G S, Rao D K, and Rao P G. 1996. Fingicidal control of turmeric leaf spot caused by *colletotrichum capsici. Andhra Agricultural Journal.* 13: 69-72.

Ghosh S P and Govind S. 1982. Yield and quality of turmeric in North Eastern Hills. *Indian Horticulture* 39: 230-232.

Gopalakrishna V, Reddy M S and Kumar T V. 1997. Response of turmeric of FYM and N fertilization. *Journal of Research ANGRAU* 24: 58-59.

Govindarajan V S. 1980. Turmeric chemistry, technology and quality. *Critical Review in Food Science and Nutrition* 12: 199-310.

Hedge G S. 1992. *Studies on the performance of turmeric (Curcuma longa L.) cultivars.* M.Sc (Ag) thesis, University of Agricultural Sciences, Dharwad.

Indiresh K M, Uthaiah B C, Reddy M J and Rao KB. 1990. Morphological, rhizome and yield character of different turmeric varieties in coastal Karnataka. *Mysore Journal of Agricultural Sciences* 24: 484-490.

Jacob S A. 1981. Biology of *Dichocrosis punctiferalis Guen* on turmeric. *Journal of Plantation crops* 9: 119-23.

Jacob S A. 1982. Biology and Bionomics of ginger and turmeric scale *Aspiodiotus hartii Green.* 131-132 pp. *In proceedings of National Seminar on ginger an turmeric*, CPCRI, Kasargad.

Jalgoankar J, Patil M M and Rajput J C. 1988. *Performance of different varieties of turmeric under konkan conditions of Maharastra.*102-105 pp. Proceedings of the National Seminar on chillies ginger and turmeric at Hyderabad

Jha R C, Sharma N N and Maurya K R. 1983. Effect of sowing dates and mulching on the yield and profitability of turmeric. *Bangladesh Horticulture* 11: 1-4.

Joseph Philip. 1984, Effects of different planting materials on growth, yield and quality of turmeric. *Indian Cocoa Arecanut Spices Journal* 7: 8-11.

Joshi L K and Sharma N D. 1980. *Diseases of ginger and turmeric.* 104-119 pp. In national symposium on spices held at Calicut.

Kandaswamy T K. 1958. Control of the leaf spot disease of turmeric caused by *Colletotrichum capsici. Madras Agricultural Journal* 45(2): 55-60.

Kotikal Y K and Kulkarni K A. 2000. Incidence of insect pests of turmeric (*Curcuma longa L.*) in northern Karnataka, India. *Journal of Spices and Aromatic crops* 5: 41-48.

Kumar B S, George J K, Zacharaiah T J, Ratnambal M J, Babu K N and Ravindran P N. 1996. IISR Prabha and IISR Prathibha two new high yielding and high quality turmeric (*Curcuma longa L.*) varieties. *Journal of spices and Aromatic crops* 5: 41-48.

Kumar B S, Ravindran P N and George J K. 1994. Breeding ginger and turmeric. *Indian Cocoa Arecanut and Spices Journal* 18: 10-12.

Kumar G V, Reddy K S, Rao M S and Ramavatharam M. 1992. Soil and plant characters influencing circumin content of turmeric. *Indian Cocoa Arecanut and spices Journal* 15: 102-104.

Kumar R and Jain B P. 1998. Evaluation of growth and rhizome characters of some turmeric (*Curcuma longa L.*) cultivars under plateau region of Bihar. 9-10 pp. Souvenir, National seminar on recent development in spices production technology. Bihar Agricultural College, Sabour.

Kurian A and Nair G S. 1996. Evaluation of turmeric germplasm for yield and quality. *Indian Journal of Genetic Resources* 9: 327-329.

Kurian, A and Valsala P A. 1996. Evaluation of turmeric types for yield and quality. *Journal of Tropical Agriculture* 33: 75-76.

Latha P, Giridharan M.P. and Naik B J. 1995. Performance of turmeric (*Curcuma longa.L*) cultivars is open and partially shaded conditions. *Journal of Spices and Aromatic crops* 4: 139-144.

Lewis Y S A. 1973. The importance of selecting the proper variety of a spice for oil and oleoresin extractions. pp 183-185. *Proceedings of the conference of spices*, T.P.L. London. .

Mathai G K. 1974. *Quality studies in cashew and spices*.156-167 pp. Annual Report, CPCRI, Kasargod.

Maurya K R. 1990. R H- 10 A promising variety of turmeric to boost farmers economy. *Indian Cocoa Arecanut Spices Journal* 13: 100-105.

Mohanbabu N and Muthuswamy S. 1984. Influence of potassium on the quality of turmeric. *South Indian Horticulture* 31: 343-46.

Muralidharan A and Balakrishna S. 1972. Studies on the performance of some varieties of turmeric (Curcuma sp.) and its fertilizer requirements. *Agricultural Research Journal Kerala* 10: 112-15.

Muralidharan A and Kutty, R H N. 1976. Performance of some selected clones of turmeric (*Curcuma longa*) in Wynad. *Agricultural Research Journal Kerala* 14: 191-193.

Murthy M K, Mathew A G, Nambudiri E S, Sivasankar S, Lewis Y S and Natarajan, C P. 1976. Oil and oleoresins of turmeric. *Tropical Science* 18: 37-45.

Murthy M K, Madhav A G, Nambudiri E S and Lewis Y S. 1972. Essential oils and oleoresins from major spices of India. *Journal of Plantation Crops* (Suppl.): 181-188.

Muthuswamy S and Shah A H. 1982. Comparative quality evaluation of Selam and Erode turmeric types. *Indian Cocoa Arecanut and Spices Journal* 5: 77.

Nagarajan M and Pillai N G. 1979. A note on the nutrient removal by ginger and turmeric rhizomes. *Madras Agricultural Journal* 66: 56-59.

Nair M K Nambiar M C and Ratnambal M J. 1982. *Cytogenetics and crop improvement of ginger and turmeric.* 15-23 pp. Proceedings of the National seminar on ginger turmeric, Calicut

Nair N C and Ramakrishna K. 1973. Effect of *Colletotrichum* leaf spot disease of *Curcuma longa L.* on the yield and quality of rhizomes. *Current science* 42: 549-550.

Nambair M C. 1972. Advances in Spices Research. *Indian Farming* 22:9-12.

Natarajan C P and Sankaracharya N B. 1974. *Processing of spices.*100 pp. (In) Proc. Symposium on Development and prospects of spice industry in India. Association of food Scientists and Technologists (India), Mysore.

Parry J W. 1962. *Spices.* 79-80 pp. Chemical Publishing Company INC, New york.

Patil D V Kuruvilla K M and Madhusoodanan K J. 1995. Performance of turmeric varieties is lower Palney hills of Tamilnadu. *Indian Cocoa Arecanut and Spices Journal* 4: 156-158.

Peter K V and Kandiannan K. 1999. *Tropical Horticulture.*pp 638-691 Edited by Bose *et al.* Naya Prakash Publishing Human, Calcutta.

Philip J and Nair P C S. 1981. Field reaction of turmeric types to important pests and diseases. *Indian Cocoa Arecanut and Spices Journal* 4(4): 107-109.

Philip J, Sivaraman Nair P C, Nybe E V and Mohan N K. 1980. Variation of yield and quality of turmeric.42-46 pp. *Proceedings of National Seminar on ginger and turmeric held at Calicut.*

Philip J. 1983. Studies on growth, yield and quality component in different turmeric types. *Indian Cocoa Arecanut and spices Journal* 6: 93-97.

Philip.1985. Effect of plant density on yield and yield components of turmeric. *Indian Cocoa, Areacut and Spices Journal* 8: 93-96.

Pillai T P K and Nambiar M C. 1975. *Germplasm collections and cataloguing of turmeric.* 138-140 pp. Annual report, CPCRI, Kasargod

Pillai T P K, Nambiar M C and Ratnambal M J. 1976. *Germplasm collections and cataloguing of turmeric.* 256-258 pp. Annual Report, CPCRI, Kasargod.

Ponnuswamy V and Muthuswamy S. 1981. Influence of spacings in yield and yield components of turmeric. *South Indian Horticulture* 29: 229-230.

Premkumar T, Sharma Y R and Gautham S S. 1980. Association of dipteran maggots in rhizome rot of ginger. pp 128-130. (In) *National symposium on spices held at Calicut.*

Pruthi J S. 1976. *Spices and condiments.* 223-227 pp. National Book Trust, New Delhi.

Pujari P P, Patil R B and Sakpal B T. 1987. Studies on growth yield quality components in different turmeric varieties. *Indian Cocoa Arecanut and Spices Journal.*11: 15-17.

Radhakrishna V V, Madhusoodhnan K J and Kuruvilla K M. 1995. Performance of different varieties of turmeric (*Curcuma longa*) in the high ranges of Iddukki district. *Indian cocoa Arecanut and Spices Journal* 29:8-10.

Radhaw K.S and Mishra. 1974. Effect of sowing date, seed size and spacing on the growth and yield of turmeric. *Punjab Horticultural Journal* 24: 53-55.

Raghava Rao D V and Swamy G S. 1978. Effect of graded doses of Farm Yard Manure on growth and yield and quality of turmeric variety Mydukur. *Journal of Research APAU* 5: 8-12.

Raghava Rao D V and Swamy G S. 1984. Studies on the effect of N, P and K on growth, yield and quality of turmeric. *South Indian Horticultture*. 32:288-291.

Ramachandran 1961. Chromosome number is the genus *Curcuma* species. *Current Science* 30: 194-196.

Ramakrishna M, Reddy R S and Padmanabham V. 1995. Studies on the performance of short duration varieties/cultures of turmeric in Southern Zone of Andhra Pradesh. *Journal of Plantation crops* 23:126-127.

Ramakrishna T S. 1954. Leaf spot diseases of turmeric (*Curcuma longa L.*) caused by *Colletotrichum capsici* (Syd.) Butler and Bisby. *Indian Phytopathology* 7: 111-117.

Ramakrishnan T S and Soumini C R.1954. Rhizome and root rot of turmeric caused by *Pythium graminicolum* Sub. *Indian Phytopathogy* 7: 152-158.

Randhawa K S, Nandpuri K S and Bajwa M S . 1973. Response of turmeric to N, P, K fertilization. *Journal of Research Punjab Agricultural University* 10: 45-48.

Rao M R and Reddy V R. 1977. Effect of different levels of Nitrogen, Phosphorus and potassium on yield of turmeric. *Journal of Plantation Crops* 5: 61-62.

Rao M R, Reddy K R C and Subbarayudu M. 1975. Promising turmeric types of Andhra Pradesh. *Indian Spices.* 12:2-5.

Rao P S and Rao T G N. 1988. Diseases of Turmeric in Andhra Pradesh. 162-167 pp. *In Proceedings of National Seminar on Chillies, Ginger and Turmeric held in Hyderabad.*

Rao A M 2000. *Genetic variability yield and quality studies in turmeric (Curcuma Longa L.).* Ph.D thesis submitted to ANGRAU, Rajendranagar, Hyderabad.

Rao A M and Reddy M L N. 1990. Population and fertilizer requirement of maize in turmeric maize intercropping systems. *Journal of Plantation crops* 18: 44-49.

Rao M R and Rao D V R. 1994. *Genetic resources of turmeric*.pp 131-148. Advances in Horticulture. Plantation and spice crops.

Rao V T. 1965. Curing and processing of turmeric in Andhra Pradesh with special reference to Duggirala. *Spices Bulletin* 4: 91-95.

Ratnambal M J, Nirmal Babu K N, Nair M K and Edison S. 1991. PCT 13 and PCT 14 – two high yielding varieties of turmeric, *Journal of Plantation crops* 20: 79-84.

Ratnambal M J. 1986. Evaluation of turmeric accessions for quality. *Plants Food Human Nutrition* 36: 243-252.

Ratnambal M J and Nair M K. 1986. High yielding turmeric selection PCT 8. *Journal of plantation crops* 14: 94-978.

Reddy E N, Padmalatha B and Reddy B K. 1990. Improvement of turmeric- present trend. *Indian Cocoa Arecanut and Spices Journal*.13: 102-103.

Reddy G S. 1957. *Studies on storage rot and leaf spot disease in turmeric*. 44 pp. M.Sc.(Ag) Thesis, submitted to Andhra University.

Reddy M L N, Rao A M, Rao D V R and Reddy S A 1989. Screening of short duration turmeric varieties *Journal of Plantation and Spices Crops*. 477-488 pp.

Rethinam P, Sivaraman K and Sushma P K. 1994. Nutrition of Turmeric. 477-478 pp. *Advances in Horticulture, Plantation and Spices Crops*, Vol 9(1).

RFRS, Anantharajupeta , Andhra Pradesh *1973, 1975, 1976, 1986 Annual reports*.

Rosengarten, F. 1969. The book of spices. 444-452 pp. Living Stone Publishing Company, Pennysylvania.

Sadanandam A.K and Hamza S. 1998. Effect of organic manures on nutrient uptake, yield and quality of turmeric (Curcuma laya L.). Developments in plantation crops research proceedings of the 12th symposium on plantation crops, Kottayam, India 175-181.

Saha A K. 1998. Note on response of turmeric to manure and source of N and P under terrace conditions of Mizoram. *Indian Journal of Horticulture* 45: 139-140.

Sekhar K and Muthuswamy S. 1985. Economics of Double Intercropping in turmeric. *Indian Cocoa, Arecanut and Spices Journal* 8: 67-69.

Shah N C.1997. Traditional uses of Turmeric (*Curcuma longa*) in India. *Journal of Medicinal and Aromatic Plant Sciences*. 19: 948-954.

Shah H A, Seemanthini R. Arumugham R Muthuswamy S and Kader J B M 1982. Co-1 turmeric A high yielding mutant turmeric. *South Indian Horticulture* 30: 276-277.

Shankaracharya N B and Natarajan C P. 1974. Technology of Spices. *Arecanut and Spices Buttletin*. 7: 27-43.

Shankaracharya N B. 1974. Spices in India. 24-37 pp. *Souvenir*. Publication of symposium on spices industry in India.

Sheshargiri K S and Uthaiah B C. 1994. Performance of turmeric (*Curcuma longa L*) varieties in the hill zone of Karnataka. *Indian Cocoa Arecanut and Spices Journal* 3: 161-163.

Sivaraman K. 1992. *Studies on the productivity of turmeric maize and onion intercropping systems under varied population and nitrogen levels*. Ph.D thesis, Tamil Nadu.

Sivaraman Nair P C. 1982. Agronomy of ginger and turmeric – Proceedings of National seminar on ginger and turmeric. pp 63-68. held in Calicut, Kerala.

Srivastava V P and Gupta J H. 1977. Fungicidal control of turmeric leaf spot incited by *Taphrina maculans*. *Indian Journal of Plant Pathology* 7:76-77.

Subbaraja K T. 1981. Studies on turmeric leaf spot caused by *Colletotrichum capsici* (syd) Batul. and Bisby, Ph.D. thesis, TNAU, Coimbatore.

Subbarayudu M, Reddy K R C and Rao M R. 1976. Studies on varietal performance of turmeric. *Andhra Agricultural Journal* 23: 195-198.

Thangevel P, Sivakumar C V and Kateem A. 1977. Control of lace wing bug *Stephanitis typicus* Distant (Tingidae: Homoptera) in turmeric. *South Indian Horticulture* 25 (4): 170-172.

Umate M G, Latchanna A and Bidigir U S. 1984. Growth and yield of turmeric varieties as influenced by varying levels of nitrogen. *Indian Cocoa Arecanut and Spices Journal* 8:23-57.

Umrani N K, Patel R B and Pawar H K. 1980. Effect of planting materials and transplanting on yield of turmeric cv. Tekuripeta. 79-82 pp. *Proceedings of the National Seminar on ginger and turmeric, Calicut.*.

Venkatesha J. 1994. Studies on the evaluation of promising cultivars and nutrient requirement of turmeric and *Curcuma domestica val.*, Ph.D. Thesis, Division of Horticulture, UAS, Bangalore.

**Underutilized and Underexploited Horticultural Crops, Vol. 2, 2007**
*K.V. Peter (ed.), pp. 353-355*
New India Publishing Agency, New Delhi (India)
E-mail : newindiapublishingagency@gmail.com
Web: www.bookfactoryindia.com

# 20 Long Coriander (*Eryngium foetidum* L.)

**P. INDIRA**
Kerala Agricultural University, P O KAU - 680 656

**Plant family** : Apiaceae

**Synonyms**

**English** : Puerto Rican Coriander, Black Benny, Saw leaf herb, Mexican coriander, Saw tooth coriander, Spiny coriander, Fitweed

**Hindi** : Bhandhania,Bhandhanya

**Sanskrit** : Phiranga

**Kannada** : Paramgi

## Origin and Distribution

The plant is native to the Caribbean islands occuring from S.Mexico to Panama, Columbia, Bolivia, Brazil and Cuba to Trinidad. It was introduced to many tropical and subtriopical countries where it is used as a substitute for coriander.

## Main Constituents

The essential oil from leaves of long coriander is rich in aliphatic aldehydes, most of which are $\alpha,\beta$ unsaturated. The impact compound is E-2-dodecenal (60%), furthermore 2,3,6-trimethylbenzaldehyde (10%), dodecanal (7%) and

E-2-tridecenal (5%) were identified. Aliphatic aldehydes appear also in other spices with coriander like scent (e.g.,Vietnamese coriander).

Another essential oil can be obtained from root; in the root oil, unsaturated alicyclic or aromatic aldehydes dominate (2,3,6-trimethylbenzaldehyde 40%, 2-formyl-1,1,5-trimethyl cyclohexa-2,5-dien-4-ol 10%, 2-formyl-1,1,5-trimethyl cyclohexa-2,4-dien-6-ol 20%, 2,3,4-trimethylbenzaldehyde 30%). The essential oil from the seed contains mainly carotol, β-farnesene, anethole and α-pinene (Wong *et al.*, 1994).

## Description

*Eryngium* is an erect, perennial,glabrous herb, about 75-80cm tall with a grooved stem that elongates before flowering. The leaves are simple, subsessile to petiolate; having a fetid odour similar to the coriander leaves. Inflorescence is terminal, flowers small, fruit a shizocarp, splitting into two globoid mericarps.The chromosome number is 2n=16. The plant can be multiplied both by seeds and vegetatively. It has become a weed in many countries because of its easy propagation . It is cultivated on a small scale in homesteads, and large scale production is not popular. Leaves are harvested as and when required and used either fresh or after drying (Ravindran *et al.*, 2005).

## Uses

Long coriander belongs to same plant family as coriander , but the plant's shape does not bear much resemblance. Yet the long, tough leaves exemanate a fragrance very much similar to coriander's aroma and thus suggest themselves as a substitute or alternative for the former. Long coriander's usage concentrates on the Far East and Central America. In Asia, it is the most popular in countries of the South East Asian peninsular. In Thailand, Malaysia and Singapore, it is commonly used together with or in place of coriander and topped over soups, noodle dishes and curries. It can also be used for Thai curry pastes especially, when coriander roots are not available.

Long coriander is of importance in the cooking of Vietnam, where fresh herbs are of main importance. Long coriander is often used as a fully equivalent substitute for the much-loved coriander leaves to decorate soups and stir-fries; occasionally, the largest leaves are used to wrap food bits in them. In Central America, long coriander is the most associated with cooking style of Puerto Rico, although it is also known in other Caribbean islands and in Eastern México. Yet Puerto Rico is the place where one is most likely to find foods common to all Central American countries enhanced with long coriander.

Salsa is a spicy sauce of varying composition that often provides extra spiciness for main courses or is used as a dip and eaten with crisp-fried tortilla chips (*tostadas*) (Gernot Katzer's spice pages, 2006). Salsa can be made from more or less everything, but most recipes are actually based on tomatoes (also *tomatillos*) or mild paprika varieties garlic, onion and more or less fiery chillies are called for by almost all recipes. In the simplest case, *salsa* is just the raw, chopped or blended ingredients, but it can also be shortly cooked or even long simmered. Also ripe tropical fruits (papaya,mango) are sometimes added. The *salsa* is then finalized with fresh herbs (oregano, coriander, epazote, parsley and others), salt and may be a dash of lime juice and sugar.

*Eryngium* is used in the traditional medicine of Central and South America. Decoction of root is taken as a sudorific , diuretic, febrifuge and stimulant. Juice is used as a laxative and as a remedy for cold and fever.; decoction of the whole plant is a potent emmenogogue and abortifacient and is used as an aphrodisiac (Van den Bergh, 1999).

## References

Gernot Katzer's spice pages. 2006. Available at http://www.uni-graz.at./~katzer/eng//eryn-foe.html.

Ravindran, P.N., Geetha,S.P. and Shylaja, M. 2005. In : Advances in Spice Research. Ravindran, P.N., Nirmalbabu, K. , Shiva, K.N. and Johny, A.K. *History and Achievements of spice research in India since independence.* Agrobios, india. P 862-863.

Van der Bergh, M.H. 1999. *Eryngium foetidum* L J. In de Guzman,C.C. and Siemonsama,J.S. (eds). *Plant resource of south east Asia.* No.13 *Spices*. Bachyv's pub; Leiden. pp121-123.

Wong, K.C. , Feng, M.C. Sam, T.W. and Tan, G.L. 1994. Composition of the leaf and root oils of *Eryngium foetidum* L. J. *Essen. Oil Res.* 6:369-374.

FSC
www.fsc.org
MIX
Papier | Fördert
gute Waldnutzung
FSC® C083411

Zeitfracht Medien GmbH
Ferdinand-Jühlke-Straße 7
99095 Erfurt, Deutschland
produktsicherheit@kolibri360.de